Sustainable Forest Management

Sustainable Forest Management

Editor:

Jaspreet Banga, Ph.D.

Delve Publishing

www.delvepublishing.com

Sustainable Forest Management

Editor: **Jaspreet Banga, Ph.D.**

708 3rd Avenue, 6th Floor
New York, NY 10017
United States of America
www.delvepublishing.com

ISBN: 978-1-68095-766-2

Library of Congress Control Number: 2016951522

About the Editor

Jaspreet Banga, Ph.D.

Jaspreet Banga has completed PhD from Central Drug Research Institute, India and Master of Technology from Indian Institute of Technology. Jaspreet's interests range from Drug development, Purification, Fermentation technology, Plant Biotechnology and Oncology.

Contents

List of Contributors

Jørgen Bo Larsen
Forest & Landscape, University of Copenhagen Denmark

Floris Dalemans
Division of Forest, Nature and Landscape, Department of Earth and Environmental Sciences, KU Leuven (University of Leuven), Celestijnenlaan 200E, BE-3001 Leuven, Belgium

Paul Jacxsens
Spatial Applications Division Leuven, Department of Earth and Environmental Sciences, KU Leuven (University of Leuven), Celestijnenlaan 200E, BE-3001 Leuven, Belgium

Jos Van Orshoven
Division of Forest, Nature and Landscape, Department of Earth and Environmental Sciences, KU Leuven (University of Leuven), Celestijnenlaan 200E, BE-3001 Leuven, Belgium
Spatial Applications Division Leuven, Department of Earth and Environmental Sciences, KU Leuven (University of Leuven), Celestijnenlaan 200E, BE-3001 Leuven, Belgium

Vincent Kint
Division of Forest, Nature and Landscape, Department of Earth and Environmental Sciences, KU Leuven (University of Leuven), Celestijnenlaan 200E, BE-3001 Leuven, Belgium
Flemish Agency for Nature and Forests, Koning Albert II-laan 20/8, BE-1000 Brussels, Belgium

Pieter Moonen
Division of Forest, Nature and Landscape, Department of Earth and Environmental Sciences, KU Leuven (University of Leuven), Celestijnenlaan 200E, BE-3001 Leuven, Belgium

Bart Muys
Division of Forest, Nature and Landscape, Department of Earth and Environmental Sciences, KU Leuven (University of Leuven), Celestijnenlaan 200E, BE-3001 Leuven, Belgium

Jeji Varghese
Sociology and Anthropology, University of Guelph, Guelph, ON, Canada N1G 2W1

Maureen G. Reed
School of Environment and Sustainability and Department of Geography and Planning, University of Saskatchewan, Saskatoon, SK, Canada S7N 5A6

Peter Dok Tindan
Institute of Distance Learning, KNUST, Kumasi, Ghana

James Boafo
Department of Geography and Rural Development, KNUST, Kumasi, Ghana

Jacob Obodai
Department of Geography and Rural Development, KNUST, Kumasi, Ghana

Walter Lintangah
International Tropical Forestry Programme, Faculty of Science and Natural Resources, Universiti Malaysia Sabah, Malaysia

Norbert Weber
Professorship of Forest Policy and Forest Resource Economics, Technische Universitat¨ Dresden, Germany

Xiaohua Wei
Department of Earth and Environmental Sciences, University of British Columbia, Kelowna, British Columbia, Canada

Juan A. Blanco
Departmento de Ciencias del Medio Natural, Universidad Pu´blica de Navarra, Pamplona, Navarra, Spain

C. H. Traynor
Discipline of Geography, School of Environmental Sciences., South Africa

T. R. Hill
Discipline of Geography, School of Environmental Sciences., South Africa

Hanna Kruk
Gdynia Maritime University, Faculty of Entrepreneurship and Quality Science, Department of Economics and Management, Morska 81–87, 81-225 Gdynia, Poland

Bożena Kornatowska
Institute of Environmental Protection-NRI, Department of Nature and Landscape Conservation, Kolektorska 4, 01-692 Warsaw, Poland

Klaus J Puettmann
Department of Forest Ecosystems and Society, Oregon State University, Corvallis, OR 97330, USA

Scott McG Wilson
Consultant Forester and Forest Ecologist

Susan C Baker
Thorngrove Crescent, Aberdeen, Scotland AB15 7FH, UK. 3 School of Biological Sciences, University of Tasmania, Private Bag 55, Hobart, TAS 7001, Australia

Pablo J Donoso
Department of Forests and Society, Faculty of Forest Sciences and Natural Resources, Universidad Austral de Chile, Isla Teja s/n, Valdivia, Chile

Lars Drössler
Southern Swedish Forest Research Centre, Swedish Agricultural University, Alnarp SE-23053, Sweden.

Girma Amente
Haramaya University, P. Box 138, Diredawa, Ethiopia

Brian D Harvey
Institut de recherche sur les forêts, Université du Québec en Abitibi-Témiscamingue, 445, Boulevard de l'Université, Rouyn--Noranda, Québec J9X 5E4, Canada

Thomas Knoke
Institute of Forest Management, TUM School of Life Sciences Weihenstephan, Technische Universität München, Hans-Carl-von-Carlowitz-Platz 2, Freising 85354, Germany

Yuanchang Lu
Department of Forest Management and Statistics IFRIT, Chinese Academy of Forestry, Dongxiaofu 2, Xiangshan Road, Beijing, Haidian 100091, China

Susanna Nocentini
Department of Agricultural, Food and Forestry Systems, University of Florence, v. S. Bonaventura 13, Florence I-50145, Italy

Francis E Putz
Department of Biology, University of Florida, Gainesville, FL 32611-8526, USA

Toshiya Yoshida
Uryu Experimental Forest, Field Science Center for Northern Biosphere, Hokkaido University, Moshiri, Horokanai 074-0741, Japan

Jürgen Bauhus
Chair of Silviculture, Faculty of Environment and Natural Resources, University of Freiburg, Tennenbacherstr. 4, Freiburg 79085, Germany

Rabiul Islam
Institute for Environment and Development (LESTARI), University Kebangsaan Malaysia, 43600 UKM Bangi, Darul Ehsan, Selangor, Malaysia

Chamhuri Siwar
Institute for Environment and Development (LESTARI), University Kebangsaan Malaysia, 43600 UKM Bangi, Darul Ehsan, Selangor, Malaysia

Shaharuddin Mohamad Ismail
Institute for Environment and Development (LESTARI), University Kebangsaan Malaysia, 43600 UKM Bangi, Darul Ehsan, Selangor, Malaysia

Nurul Hidayah Chamhuri

Institute for Environment and Development (LESTARI), University Kebangsaan Malaysia, 43600 UKM Bangi, Darul Ehsan, Selangor, Malaysia

Moritz Albrecht
Department of Geographical and Historical Studies, University of Eastern Finland

Tiemo Kahl
Chair of Silviculture, Faculty of Environment and Natural Resources, University of Freiburg, Germany

Jürgen Bauhus
Chair of Silviculture, Faculty of Environment and Natural Resources, University of Freiburg, Germany

AbhishekChaudhary
Institute of Environmental Engineering, ETH Zurich, 8093 Zurich, Switzerland

Zuzana Burivalova
Woodrow Wilson School of Public and International Affairs, Princeton University, 08540 Princeton NJ, USA

Lian Pin Koh
Environment Institute, and School of Biological Sciences, University of Adelaide, 5005 Adelaide, Australia

Stefanie Hellweg
Institute of Environmental Engineering, ETH Zurich, 8093 Zurich, Switzerland

Mauro Maesano
Department of Biosciences and Territory, University of Molise, Italy

Bruno Lasserre
Department of Biosciences and Territory, University of Molise, Italy

Marco Marchetti
Department of Biosciences and Territory, University of Molise, Italy

Yayuk Siswiyanti
Post Graduate School on Forest Policy, Forest Management Science, Bogor Agricultural University Academic Ring Road Campus IPB Dramaga, Bogor 16680 Indonesia

Dudung Darusman
Forest Management Science, Bogor Agricultural University, Academic Ring Road Campus IPB Dramaga, Bogor 16680 Indonesia

Hariadi Kartodihardjo
Forest Management Science, Bogor Agricultural University, Academic Ring Road Campus IPB Dramaga, Bogor 16680 Indonesia

Iin Ichwandid
Forest Management Science, Bogor Agricultural University, Academic Ring Road Campus IPB Dramaga, Bogor 16680 Indonesia

Preface

Sustainable forest management deals with the increasing socio-economic demands of growing human populations while maintaining ecological function of forest ecosystems. Forest management deals not only with conservation of biodiversity of forests, but also its role in carbon recycling, climate change and ecosystem based management of broader landscapes, which can be well beyond the confines of the forests. The conservation of biological diversity of forests in an ecologically sustainable way requires successful design of management strategies, especially when the relationships between potential indicator species and biodiversity are not well established. The strategies required for sustainable management of forests are based on the fact that human interactions with the environment are required to be modified to ensure the sustenance of healthy ecosystems. This book presents the studies for sustainable biological diversity considerations in forestry and in consequence strategies applicable for future forest management

Editor
Jaspreet Banga, Ph.D.

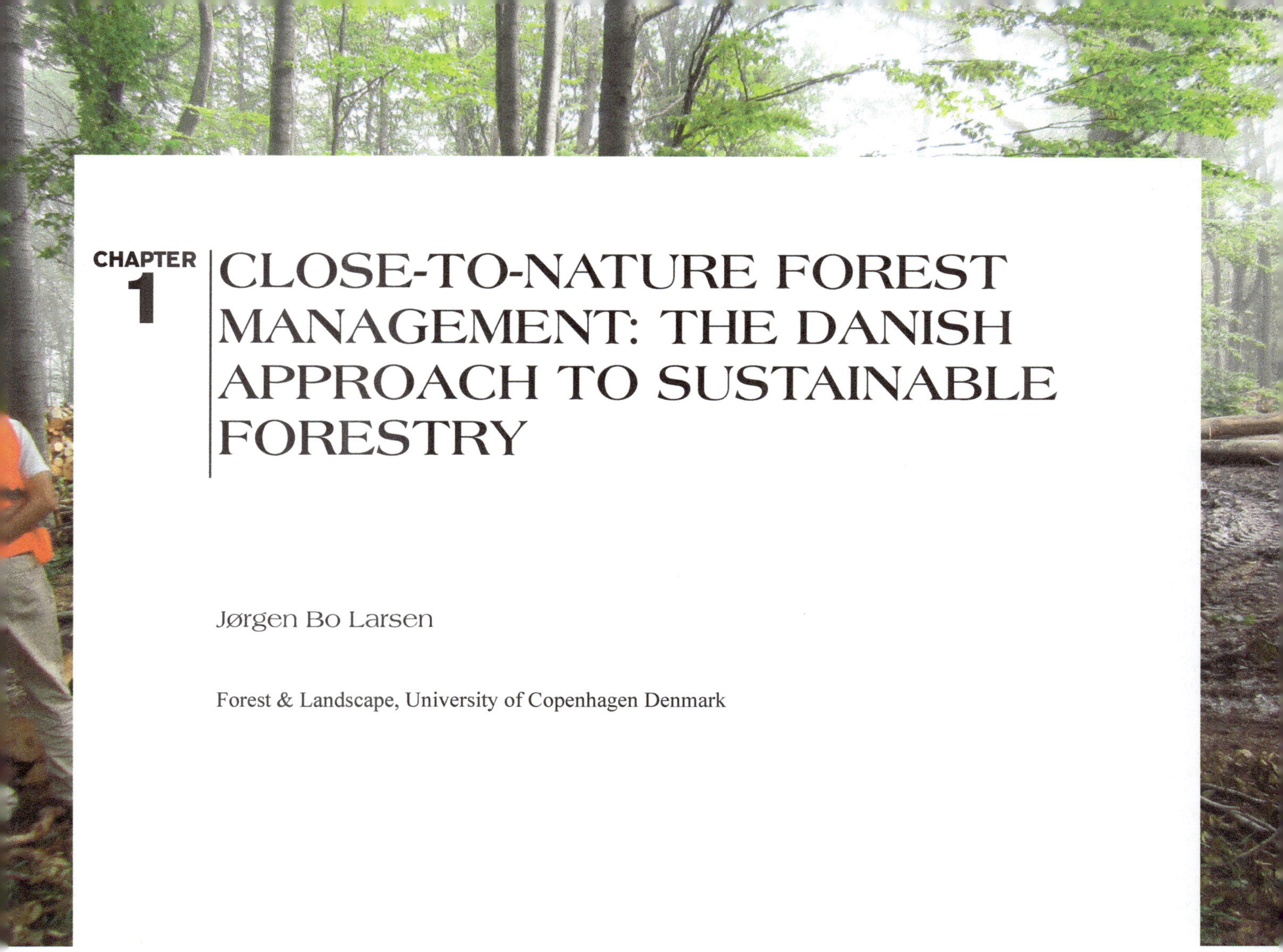

CHAPTER 1

CLOSE-TO-NATURE FOREST MANAGEMENT: THE DANISH APPROACH TO SUSTAINABLE FORESTRY

Jørgen Bo Larsen

Forest & Landscape, University of Copenhagen Denmark

INTRODUCTION

How should tomorrow's forests look and which future climatic conditions should they prevail? What kind of goods, services and experiences should they be able to provide; what kind of functions should they be able to perform? These are some of the multifaceted questions forest management faces today.

Forestry policy objectives have grown into a broad range of benefit provisions, other than serving exclusively as the traditional timber suppliers. Today we thus address multiple-use forestry. Production of wood commodities and securing carbon storage is central, but does not necessarily rate above the creation of non-timber forest products. Increasingly highly esteemed qualities, such as protecting landscape amenities and cultural heritage, nature conservation and environmental protection, as well as the entire chapter of recreational interests are considered. Consequently, economic and technical efficiency is still prioritized, but ecological and social parameters are progressively taken into account to ensure the multiple use.

Citation: Jørgen Bo Larsen (2012). Close-to-Nature Forest Management: The Danish Approach to Sustainable Forestry, Sustainable Forest Management - Current Research, Dr. Julio J. Diez (Ed.), InTech, DOI: 10.5772/30354.

For these reasons, silvicultural strategies are required to develop economically productive forests with a high potential for nature conservation, ecosystem protection, and social values. One promising management strategy is to incorporate structural qualities and functional features of natural forest ecosystems – "to follow and assist nature in her development" as already stated 230 years ago by the Danish forester von Warnstedt (Decree of 1781 regarding the management of the Royal forests). This approach can be summarized by the term "nature-based silviculture" or "close-to-nature management" (Gamborg & Larsen, 2003). In North America, on a more general forest management level, 'ecosystem management' and 'adaptive management' can be recognized as part of this trend (Franklin et al., 2002). The aim is to reform current practices so that they are still profitable, but more environmentally benign and even more sensitive to the complexities of nature conservation and the multiple, varying and steadily increasing demands of society by mimicking natural forest structures, their processes as well as their dynamics (Angelstam et al., 2004; Lindenmayer, et al., 2006; Hahn et al., 2007; Larsen et al. 2010).

The disturbances and regeneration processes in natural forest ecosystems, which cause structural heterogeneity at both large and small scale levels are linked to regional characteristics of climate, soil, and species compositions. These processes are being expressed as e.g. infrequent, large-scale storm and fire-driven disturbances in boreal ecosystems and as frequent, small-scale disturbances in Central-European forests. Hence models, describing the region-specific disturbance patterns, such as the forest cycle model, should be used in the development of applied silvicultural methods in such natural ecosystems (Hahn et al., 2005).

In central and western Europe the forest cycle models have been successfully applied to describe the temporal and spatial dynamics and cyclic preoccupation of a specific forest type in natural forest reserves (Leibundgut, 1984; Christensen et al., 2007; Larsen et al., 2010). Such models could serve as an adequate basis for close-to-nature forest management.

The use of natural disturbance regimes to guide human management (i.e. thinning and cutting systems) must, however, be complemented with other measures to restore naturalness in forest management. Lindenmayer et al. (2006) emphasize the importance of maintaining aquatic ecosystem integrity for biodiversity protection in managed forests. Hence, maintaining and restoring natural hydrology in forests previously subjected to stand management operations (such as drainage) is important. Therefore promoting species and forest structures that reflect and emphasize the variation in hydrology is an integral part of close-to-nature management, thus contributing to habitat richness in forest landscapes.

One of the basic axioms of nature-based forestry is the mimicking of natural structures and processes in order to obtain a high degree of stability within the ecosystem and thus a high degree of flexibility. All of this necessary to opening up for possible future demands and needs from various players, such as landowners, interest groups and society in general. The logic of this assumption might be best illustrated by considering the contrary position: *without stability – which functions will we be able to sustain in the future.*

One of the major problems experienced with the classical forestry approaches, is the lack of ecological and structural stability and the limited flexibility, towards addressing various at present unknown future demands (who would 20 years ago have predicted the present focus on biodiversity in forest management?). An approach, which has alerted us of the necessity to search for better management systems aiming at increasing functionality as well as flexibility in forest ecosystems; both in relation to multiple uses. In other words: We seem to face a high potential for both ecological adaptability (resilience) and functional flexibility of forest ecosystems, when opening up for greater functional integration – a central aspect of close-to-nature management.

To illustrate the differences between the traditional plantation approach, the close-to-nature approach, and a strict nature protection approach, a "goal-fulfilment assessment" and comparison of the three different management approaches is shown in table 1.

Table 1. Different management approaches and their respective fulfilment of different specific management goals. The scale from 1 to 5 plusses, and '+' = low goal fulfilment, whereas '+++++' = high goal fulfilment

Management approach	**Plantation (production) approach**	**Nature-based (integrative) approach**	**Nature protection (conservation) approach**
Specific management goals	Focus on timber production and direct economic outcome	Flexible wood production, nature protection and recreation	Strict forest reserves following natural structures and processes
Production of timber	+++++	++++	+
Economic outcome, long term	+++	+++++	+
Economic outcome, short term	+++++	+++	+
Production of quality timber	++++	++++	+
Biodiversity protection	+	+++	+++++
Protection of wetlands	+	+++	+++++
Ecosystem integrity	+	++++	+++++
Aesthetic qualities	+	+++++	+++++
Landscape integration	++	++++	+++++
Historical and cultural values	+	++++	+++
Space for public recreation	++	++++	++
Place of quietness/meditation	+	+++	+++++
Hunting qualities	+++	++++	+
Robust and resilient forests	+	++++	+++++
Flexibility to changing goals	+	+++++	+

Table 1. Different management approaches and their respective fulfilment of different specific management goals. The scale from 1 to 5 plusses, and '+' = low goal fulfilment, whereas '+++++' = high goal fulfilment.

Table 1 indicates that the plantation approach and the conservation approach both are rather narrow and inflexible in their goal-fulfilment, while the nature-based wood production approach is broad and flexible in its goal-fulfilment. The weaknesses of the plantation approach focusing on timber in short rotations and neglecting most natural, cultural and social values are clearly reflected in the table. Further, the plantation approach often leads to less robust forests stands. The conservation approach obviously performs strongly in all nature protection goals but is consequently not, or less able, to deliver on socio-economic goals thus scoring rather low in terms of flexibility to changing goals.

The maintained focus on production economy in combination with relative high scores in ecological as well as social values addressing the needs for stability, explains why the nature-based approach to sustainable forest management has been chosen in many countries. The integrative ability and flexibility of the nature-based approach to fulfil different management goals is a key feature of this management type. Because of this feature, it is possible to gradually adjust the course of management to address the ever-changing objectives and aspirations of society.

THE HISTORY OF NATURE-BASED FOREST MANAGEMENT – IN SHORT

In Europe there have been attempts and local traditions to literally follow nature-near principle, to follow and steer the natural development in order to meet some more or less specific goals (Leibundgut, 1984; Schütz, 1990; Otto, 1993). However, the main trend in European forestry has followed the principles of organized forestry with a strong emphasis on clear-cutting, planting, thinning, homogenization of structures, as well as rationalization of working procedures. Organized forestry has had longstanding strong advantages in terms of overview, planning, standardization, prediction and control.

While organized forestry has become the dominating concept in most parts of Europe, the more nature-inspired forestry approaches have been left to survive in the shade. Such concepts have not been given much attention, nor has much research been carried out to highlight possible advantages of this branch of silviculture. The ideal "to follow and assist nature in her development" has often been cited – but in reality rarely been followed in practice. Until recently, most attempts to apply nature-based forestry have been mainly exceptions from the rule. They have been carried out under special conditions and have been conducted by individuals mainly driven by conviction. A belief which has led to the assumption that nature-base forest management could turn out to be a more promising approach than traditional plantation forestry. People practicing nature-based forestry have thus in the recent past often been given the image of being some kind of "religious freak" (Heyder, 1986).

Close-to-nature forestry is unquestionably focused around the idea of selection forest. The single tree and group selection system represent a clear contrast to the even-aged forests of organized forestry. Many foresters have tried to develop such uneven-aged mixed forests and have searched for appropriate methods to evaluate management successes, in order to compare them objectively to even-aged systems. The French forester Adolphe Gurnaud (1825-1898) once succeeded with the French *Méthode de contrôle*. His method based on regular inventories of forests parameters, especially diameter distribution and increment. Although not successful in implementation of his ideas, Henri Biolley (1858-1939) later succeeded in managing the community forest of Couvet with this "modern" selection system (Biolley, 1920).

Another important source of nature-based forestry started around the ideas of Karl Gayer (1822-1907), a silviculture professor in Munich. At that time, organized forestry with clear-cut systems and introduction of conifers had already expanded over large forest areas. Consequently, following this process, soil degradation, fungi and insect outbreaks, as well as frequent windbreaks had been observed in those areas. As a reaction, Gayer then developed his idea of mixed forests, which were about to be achieved merely through natural regeneration (Gayer, 1886), often in combination with the irregular shelterwood system. Using irregular regeneration over a longer time-span would thereby enable various different species to establish and thereby creating mixed forest structures.

His ideas were further developed in Switzerland. At that time Swiss forests suffered severely from torrents, landslides and windbreaks, as a result of spruce monocultures and clear-cut management systems. Arnold Engler (1858-1923) succeeded in gradual change of the Swiss forestry paradigm, which was untied from the regeneration scheme of organized forestry.

Today, variations of the Swiss irregular shelterwood systems are the most widely applied nature-based silvicultural systems all over Central and Eastern Europe. This is mainly thanks to the great flexibility of the system, which is based on the principles of adapting the felling temporally and spatially to the regeneration ecology of various tree species. Apart from the selection system, the irregular shelterwood system for nature-based forestry and the "free-style silvicultural technique" are significant as well; especially when it comes to managing degraded forests or transforming uniform and even-aged forests into mixed uneven-aged forests.

The third "wave" of nature-based forestry has developed arund 1920 in northern Germany when Alfred Möller published the book "Der Dauerwaldgedanke" (Möller, 1922). His paradigm of a continuation forest differs essentially from other nature-based concepts. Möller's approach is based on an organismic and holistic conception of forests and it follows stricter felling rules. His ideas had been shaped in forests where careful, continuous forest cover forest management had been applied for many years (Bärenthoren in eastern Germany). Möller carries out different inventories and publishes his results in favour of continuous forest cover management (German: Dauerwald), which, according to his conviction, offer improved forest sites, abundant regeneration as well as increased wood production.

Möller's forest approach was welcomed with great sympathy during the first years after his book had been published. 'The Dauerwald concept' was embraced with great enthusiasm all over Germany. When Möller died soon after publishing his book and his ideas proved unable to deliver the hoped success in the field, his approach became increasingly questioned and in the end even doubts about his scientific credibility ended this chapter of nature-based forestry in Germany during the 1930's.

With the foundation of a working group for close-to-nature forestry (Ger. Arbeitsgemein-schaft naturgemäße Waldwirtschaft - ANW), in 1950 yet another force steps onto the forest management scene. The ANW was rooted in the Dauerwald movement, and the groups ideas based on a set of principles rather than a management system. The group's members are mainly practicing foresters and forest owners. Decisions on how to manage forests and strategies are empirical and often intuition based.

The call for for expanding the ANW-movement outside Germany resulted in the foundation of Pro Silva Europe in Slovenia in 1989. Pro Silva advocates close-to-nature forest management based on natural processes. Most European countries (at present 24) have joined and established national, independent Pro Silva sub-organizations. Their common ground on the national level is to develop and promote the principles of sustainability. These principles are considered to allow for the full development of the forests ecological and social roles, while a simultaneous economic production of high quality forest products can take place - all by mimicking natural processes. Members are forest owners, foresters, students and others who wish to practice and learn more about nature-near forestry.

The Toolbox of Classical and Nature-Based Forestry

Basically classical plantation silviculture and close-to-nature approaches make use of the same toolbox in managing forests. However, the importance of single tools differs between the two concepts. Table 2 illustrates how different silviculture tools can be applied and combined under different management approaches. The plantation approach is displayed in two versions: traditional and modified (to achieve a higher degree of sustainability). Accordingly, the nature-based approach is displayed in a more economic, and a conservation focused version.

Table 2 shows how management approaches determine what tools might be appropriate and most widely used. It further shows that although it is meaningful to differentiate between the various management approaches, it is neither possible, nor is it meaningful to draw a strict watershed line between those definition categories. Naturally transgression corridors occur. For each strategy however, it is possible to provide a set of relevant silviculture tools. Depending on management styles and aims within plantation, respectively nature-based management, the relative importance of the different tools can be adjusted. Each forest owner and each policy maker must critically choose his or her favorite tools for the situation and objectives which are being focused upon.

Table 2. Examples of silviculture tools and anticipated stand structure and their relative importance in nature-based as well as classic (plantation) forest management: +++++ greatly used, ++++ frequently used, +++ regularly used, ++ rarely used; + hardly ever used

Silviculture tool / Anticipated stand structure	Traditional plantation (production) approach	Modified (sustainable) plantation approach	Nature-based economic production approach	Nature-based nature conservation approach
Clear cutting at rotation age	+++++	++++	++	+
Single tree/group cutting	+	++	+++++	+++++
Planting or sowing	+++++	++++	++	+
Natural regeneration	+	++	++++	+++++
Use of soil preparation	+++++	++++	++	+
No soil preparation	+	++	++++	+++++
Use of pesticides	+++++	++	+	+
Ban of pesticides	+	++++	+++++	+++++
Use of exotic species	+++++	++++	+++	+
Use of native species	+	++	+++	+++++
Stand management	+++++	++++	++	+
Single tree management	+	++	++++	+++++
Harvest when ripe	+++++	++++	++++	+
Preserving old trees	+	++	++	+++++
Wood salvage	+++++	++++	++++	+
Leaving dead wood	+	++	++	+++++
Draining for production	+++++	++++	+++	+
Maintain wet habitats	+	++	+++	+++++
Monoculture	+++++	++++	++	+
Species mixtures	+	++	++++	+++++
Even-aged stands	+++++	++++	++	+
Uneven-aged stands	+	++	++++	+++++

Table 2. Examples of silviculture tools and anticipated stand structure and their relative importance in nature-based as well as classic (plantation) forest management:
+++++ greatly used, ++++ frequently used, +++ regularly used, ++ rarely used;
+ hardly ever used

Nature-based approaches in general refrain from larger clear cuts, but in specific cases - often in order to promote light demanding (pioneer) species - clear-cuts can be applied. Nature-based management relies heavily on natural regeneration but includes planting or direct seeding if natural regeneration is insufficient and/or if desired species are missing (enrichment planting). Nature-based approaches often make use of single tree selection based on target-diameter cutting, which should not be misinterpreted as "high grading" known from overexploitation of natural stands. Thus we here focus on a system to provide a sustained yield by making thinning among the various age classes in order to ensure their desired proportions and to maintain a suitable mixture of species. It should further be stressed that the heterogeneity of the stands is not just an end in itself, but rather a way of allocating species to various soil conditions and creating good forest floor conditions for natural regeneration.

The toolbox concept implies the refrain from any specific (religious) interpretation of what nature-based forest management is or should be - rather, the toolbox should be open to anyone finding the tools appropriate for any use he or she might wish for. The tools from this nature-based toolbox can be used for nature-protection, for wood-production or to develop new types of urban forest (Larsen and Nielsen, 2011).

However, the main prerequisite for defining an approach "nature-near" or "close-to-nature" should be *that the practices are founded in, or inspired by, the structures and processes that occur in natural forests of a specific (reference) region.* This principle can be used to achieve all kinds of different management goals and objectives, including timber production, nature protection and social values.

CLOSE-TO-NATURE FOREST MANAGEMENT IN DENMARK

The forests in Denmark amount to a total area of 570.800 ha, equivalent to some 13 percent of the total land cover. Originally, most of the land has been forested, but after centuries of uncontrolled logging and deforestation for agriculture, forest areas begun to decline drastically and consequently collapsed to a mere 2 to 3 percent around the 1820's. Since then the forest area is increasing due to large forestation efforts from 1860 and onwards and expected to reach around 20 % within this century.

Originally the Danish forest consisted mainly of deciduous trees - especially beech and oak. Over the past 200 years of forest management - including the large forest plantings in Central and West Jutland – the species distribution changed radically. Today, more than 50 % of forested areas are covered with non native conifers such as Norway and Sitka spruces, Douglas fir, as well as different Abies-species. Deciduous forests cover not more than 44 %, with beech and oak as the most common species, and ash, sycamore maple, Norway maple, birch, alder, wild cherry and lime as minor species.

As a general trend, forestry in Denmark has followed the overall European development when focussing on timber production in mostly plantation like structures. As a result, highly productive forests have been promoted, a process, which simultaneously created the matrix for increasingly intense conflicts with nature protection interests. First and foremost, the stability of the forests suffered through the development of even-aged monocultures. During the last 40 years, in 4 storms (1967, 1981, 1999, and 2005) a total of 15 million m^3 were blown down, whereas "only" 1 million m^3 fell down during the first 60 years of the past century. Hence, a major reason for the increasing impact of storms in Danish forests is the increasing use of storm sensitive conifer species.

In order to realize sustainable forestry at the management unit level (to achieve a proper balance between economic, ecological and social functions), a set of overall aims and operational guidelines has been developed in a stakeholder driven process during 2001. The National Forest Programme (Skov- og Naturstyrelsen, 2002) now consequently prescribes that Danish public forests should be managed in accordance with close-to-nature principles. The essence in these close-to-nature principles can be summarized as follows: Increase the stability and prepare the forests for an unknown future of changing climate, changing values and a variety of goals.

This close-to-nature approach is in particular focused on:

- Creating optimal conditions for natural regeneration by maintaining the permanent forest climate by refraining from clear-felling.
- Stability improvement and risk diversification (resilience) through the creation of uneven aged mixed forest stands of site-adapted tree species.
- Active stand improvement through frequent and weak thinning.
- Protection of natural equilibriums among forest organisms, including pests, with the aim of promoting biodiversity and avoid the use of pesticides.

The close-to-nature forest management, combined with an increased use of climate robust deciduous and coniferous species and the reduction of climate change intolerant conifers (i.e. Norway spruce and Sitka spruce), are here identified as the overarching principles to secure sustainability, safeguard stability, and prevent the negative effects of climate change. Consequently, The Forest Act from 2004 supports the change from classical mono-species and even-aged management of stands into close-to-nature management characterized by more single tree and group management, incorporating and supporting natural regeneration and structural differentiation.

This decision to transform "classical" age-class forests (plantation forestry) towards nature-based forest stand structures implied no less than a paradigm shift in the management of state owned forests. Realizing that the complex character of these near-natural forest structures and dynamics require integrative and flexible management frameworks, as well as tools, a two step process was established: Firstly, the need for defining and describing long term goals for nature-near stand structure and dynamics was recognized and taken into the picture (where are we going?). Secondly, methods for transformation from plantation to nature-near structures were specified (how do we get there?).

The Long-Term Goals – Creating Forest Development Types (FDT)

The concept "Forest Development Type" (FDT) was considered as an adequate framework for advancing and describing long-term goals for stand structures and dynamics in stands subjected to close-to-nature management (Larsen and Nielsen, 2007). An

FDT describes the direction for forest development on a given locality (climate and soil conditions) in order to accomplish specific long-term aims of functionality (ecological-protective, economical-productive, and social-/cultural functions). It is based upon an analysis of the silvicultural possibilities on a given site in combination with the aspirations of future forest functions. It will serve as a guide for future silvicultural activities in order to "channel" the actual forest stand into the desired direction. Such a common understanding and agreement upon the desired development is crucial, since the conversion from age-class to nature-based stand structures is a continuous process.

In Denmark, a participatory process lead and described by Larsen and Nielsen (2007) resulted in the creation of 19 FDT's, which can be grouped into 9 broadleaved dominated, 6 conifer dominated, and an additional 4 "historic" types (Table 3). Whereas all "nature-based" FDT encompass a balance between productive, protective and recreational/social functions, the other four "historical" types mainly serve to protect recreational, natural and cultural functions. Especially the historical Forest Pasture (FDT No. 92) and Forest Meadow (FDT No. 93) can be actively used to create habitat diversity and experiential richness in forest landscapes.

Each FDT is described as follows (See also Figure 2, describing FDT No. 12 "Beech with ash and sycamore"):

- Name: The name encompasses the dominating and co-dominating species. The first digit in the FDT-number indicates the main species (1 = beech, 2 = oak, 3 = ash, 4 = birch, 5 = spruce, 6 = Douglas fir, 7 = true fir, 8 = pine, and 9 indicating a "historic" FDT). The second digit is numbered at random.
- Structure: A description of how the forest structure could appear when fully developed. This description is supplied with a profile diagram depicting a 120 m transect of the anticipated forest structure at "maturity" (In Figure 1 profile diagrams of all 19 FDT´s are displayed and in Figure 4 the profile diagrams of four FDT´s are with different forest-edge types shown: No. 11-Beech, No. 21-Oak with ash and hornbeam, No. 71-Silver fir and beech, and No. 92-Forest pasture).
- Species distribution: The long-term distribution of species and their relative importance.
- Dynamics: The regeneration dynamics described in relation to the expected succession and spatial patterns (species, size).
- Functionality: Indication of the forest functionality (economic-production, ecologic-protection, and social/cultural functions).
- Occurrence: Suggested application in relation to climate and soil. For this purpose the country is divided into 4 sub-regions each with their typical climatic characteristics. Further, the application of the specific FDT in terms of soil conditions is stated in relation to nutrient and water supply.

Table 3. The 19 Danish Forest Development Types

Broadleaved dominated:		Conifer dominated:	
11	Beech	51	Spruce with beech and sycamore
12	Beech with ash and sycamore	52	Sitka spruce with pine and broadleaves
13	Beech with Douglas fir and larch	61	Douglas fir, Norway spruce and beech
14	Beech with spruce	71	Silver fir and beech
21	Oak with ash and hornbeam	81	Scots pine with birch and Norway spruce
22	Oak with lime and beech	82	Mountain pine
23	Oak with Scots pine and larch	**"Historic" forest types:**	
31	Ash with alder	91	Coppice forest
41	Birch with Scots pine and spruce	92	Forest pasture
		93	Forest meadow
		94	Unmanaged forest

Matching Forest Development Types to Site

While different forest development types possess different site requirements it is possible to address and utilize potential variation in site conditions by matching FDT to site. This requires a thorough site survey, in which analyzing the basic growth conditions such as geology and soil types, nutrient and water supply, as well as specific site factors (such as compact layers and insufficient drainage) are taken into account. A hydrological status analysis on site is necessary, and it should include a survey of existing drainage systems, in combination with a plan of the historic landscape with former wet-lands, prior to any draining process. This hydrological analysis will provide an important tool and inspiration for delineating the landscape into ecological functional units. The site classification map works correspondingly as a frame for applying FDT to the site, thus facilitating the creation of forested landscape where site adapted forest and nature types reflect and emphasize variations within landscape. Further, different FDTs possess different combinations of goal fulfilment - some are more production oriented, some more oriented towards nature/biodiversity protection, while others focus on enhancing landscape and recreational values. This variation in goal achievement can correspondingly be used to select FDTs - all according to specific functional requirements defined by the forest owner and - in case of public forest - by society/interest groups.

Figure 1. Profile diagrams of the 19 Danish Forest Development Types.

At present the Nature Agency, responsible for the management of the Danish state forests, is laying out a grid system of forest development types on all public forests. This grid system will provide the local forest manager with information about the long-term goals he should aim at in each and every part of "his" forest. The managers job as local silviculturist will consequently be to observe the natural development and only then, after having conducted his observational research, to start making adjustments (cutting, planting, weeding, fencing, soil scarification etc.) in case the stand is due for short-term economic intervention (commercial thinning) and/or the actual development compromises the long term goal, as described in the attributed forest development type.

As mentioned above, the process of marking out FDTs on a management unit level is at present ongoing in Denmark. To illustrate this process, as well as the outcome, an example will be shown below. This example inspects the FDT-plan for the eastern part of Vestskoven as proposed by a group of students attending the international master course in Urban Woodland Design and Management (plan described in detail in Larsen & Nielsen, 2011).

Vestskoven was established in the 1960´s west of Copenhagen to create a large recreational forest that could separate and structure the intense and rapid urban sprawl, and provide for important recreational qualities for the 300.000 new citizens in the western parts of Copenhagen. Fields were planted successively as they were purchased; little consideration was given to the overall composition and interlock zones between stands, or those parts dividing forested from open areas. The fields were planted according to traditional manuals with monoculture stands or simple species mixtures, using the species that were available at nurseries. The forest thus consists of small stands with abrupt species transitions and edges, all together lacking valuable interlock zones between the forested and more open areas. Today the area functions as a traditional Danish timber production forest with some large open spaces for recreation sprinkled onto it (Figure 3).

Forest Development Type 12: Beech with ash and sycamore

Structure: Species rich, well structured forest with beech as dominating element mixed with ash and cherry and in south-eastern Denmark additionally with hornbeam and lime. The in-mixed species occur mainly in groups. The horizontal structures arise between groups of varying size and age. Where the light demanding species such as ash, sycamore and cherry dominate, vertical structures occur periodically with shade trees (beech, hornbeam, elm, and others) in sub-canopy strata.

Species: Beech. 40 – 60 %, ash and sycamore: 30 – 50 %,
cherry, hornbeam, oak, lime, and others up to 20 %

Dynamics: Beech regenerates mainly in groups and smaller stands. Ash and sycamore as gap specialists regenerate in openings later followed by beech. Hornbeam belongs to the sub-canopy stratum and regenerates under shade, whereas the pioneer species (cherry and oak) only regenerate after lager openings and/or in relation to forest edges.

Functions: Productive: The forest development type has a high potential for production of hardwood in larger dimensions and of good quality.

Protective: In most parts of the country the beech dominated forest represents the potential natural vegetation; consequently, many indigenous species are connected to this forest development type. It has a great potential for conserving biodiversity connected to the NATURA 2000 habitat type 9139 and 9150.

Recreational: Through ist mixture of (indigenous) species in combination with pronounced variation in size the forest development type gives a multitude of recreational experiences and intimacy.

Occurrence: The forest development type belongs on protected sties in the eastern and northern parts of Denmark on rich, well drained soils with good water supply as illustrated below.

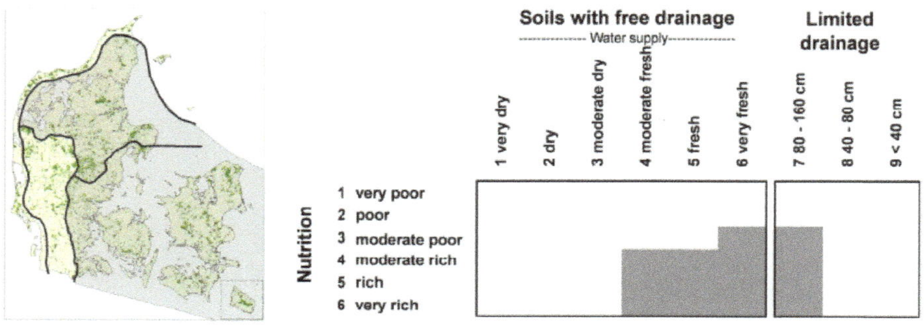

Figure 2. Description and illustration of Forest development Type 12: Beech (*Fagus sylvatica*) with ash (*Fraxinus excelsior*) and sycamore (*Acer pseudoplatanus*).

Figure 3. Photo (from east towards west) of the eastern part of Vestskoven, showing the fragmented composition of uniform blocks of geometrically shaped stands and open spaces.

The above description demonstrates that Vestskoven incorporates most of the potentials, but even many problems, which urban woodlands inherited from the commercial forest management tradition with its uniform stand structures and its fragmented blocks of geometrically formed stands and open areas. The absence of smaller openings and glades, and the lack of valuable wetlands thus mould a fragmented, disconnected forest landscape.

Since Vestskoven is a public forest it will be managed according to close-to-nature principles and it is currently in the process of being charted into the FDT grid. Figure 4 presents a conversion/restoration plan where four Forest Development Types (FDT´s) have been laid out in respect of existing values in the young plantations and adjacent plains. The four selected FDT´s (FDT 11, Beech; FDT 71, Silver fir with beech and spruce; FDT 21, Oak with ash and hornbeam; FDT 92, Grazing forest), each with distinct experiential and ecological characteristics, unify the many small stands within larger units. The variety of size in open areas is increased by adding small, intimate glades in the forested parts. Some of the open areas have been linked to add further spatial variation and to increase coherence.

Parts of the forested, as well as the open areas have been converted into grazing forest through heavy thinning and some additional planting of trees. The borders between forested parts and open areas have been re-shaped organically by cutting out some of the existing stands, and instead giving room for edge species in those corridors. Thereby important interlock zones are being shaped between the denser forested and the more open areas, allowing for more diverse and complex edge structures. Ponds have been restored at emerging wetlands to render valuable landscape attractions, both in regard to landscape interpretation by visitors, as well as in regard to biodiversity in general. This landscape re-shaping takes place in the vicinity of small glades, at forested edges and in larger plains.

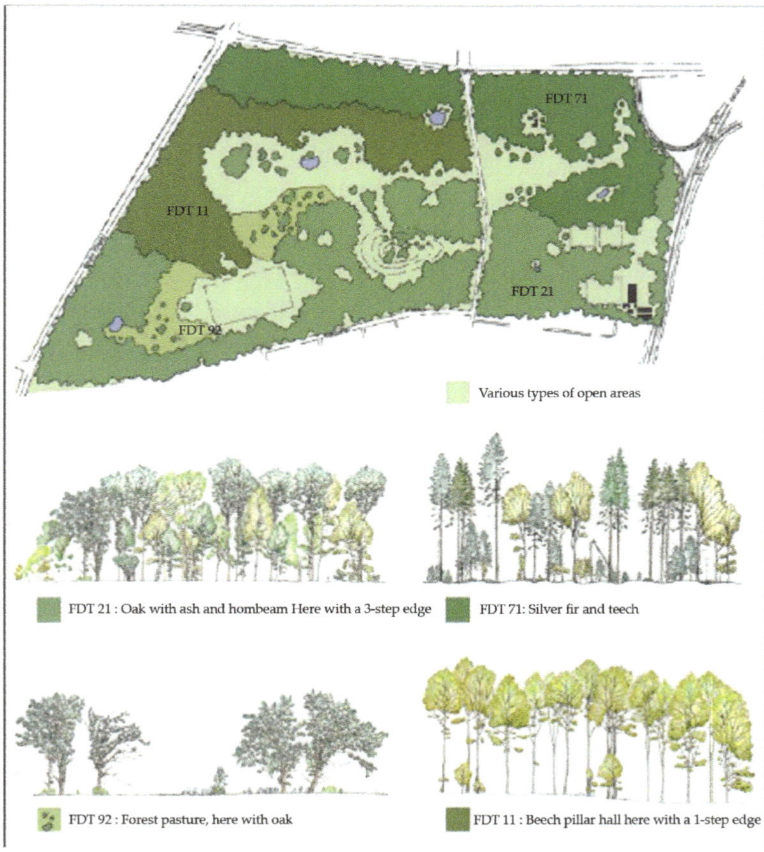

Figure 4. Restoration plan for the eastern part of Vestskoven:

This plan was developed by a group of students attending the international master course in Urban Woodland Design and Management (Larsen & Nielsen, 2011). The chart, in combination with the profile diagrams of the four FDT´s, including examples of different edge-types, gives an instant impression of the anticipated urban forest landscape goals.

Furthermore, it provides an outline for appropriate developments in different parts of the forest. Such a developed and augmented plan, in combination with an FDT-map and profile diagrams of the different forest development types applied, can be used in multiple participatory planning processes.

Conversion Principles and Methods

Having defined the long-term goal at each part of the forest, the practitioners' principal task is to "guide" the forest from the current structure toward the targeted FDT. To help the local manager in this new endeavor, a number of conversion models haven been developed through a participatory process with local practitioners, forest workers and entrepreneurs. The primary purpose of this process is to come up with ideas as to how the conversion of a number of typical output models toward the desired forest type of development can take place. Since the conversion of uniform stands of spruce and beech are the main challenges in the transition to close-to-nature forest management in Denmark, the emphasis is on models for these species. Therefore, it is important to emphasize that these models are intended only to be used as inspiration, and they will always have to be adapted to any local situation, as well as to the concrete economic and technical possibilities. Especially the pace, at which the conversion is to be preformed, must be thoroughly analyzed in regard to any economic aspects, paying special attention not to compromise expectation values for wood production in the transition phase. Therefore, in most cases, the full transformation to nature-near structures might take up to one or two tree generations.

Deciding on conversion strategy and tools there are two fundamentals, which must be kept in mind: Firstly, stand stability must be ensured and natural regeneration conditions must be improved. Thus creating various options and "freedom," timely to initiate rejuvenation (including bringing in new species), if required. Secondly, it is essential to initiate these elements at the ecologically and economically right moment in time. Thus, we speak of 1) a preparation phase, where the forest is stabilized and prepared for regeneration - mainly through selective thinning operations, and 2) a transformation phase, characterized by passive or active initiated regeneration, respectively by introduction of new species (and the procedure of ensuring their development). The preparation phase is usually associated with income (or at least cost neutrally implemented); whereas the transformation phase often entails costs (investments). Although, according to the principles of biological rationalization (a central economic aspect of close-to-nature-management), these costs could be kept on minimized levels by letting nature itself do as much of the "work" as possible.

If the forest development type prescribes species which are not present, or their genetic constitution (provenance) is not acceptable, additional seeding or planting (enrichment) in groups (typically, beech, ash, maple, birch, bird cherry, fir, larch, Douglas fir, etc.) is foreseen. These groups can later on contribute to a more widespread distribution of the species (done through seed dispersal). In order to allow rejuvenation of stable, but often frost-sensitive species (beech, firs including Douglas fir, etc.), a continued forest climate is regarded vital. Under such circumstances a stable forest canopy is paramount; especially in critically exposed, storm sensitive spruce monocultures. If a more complete conversion to new main tree species is aimed at already in the first generation, an extra widespread planting or seeding is envisaged, but often at higher cost. However, the close-to-nature approach is in general more inclined to exploit cheap regeneration methods, thereby accepting a longer conversion phase.

Generally, we distinguish between passive and active conversion strategies. The passive strategies are primarily based on existing vegetation, in order to convert as economically efficiently as possible. This implies mostly long conversion periods (up to several tree generations). The active approach is used where stability does not allow a slow (pending) conversion and/or there are other motives (ecological, aesthetical, and recreational) that advocate for a fast conversion.

Passive Strategies

The purpose of the passive strategies is to implement as low-cost rejuvenation as possible, while maintaining optimum production in the upper canopy and the area as a whole. Exhibit stands a high degree of stability; a passive strategy can be used that largely exploits the stand productive potential for transition to target diameter cutting without losing the possibility of a conversion. Transition phase can likewise extend over a long period of time, utilizing the system's own forces (natural regeneration), supplemented with scattered introduction of "new" species, if needed in the emerging gaps. Under such conditions, there are usually no major conflicts between the long-term objectives and the operating economy of the conversion phase. Gap size, and thus the potential light radiation, plays a crucial role in the choice of implanted species where they do not appear spontaneously. Thus, light demanding species such as larch, Douglas fir, oak, birch etc. require larger gaps (above 0.4 ha), while in the smaller gaps (0.1 - 0.2 ha), more shade tolerant species such as beech, maple and fir will be suitable.

Figure 5. Passive approach; Spontaneous regeneration of fir, spruce, birch, larch and Mountain ash in wind-throw gaps in a Norway spruce stand (group regeneration), Klosterheden Statsskovdistrikt. Photo: J.B. Larsen

Active Strategies

The active conversion approach is used under conditions, where lack of stability does not allow a passive conversion. Active strategies are used in unstable stands primarily of spruce. In potentially unstable stands which have not yet reached a height that makes them storm exposed (below approx. 14 m), it is important to conduct an active thinning to promote stability and structural variation. This can happen partly through an early shelterwood formation or by liberating a number of future trees, thereby creating stable single trees (anchor trees). Important is that the thinning is conducted "from above" (removing dominating and co-dominating trees) thereby promoting variation in tree size (diameter, height) and a more heterogeneous stand structure. Group felling, in combination with early introduction of regeneration are also examples of active strategies. It is common to these approaches that a portion of the potential production in the stand will be sacrificed to safeguard the success of regeneration. In some cases the only economically realistic approach for regeneration/conversion of unstable spruce stands will be a clear cut; a measure, which also can be considered as an active strategy. In situations, when clear-cutting is the only way to regenerate the stand, frost hardy pioneer species such as Scots pine, oak, larch and birch will be introduced by planting/sowing to supplement, to improve the frequent natural regeneration of spruce and birch, thereby increasing future silvicultural options and thereby successively moving towards the planned long term goal – the FDT.

Figure 6. Active approach; 9-year old beech planted under a canopy of Norway spruce (shelter-wood regeneration). Klosterheden Statsskovdistrikt. Photo: J.B. Larsen.

The choice of conversion strategy depend on the starting point including the potential stability of the concrete stand, the objective defined by the FDT, and the time available for the conversion according to the economic perspective of net-present values of anticipated functions, together with the conversion costs. In total 10 different conversion and regeneration models have been developed for converting monocultures of beech, spruce and oak into nature-near structures. In Figure 7 such two models are displayed by means of profile diagrams, depicting a possible development from a uniform plantation like structure towards the decided nature-near forest development type.

Figure 7. Conversion models displayed with profile diagrams. Left: Passive approach showing the conversion of a 54-year old even-aged beech stand to FDT 12 – Beech with ash and sycamore maple. The so-called "qualitative group cutting" is applied. Thinning is preformed by cut-

ting trees according to their quality disregarding an even distribution of the remaining trees. This will create openings in the closed beech stand, where ash and maple is introduced. The regeneration is completed by natural regeneration slowly creating a group-wise structure of beech, ash and maple.

Right: Active approach showing the conversion of a 24-year old Norway spruce plantation to FDT 61 – Douglas fir, Norway spruce and beech. The thinning regime aims at creating variation in the overall thinning density in the area. It is done to open up for creation gaps, to be filled with Douglas fir and beech. The rest of the area is regenerated naturally with spruce and birch.

CONCLUSION

The management of forests "closer to nature" has increased significantly in recent decades, simultaneously accompanied by ever more reliable and refined models, promoting its efficient implementation. The basic idea is to reach a better balance between productive, protective and social functions. Other important goals are to increase economic competitiveness by cost reduction and increase robustness to climate change.

In Denmark, the Nature Agency started to manage all public forest according to close-to-nature principles in 2005. To facilitate the transition from classical even-aged plantation forestry to close-to-nature silviculture a total of 19 Forest Development Types (FDTs) and different conversion models have been developed in a participatory process with forest practitioners, scientists, forest workers, contractors and other stakeholders.

Now, almost 10 years after the political initiation, and 6 years after the state forest once started to be managed according to close-to-nature principles, the picture is multifaceted: The conversion process in the state forests is continuing with special focus on developing nature rich recreational forest landscapes, by means of the FDT planning scheme. A massive effort to restore natural hydrology is one of the most significant ingredients in the process; as well as the integration of permanent open spaces in the forest (forest meadows – FDT 93), the introduction of grazing animals (forest pasture – FDT 92), and the delineation of larger reserves (unmanaged forest - FDT 94). Furthermore, different methods and models for converting spruce plantations have been used. Still, it seems too early to draw any final conclusions in regard to his last aspect. The lack of funding for a scientific follow-up is a potentially jeopardising aspect.

Many forests belonging to municipalities have also changed management strategies fundamentally and they now apply the close-to-nature silviculture guidelines. Especially the FDT planning tool-box has proven highly effective to generate discussion platforms to define goals and ways of forest management among various stakeholders in urban forests.

The private forest sector is still rather reluctant in applying close-to-nature management. Some forest owners are doing it with great enthusiasm, while a majority still sticks to the classical age-class plantation system. However, the running debate about the pros and cons has had its effect on the size of clear-cuts and the use of natural regeneration.

We are learning by doing: Some of the pending issues are: How much reduction in professional input/contribution is possible without losing the advantages of close-to-nature management? To what extend is it possible to educate private forest contractors to apply close-to-nature silviculture with their big machines? Is it possible to create the same high wood quality in un-even aged forest systems as in plantation like structures – and to what costs? How can the close-to-nature managed forest cope with the increased need for bio-energy production?

ACKNOWLEDGMENTS

The author wants to thank Dagmar Nordberg for valuable contributions and Alan & Jane Newbury for proof reading.

REFERENCES

1. Angelstam, P.; Boutin, S.; Schmiegelow, F.; Villard, M.-A.; Drapeau, P.; Host, G.; Innes, J.; Isachenko, G.; Kuuluvainen, T.; Mönkkönen, M.; Niemelä, J.; Niemi, G.; Roberge, J.-M.; Spence, J. & Stone, D. (2004). Targets for boreal forest biodiversity conservation – a rationale for adaptive management. *Ecoogical Bulletins.* Vol. 51: pp 487–509.
2. Biolley, H. (1920). L'aménagement des forets par la méthode expérimentale et spécialement la méthode du contrôle. *Beiheft Schweizerischer Forstverein,* Vol. 66, 1980, pp 51-134.
3. Christensen, M.; Emborg, J. & Nielsen, A.B., (2007). The forest cycle of Suserup Skov - revisited and revised. *Ecological Bulletins,* Vol. 52, pp 33-42.
4. Franklin, J.F.; Spies, T.A.; Van Pelt, R.; Carey, A.B.; Thornburgh, D.A.; Berg, D.R.; Lindenmayer, D.B.; Harmon, M.E.; Keeton, W.S.; Shaw D.C.; Bible, K. & Chen, J. (2002). Disturbances and structural development of natural forest ecosystems with silvicultural implications, using Douglas-fir forests as an example. *Forest Ecology and Management,* Vol. 155; pp 399-423.
5. Gamborg, C. & Larsen, J.B. (2003). 'Back to nature' a sustainable future for forestry? *Forest Ecology and Management*, Vol. 179, pp. 559-571.
6. Gayer, K. (1886). *Der gemischte Wald.* Verlag von Paul Parey, Berlin, 168 pp.
7. Hahn, K.; Emborg, J.; Larsen J.B. & Madsen, P. (2005). Forest rehabilitation in Denmark using nature-based forestry. In Stanturf J.A., and Madsen P. (eds.): *Restoration of boreal and temperate forests*. CRC Press, 299-317.
8. Hahn, K., Emborg J.; Vesterdal L.; Christensen S.; Bradshaw R.H.W.; Raulund-Rasmussen K. & Larsen J.B. (2007): Natural forest stand dynamics in time and space - synthesis of research in Suserup Skov, Denmark and perspectives for forest mangement. *Ecological Bulletins*. Vol. 52, pp. 183-194.
9. Heyder, J.C. (1986). *Waldbau im Wandel.* J.D. Sauerländer's Verlag, Frankfurt am Main. Larsen, J.B. & Nielsen, A.B. (2007): Nature-based forest management – where are we going? Elaboration forest development types in and with practice. *Forest Ecology and management,* 238, 107-117.

10. Larsen, J.B.; Hahn, K. & Emborg J. (2010). Forest reserve studies as inspiration for sustainable forest management – Lesson learned from Suserup Skov in Denmark. *Forstarchiv,* Vol. 81, pp 28-33.

11. Larsen, J.B. & Nielsen, A.B. (2011): Urban forest landscape restoration - Applying Forest Development Types in design and planning. In: *Forest Landscape Restoration: Integrating Natural and Social Sciences.* Springer Publishing Company. Accepted

12. Leibundgut, H., 1984: *Die Waldpflege.* Verlag Paul Haupt Bern, Stuttgart, 216 pp. Lindenmayer, D.B.; Franklin, J.F.; & Fischer, J. (2006). Conserving forest biodiversity: A checklist for forest managers. *Biological Conservation,* Vol. 129, pp 511-518.

13. Möller, A., 1922: *Der Dauerwaldgedanke: Sein Sinn und seine Bedeutung.* Reprint of the original 1922 publication. Erich Degreif Verlag, Oberteuringen, 134 pp.

14. Otto, H.-J. (1993). Waldbau in Europa - seine Schwächen und Vorzüge - in historischer Perspektive. *Forst und Holz,* Vol. 48, pp. 235-237.

15. Schütz, J-P. (1990). Heutige Bedeutung und Charakterisierung des naturnahen Waldbaus. *Schweizerische Zeitschrift für Forstwesen,* Vol. 141, pp. 609-614.

16. Skov- og Naturstyrelsen. (2002). *The Danish National Forest Programme in an International Perspective.* http://www.naturstyrelsen.dk/NR/rdonlyres/6BA78078-1188-494B-841E-EF89ECF0C064/13461/dnf_eng.pdf

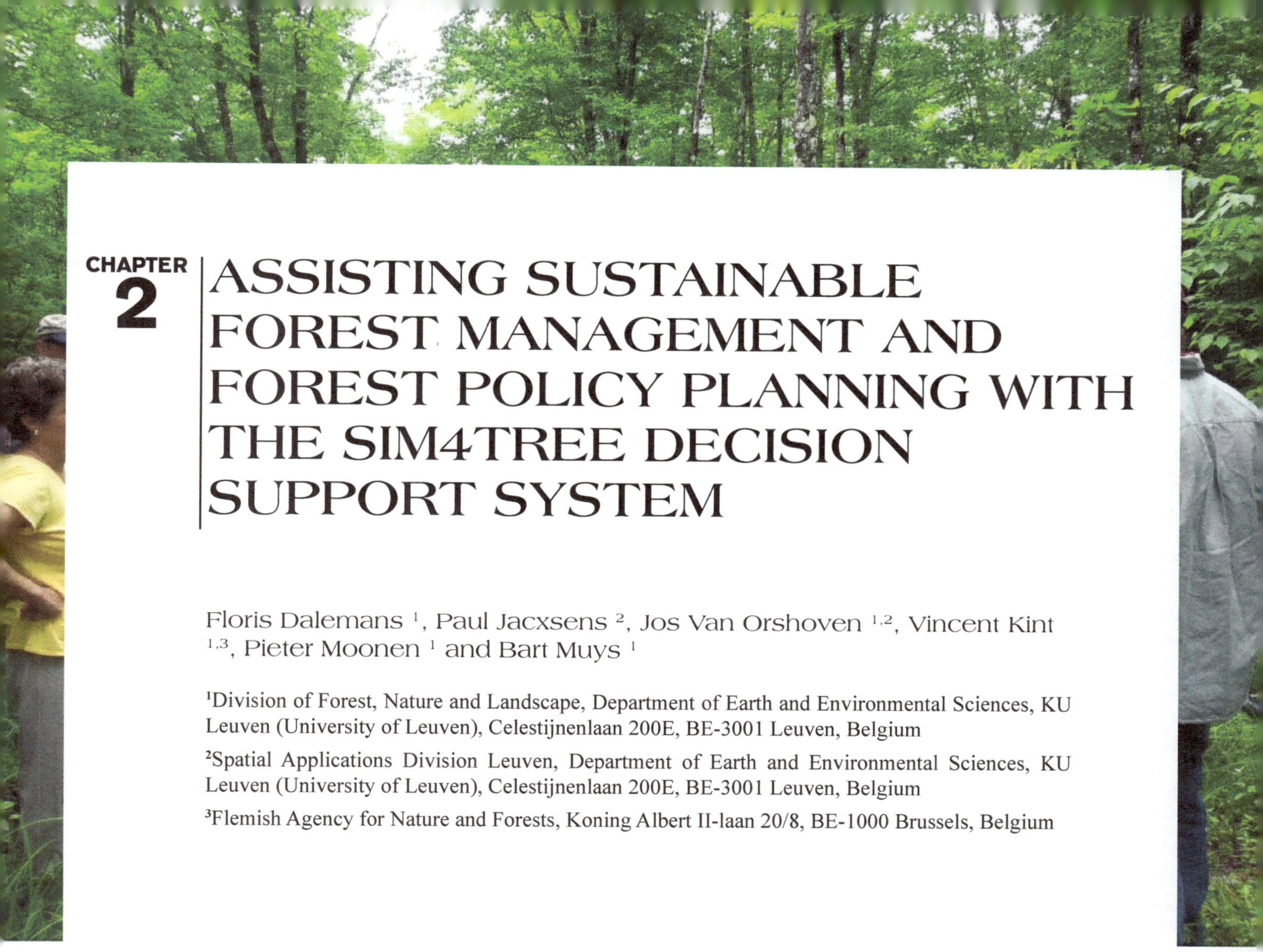

CHAPTER 2

ASSISTING SUSTAINABLE FOREST MANAGEMENT AND FOREST POLICY PLANNING WITH THE SIM4TREE DECISION SUPPORT SYSTEM

Floris Dalemans [1], Paul Jacxsens [2], Jos Van Orshoven [1,2], Vincent Kint [1,3], Pieter Moonen [1] and Bart Muys [1]

[1]Division of Forest, Nature and Landscape, Department of Earth and Environmental Sciences, KU Leuven (University of Leuven), Celestijnenlaan 200E, BE-3001 Leuven, Belgium

[2]Spatial Applications Division Leuven, Department of Earth and Environmental Sciences, KU Leuven (University of Leuven), Celestijnenlaan 200E, BE-3001 Leuven, Belgium

[3]Flemish Agency for Nature and Forests, Koning Albert II-laan 20/8, BE-1000 Brussels, Belgium

ABSTRACT

As European forest policy increasingly focuses on multiple ecosystem services and participatory decision making, forest managers and policy planners have a need for integrated, user-friendly, broad spectrum decision support systems (DSS) that address risks and uncertainties, such as climate change, in a robust way and that provide credible advice in a transparent manner, enabling effective stakeholder involvement. The Sim4Tree DSS has been accordingly developed as a user-oriented, modular and multipurpose toolbox. Sim4Tree supports strategic and tactical forestry planning by providing simulations of forest development, ecosystem services potential and economic performance through time, from a regional to a stand scale, under various management and climate regimes. Sim4Tree allows comparing the performance of different scenarios with regard to diverse criteria so as to optimize management choices. This paper explains the concept, characteristics, functionalities, components and use of the current Sim4Tree DSS v2.5, which was parameterized for the region of Flanders, Belgium, but can be flexibly adapted to allow a broader use. When considering the current challenges for forestry DSS, an effort has been made towards the participatory component and towards integration, while the lack of robustness remains Sim4Tree's weakest point. However, its structural flexibility allows many possibilities for future improvement and extension.

Keywords

Decision support systems (DSS); forest management and policy planning; ecosystem services; multi-criteria decision analysis (MCDA); Flanders

---

Citation: Floris Dalemans, Paul Jacxsens, Jos Van Orshoven, Vincent Kint, Pieter Moonen and Bart Muys, Assisting Sustainable Forest Management and Forest Policy Planning with the Sim4Tree Decision Support System. Forests 2015, 6, 859-878. doi:10.3390/f6040859.

INTRODUCTION

In recent decades, European forest management has increasingly focused on integrating the multitude of forest ecosystem services within the context of sustainability [1,2,3,4]. Forest management planning aims at maintaining a sustainable provision of multiple goods and services, such as timber and non-timber products, carbon sequestration, air and water purification, buffer to natural hazards, climate regulation, biodiversity conservation or recreation [5,6,7], at the level of the forest management unit [8]. However, the consideration of multiple ecological, economic and social objectives [1] often implies trade-offs and conflicts [9], along with complex interactions on various spatial and temporal scales [10]. The more diverse the demands on forest ecosystem services, the more complex and challenging planning and decision making is [8,11]. Simultaneously, sustainable forest management goes hand in hand with a demand for the increased involvement of various stakeholders (land owners, forest managers, land managers, forest industry, forest-dependent communities, policy makers) and the general public in the planning and decision making process [12,13,14]. This increasing technical and social complexity of forest decision making poses considerable challenges for forest managers and policy makers [15,16], whose choices have a direct and long-term influence on the quality and versatility of the ecosystem services provided by forests.

To accommodate multiple objectives and facilitate participatory decision making in forest management, various decision support systems (DSS) have been developed, mainly in the past three decades [10,17]. These computer-based tools typically combine a simulator, which calculates the outcome of management scenarios using one or several models, with a user interface, a database management system and an optimization algorithm [11]. They may be coupled to a geographic information system (GIS), include visualization technology and allow for stakeholder preferences. By structuring decision problems and analyzing the outcome and implications of alternative management scenarios, DSS greatly facilitate reasoned decision making for stakeholders [10,17]. Available DSS account for a broad range of forest ecosystems [4] and vary widely in terms of functionality (temporal scale, spatial scale, simulated variables, spatial analysis, optimization, visualization), typology (scope and objectives, user involvement, models and methods) and architectural set-up (user interface, database, modules) [18,19]. Comprehensive overviews of available DSS for forest management planning have been compiled within the framework of the European COST Action FP0804 forest management decision support systems (FORSYS); see [4,20].

In the 1980s and 1990s, DSS were addressing relatively narrow, well-defined problems, leading to an abundant collection of specific DSS [10]. These systems were mainly used by scientific researchers themselves for case studies and demonstrations [18]. Later on, driven by the increasing demand for integration and participation along with rapid advances in computing hardware and software systems, there was a clear trend towards developing fewer, yet more general-purpose DSS, with a broader scope, more functionalities (e.g., 3D-visualization, optimization and choice algorithms), a modular architecture with linkages to other systems (e.g., GIS tools, web services), larger stakeholder involvement and increasing user-friendliness [10,11,18]. This evolution is not at an end, and there still exists a need for further improvement and tailoring of DSS [16]. In this respect, the three following key targets of future DSS development could be synthesized from recent literature. Firstly, there is a need for further integration, referring to the consideration of multiple spatio-temporal scales and a broader array of services [17,21,22], in particular regulating and socio-cultural ones [19]. Several authors recommend a toolbox approach, which refers to a modular software architecture flexibly combining different models and systems [8,23,24] and enabling linkages with different (web-based) data sources and/or other DSS [24,25,26]. With regard to the current lack of integration, the fact that most DSS are case-specifically developed and are not systematically recorded in a meta-database of DSS [20] is leading to inefficiencies and overlap in development efforts [19]. Secondly, there is a need for more robust outputs, which account for the risks and uncertainties in forest management and policy, related to the increasing exposure of European forests to climate change and natural disturbances (heat waves, drought, fire, storm, flooding, biotic risks) [27,28,29] or to the future development of the wood market and other human needs. In this context, there is a need to generate outputs that consider factors of resilience, such as the genetic variation between provenances [30,31] and the occurrence of mixed and uneven-aged stands within the forest [11]. Lastly, the success of approaches for participatory decision making has been limited so far [13,15]. Although multiple obstacles for successful stakeholder involvement might exist in the entire planning cycle [14], with regard to DSS design, emphasis should be put on the inclusion of tools that help with assessing options, such as multi-criteria analyses, voting models, cognitive mapping, visualization techniques and web-based tools [13,26], as well as on the customization of the DSS (interface, support) towards the specific type(s) of users, in order to obtain user-friendliness, transparency, credibility and, ultimately, effective adoption [13,15,32]. The identification of the users and their needs, as well as user involvement in DSS development and testing are crucial with regard to the latter [13].

In addition to these three key targets, an elaborate and more practically-oriented set of empirical guidelines for the development and implementation of DSS has been synthesized by [16] from past experiences and lessons-learned of the experts involved in the FORSYS action. Similarly, a lot of these are obvious, yet lacking conditions for generalized use of DSS [11].

Future prospects indicate a continued increase in the demand for integration and participation, in the labor costs for inventories and field observations and in computer power. Therefore, the development of multifunctional, reliable and user-friendly DSS will become an even more essential and economically relevant challenge for flexible and interactive forest management and forest policy planning [11,19]. The current shortcomings, as well as newly emerging decision problem domains propel the demand for new DSS from audiences ranging from individual forest managers to landscape planners to (supra-)national policy makers and administrations [9,10,17].

This paper presents the Sim4Tree DSS, which has been developed taking into account several of these issues. The concept and characteristics of Sim4Tree, its current functionalities and its software components will be elucidated in Section 2.Section 3 will illustrate Sim4Tree's practical use and applicability by means of a case study for Flanders, Belgium, the region where it was developed. The strengths, weaknesses and opportunities of Sim4Tree will be evaluated against the three aforementioned key targets in Section 4.

THE SIM4TREE DECISION SUPPORT SYSTEM

Concept and Characteristics

Sim4Tree supports strategic (long term) and tactical (medium term) forestry planning by providing simulations of forest development, ecosystem services potential and economic performance through time, from a stand to a regional scale, under various management and climate regimes. Sim4Tree allows comparing the performance of different scenarios with regard to diverse objectives so as to optimize the management choices. Its main characteristics are seven-fold.

Decision Support Based on Best Available Models and Data

Sim4Tree is not a model or simulator in its own right, but a decision support tool in which modelling outputs from very different models or simulators (e.g., yield tables, mechanistic ecosystem models) can be flexibly plugged in and connected to an intuitive user interface and management optimization routines. Models are not coupled to the DSS in real time; rather, model runs are taken in a tabular format following a predefined structure. The actual Version 2.5 is designed for use in Flanders, using best available forest data and models for the region, but the flexible database structure of Sim4Tree and the open source design allow for an easy adaptation for use in other geographical locations with a different decision environment.

Three Decision Levels: N1 to N3

Sim4Tree is conceived as a toolbox, rather than as one overdesigned, all-round tool, which would have difficulty matching the needs of all users. This is reflected in separate decision levels for targeted use at three spatio-temporal scales. The first level (N1) (The letter "N" in each level designation refers to the Dutch equivalent for level, *i.e.*, "Niveau") can tackle strategic policy questions at a regional extent, the second level (N2) strategic management questions at the forest management unit level and the third level (N3) tactical management questions at the stand level. In the current v2.5, only N1 and N2 are operational, while for N3, the conceptual and technical basis is provided.

Spatial Characterization of the Forest by a Set of Pixels

At all three decision levels, the forest area is represented by a grid of independent square spatial units (pixels) and correspondingly stored in the database. Each pixel is characterized by one soil type, one tree species category (or other forest-related land use type, e.g., forest gap, heathland) and one age class, and no within-pixel variation is considered. Forest development comes down to each individual pixel delivering ecosystem services and producing certain costs and benefits through time. The spatial resolution of the pixel grid varies for the three decision levels, in line with the spatial character of the corresponding problem type: pixels are 10,000 m^2 in N1, 1000 m^2 in N2 and 100 m^2 in N3. The time resolution equals five years in N1 and N2 and one year in N3, while forest development can be simulated for 150 years in N1, 90 years in N2 and 20 years in N3.

As a consequence of these conceptual choices, Sim4Tree does not account for interactions between pixels. For example, although mixed and/or uneven-aged stands can be initialized with an appropriate pixel set, no effects of species or age interactions will be simulated. Furthermore, Sim4Tree is not suitable for simulation of ecosystem services that cannot be easily attributed to individual pixels (e.g., recreational value is often attributed to a larger extent) or that require neighborhood analyses between pixels (e.g., air pollution filtering depending on canopy surface roughness).

Forest Management Scenarios

Forest development is directly driven by a user-defined management scenario, summarizing choices on afforestation and deforestation, regeneration area and management of current and future generation stands and habitats. Species-specific management choices essentially comprise the silvicultural regime, the timing of final cut and regeneration options. Choices on silvicultural regime are directly determined by the extensiveness and flexibility of the available growth models. All management choices can be differentiated based on geographic location, site suitability, owner category and/or legal status. The defined management choices are correspondingly assigned to the individual pixels.

Climate Change Scenarios

For each simulation of forest development, a climate change scenario has to be selected. In Sim4Tree v2.5, these are either a baseline scenario (no climate change), a weak climate change scenario or a strong climate change scenario. The latter two correspond to the two extreme scenarios in the Special Report on Emissions Scenarios (SRES) report by the Intergovernmental Panel on Climate Change (IPCC), respectively the B1- and the A1F1-scenario [33]. Depending on the used model type, Sim4Tree can either directly use model output (restricted to mechanistic models that allow simulation of climate change) or assume a climate-driven change in site suitability (e.g., for climate-invariant empirical yield tables).

Forest Development

At the start of the simulation, each individual pixel is assigned a tree species and age class (or forest-related land use type), a corresponding site suitability and a silvicultural regime. Site suitability is derived from soil type and prevalent climate in terms of yield potential (volumetric increment) and discretized into five ordinal yield classes for each species. User-defined management choices are translated into changes of the pixel composition and attributes throughout time. Individual pixel development is then correspondingly derived from the user-chosen model, determining its ecosystem services provision and cost-benefit structure at each time step. Forest development is derived by integration of individual pixel development. The whole procedure is implemented in an open source relational database management system (RDBMS) through SQL statements.

Types of Questions

Sim4Tree can be used to solve different types of questions [34]. The first type of question ("what if") has the general structure "What is the outcome, if certain parameters are chosen?", where the outcome refers to ecosystem services and/or realized costs and benefits, while the parameters refer to all user-defined options. In the context of decision making, however, managers and policy makers usually aim at optimizing management choices. This concerns other question types, such as the "how" question ("What should management be like to obtain an optimal outcome?"), the "where" question ("where to apply a certain management to obtain an optimal outcome"), the "how long" question ("how long to apply a certain management to obtain an optimal outcome") or a combination of these. These question types require an optimization module.

Sim4Tree v2.5 Functionality and Flexibility

Sim4Tree v2.5 (January, 2015) is designed for and implemented for use in Flanders (northern Belgium). Nevertheless, most functionality is extendable to other regions or site conditions, given appropriate adaptation of the software. In this section, the current functionality is described, as well as how it can be improved and extended.

Initialization

Because of the large spatial extent, on decision level N1, the forest composition at time step zero (reference year 2010) is pre-initialized and cannot be altered. Conversely, on levels N2 and N3, the initialization should be provided by the user. In both cases, it is possible to subdivide the area into different geographic units. On N1, various options are included for Flanders: e.g., provinces, ecoregions, management regions, special protection/conservation areas. Areas on N2 and N3 can be subdivided into zones and stands. Similarly, the pixels can be assigned a legal status and an owner category. The geographic units, legal status and owner categories are equally pre-initialized and unchangeable on N1, while the user should define them on levels N2 and N3. They enable two important functionalities. Firstly, simulations can be run for only one or several units (or statuses or categories) instead of for the entire initialized area, which might be useful for specific studies wherein only a specific unit is of interest. Secondly, even when the simulation is done for the entire initialized area, the subdivision allows management to be separately defined for and differentiated between the constituting units (or statuses or categories).

The initialization includes the characterization of the pixels by one soil type, using the Belgian soil classification system, by one tree species category (or other forest-related land use type) and one age class. Sim4Tree v2.5 is mainly operational for high forests, and while other forest-related land use types (coppice, forest edges, forest gaps, heathland, water and other non-forest areas) can be assigned to a pixel, their development and corresponding ecosystem services provision cannot be simulated in Sim4Tree v2.5, due to a lack of appropriate models for Flanders. Correspondingly, these pixels are only accounted for by the area statistics.

The software easily allows other regions and countries to be added on N1. This is conditional on the availability of 10,000 m^2-scale forest maps containing the necessary information on land use, species, age class and soil type. Effective implementation might also require minor changes to the graphical user interface (GUI), if geographic subdivisions, legal statuses or owner categories do not match those for Flanders. Analogously, the development of land use types other than high forest could be simulated, given the availability of appropriate models.

Model Choice

Pixel development through time is derived from best available models for Flanders, which are stored in the database in tabular format following a predefined structure. Models are of crucial importance, since they determine which tree species and ages, land use types, silvicultural regimes and ecosystem services can be accommodated. In Sim4Tree v2.5, a set of empirical growth models is used, based on Dutch [35] and Flemish [36] yield tables. These describe the evolution of even-aged, monospecific stands with optimal tree density, in terms of stand and harvest dimensions, with a temporal resolution of five years. Data are available for all common species of the region, while less common species are set equal to a common species based on expert knowledge (see Supplementary Materials, Table S1). These yield tables have been extrapolated towards higher age categories and reclassified into five yield classes, linked to five site suitability levels. The main drawbacks of this empirical model are the limited set of silvicultural regimes for high forest that it allows, the limited set of ecosystem services that it predicts and its inability to model the development of stands with suboptimal tree density and of other land use types.

A special note should be added on the specification of the climate change effect in the model. Since empirical models, such as yield tables, predict growth assuming a fixed site suitability, they cannot be used for the climate change scenarios. Still, to account for climate change using these models, a hybrid approach has been adopted [11]. While in the baseline scenario (no climate change), site suitability is considered invariable for a pixel, the climate change scenarios might imply changes in time of site suitability for a pixel. These changes are based on estimated species-specific climatic responses, species ecology, expert knowledge and climatological projections for Flanders. More specifically, for each combination of species and soil type, a corresponding discrete site suitability has been modeled as a step function of time. At each new generation, this function determines whether a shift of site suitability/yield class should occur for a given species, as well as the magnitude and direction of this shift. Obviously, this approach comprises certain shortcomings. Firstly, it is an artificial and discretized estimation of the ecophysiological and multi-directional effect of climate change. Secondly, it only includes direct temperature and precipitation effects, while not accounting for other effects, such as atmospheric CO_2-enrichment and natural hazards. Thirdly, since the SRES-scenarios do not serve as a direct model input, the accuracy of the estimation is directly dependent on the expert interpretation. These shortcomings would largely be resolved by coupling a mechanistic (e.g., ecophysiological) model into Sim4Tree, which can account for a varying climate in modeling growth paths.

In general, while the current models for Flanders imply certain limitations, a major part of the flexibility of Sim4Tree is owed to the fact that it has not been grafted on this or any other particular model and that its design enables one to accommodate a wide variety of models. These can, for instance, account for other regions, species, land use types, management options, climate changes scenarios and ecosystem services, as long as they can be added in a pre-calculated tabulated format. The procedure of plugging-in models is however not yet automated and may still require minor adaptations to the software according to the specific features, requirements and outcomes of the model.

Forest Management Options

To influence the overall forest structure and composition, the user can define targeted land use changes for pixels, as well as introduce 'new pixels' (which were not assigned a species or forest-related land use types during the initialization) to enable afforestation and nature expansion. With regard to specific land use management, options are only available for high forests in Sim4Tree v2.5. The user can first designate for each species a fraction of the high forest as non-productive, for which, analogously to the other land uses types, development is not simulated. For the remaining productive high forest, species-specific options for the silvicultural regime, timing of final cut and regeneration are available:

- The availability of choices for silvicultural regime depends on the model used. In Sim4Tree v2.5, based on yield models for Flanders, the choice is limited to planting density (e.g., poplar) or thinning intensity (e.g., Corsican pine).
- The time of final cut it is either fixed at a certain age or target diameter or determined by gradual conversion with settings for the conversion period and the age range where conversion should be avoided.
- Regeneration is defined by species choice. Users may either select one species or define a set of species. In the latter case, a probabilistic algorithm distributes the species over the pixels according to an assigned priority and in such a way that species can be put either on the most suitable locations or that overall productivity is maximized.
- Sim4Tree simulates final cut as a clearcut and regeneration as planting of individual pixels. However, by appropriate pixel selection and timing, also group selection or shelterwood cuttings with gradual regeneration can be simulated, especially at N2.

Finally, the overall annual area of final cuts can be limited to a user-defined maximal value, for reasons of feasibility of workload or to meet sustainability standards. If this limit is exceeded in a certain time step, the clearcut in excess is automatically postponed, according to an algorithm that assigns a priority to each pixel. As such, all management rules are ultimately executed, but their initial timing may be overruled. The software assists the user in setting this upper limit by providing a measure of the mean expected clearcut area assuming a normal forest.

Cost-Benefit Options

Besides simulations of forest composition and ecosystem services, Sim4Tree can also provide an economic analysis of management scenarios. It allows one to take various costs and benefits into account: the software includes pre-defined settings for species-specific management costs (e.g., planting, pruning, tending) and benefits (timber prices, differentiated by diameter class), yearly costs (e.g., overhead, transaction costs), subsidies (specifically for Flanders) and various other benefits (yearly or at a specific moment in time). The user can decide to either exclude or add each of these to the simulation, modify their values and timing and add additional costs and benefits, which are not included by default. To account for the temporal dimension of costs and benefits, it is possible to define a rate of return for discounting towards the net present value (NPV).

In terms of database functioning, the costs and benefits are calculated for each time step and for each pixel according to its assigned silvicultural regime and development. If new models would allow for additional silvicultural regimes, the corresponding cost and benefit items can be added to the economic module.

Output

Composition, ecosystem services and cost-benefit structures are calculated for each pixel and for each time step in a tabular output. This full detailed output can be exported from the software for subsequent analysis or visualization in external software. Within Sim4Tree itself, the results can be aggregated and/or detailed both spatially (per species, geographic unit, legal status and/or owner category) and temporally (per time step or for the entire simulation period) and can be easily visualized in tables, graphs and maps (see Supplementary Materials, Figure S1).

Sim4Tree provides standardized output on forest composition and structure (e.g., area, age structure), diversity (e.g., species diversity, fraction of native/hardwood species, age diversity) and costs-benefits throughout time. The ecosystem services that can be calculated, are highly dependent on the model that is used, more specifically on the nature, temporal resolution and extensiveness of the development path provided by the model. Based on yield tables, Sim4Tree v2.5 derives output on timber production (e.g., increment, roundwood yield, regeneration) and, via expansion and conversion factors, on biomass yield and C-sequestration in trees.

The database structure is designed to allow the addition of extra variables to the DSS output by programming the appropriate SQL statements. In this way, also other ecosystem services for high forests or for other land use types (e.g., related to soil C, water, timber quality) can be added, conditional on the availability of appropriate source data for the models that are added to the software.

Optimization Options

Sim4Tree v2.5 is able to answer the "what if" and the "how" questions (see earlier). To answer the "what if" question, the user should solely interpret the Sim4Tree output of the simulated scenario based on the particular parameters. In addition, an optimization module is included for the "how" question. It enables users to compare the outcome of different management scenarios for a forest area, based on identical model, climate and simulation time choices. Importantly, this optimization is relative rather than absolute, meaning that Sim4Tree can only compare predefined scenarios, while it is not able to identify the most optimal approach out of every possible set of applicable management options [37]. Optimization is achieved by means of a multi-criteria decision analysis (MCDA) approach [38]: scenarios are compared based on their values for certain ecosystem services and/or economic variables (= multi-criteria), either at a given point in time or cumulatively over the simulation period, which allows stakeholder preferences to be included in terms of the criteria that are selected and their weights. In addition, the user can opt to exclude certain scenarios from the optimization by defining threshold values for ecosystem services provision and/or economic performance that should be met. Non-excluded scenarios are ranked by means of the iterative ideal point thresholding (IIPT) algorithm [39,40], after which the optimal scenario(s) can be identified. Note that the optimization module can be analogously used to investigate the sensitivity of a management approach to the cost-benefit parameters; in this case, these parameters are varied instead of the management options, and optimization occurs only with regard to economic performance.

Language Choice

In accordance with the current application of the DSS in Flanders, Sim4Tree v2.5 and all of its documentation are available in Dutch. Translation to other languages is straightforward.

The Sim4Tree Components

As a DSS, Sim4Tree flexibly integrates various components in a logical workflow (see Figure 1). Several of the features in this workflow have already been discussed; this section will describe the remaining ones.

Database Management System

The database management system links the models, initialization, user choices, output and optimization components with each other in a modular structure (see Supplementary Materials, Figure S2). It is implemented in the free software package, PostgreSQL. In practice, a client-server model is being used, where clients locally install a software and a database component, which are automatically updated according to the central database, which is hosted and managed on a server. Sim4Tree is only developed for use with the Microsoft Windows (XP Service Pack 3 or higher) operating system.

Graphical User Interface

The graphical user interface (GUI) adds functionality for the users, by guiding them in a step-by-step approach through the different DSS settings (initialization, climate and management scenarios, cost-benefit parameters and optimization) and by displaying the output (see Supplementary Materials, Figure S3), while translating correspondingly to and from the database via queries throughout. The GUI has been developed in an attempt to optimally combine intuition and performance, with strong feedback from and interaction with the users. It is a stand-alone application that has been developed using the free software framework, C#.NET 4.

User Manual and Support

The GUI provides access to three user support materials. Firstly, the extensive on-line user manual explains the concept and all functionalities of Sim4Tree and provides examples, useful links and background information. Secondly, the Sim4Tree final report contains the contextual and technical information. Lastly, a feedback form is provided in which specific questions or remarks to the software administrator can be filed. All of these are hosted and managed on a server, but are currently only available in Dutch. In addition, interactive training sessions, including follow-up, have been designed and organized for a variety of stakeholders and potential users.

(a)

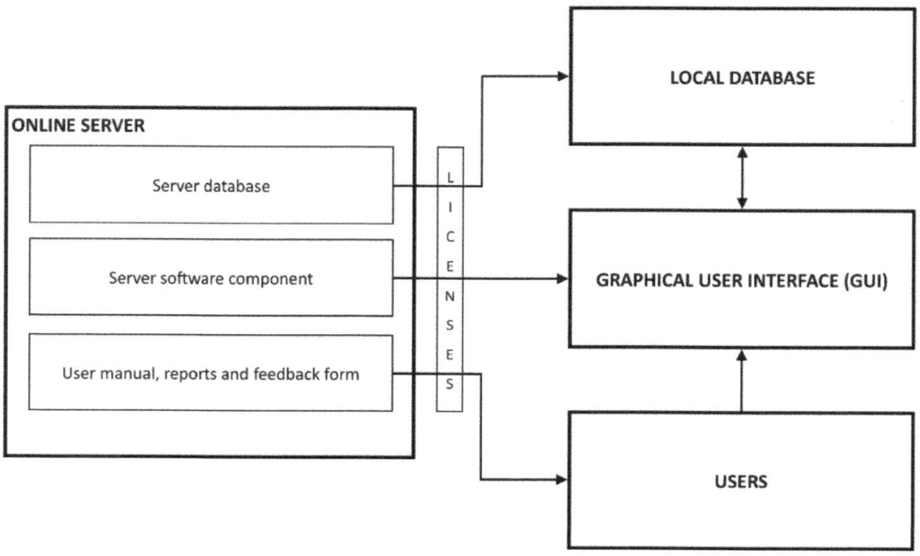

(b)

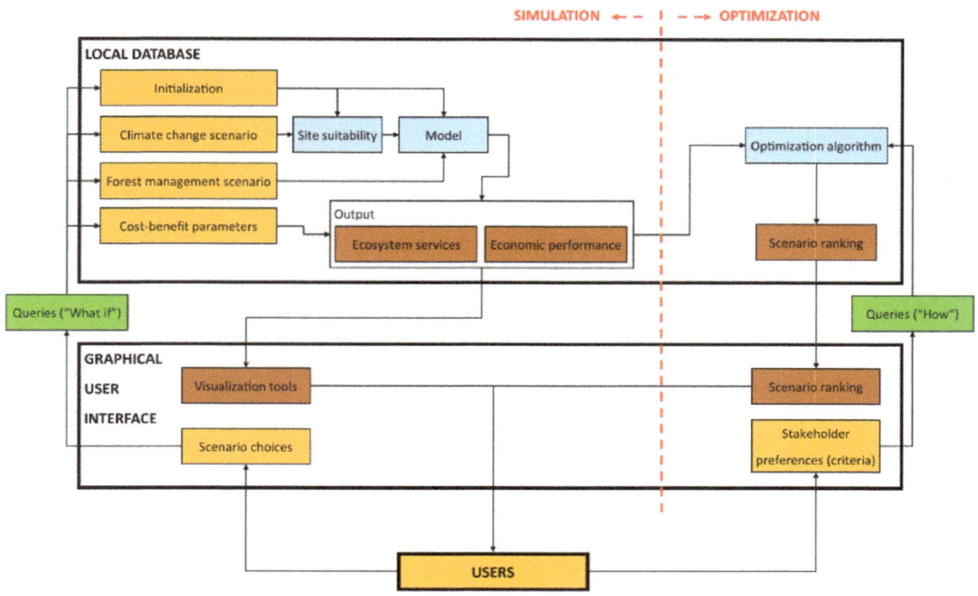

Figure 1. Sim4Tree components and workflow. (**a**) Communication between the server and the local components; (**b**) workflow of Sim4Tree.

Licenses

Sim4Tree combines different existing datasets and tools and adds functionality within a DSS. The licenses and restrictions for use of these datasets and tools apply when using the software. The DSS itself is owned by the Flemish government, which has the exclusive right to issue licenses for third party use. Sim4Tree v2.5 is licensed free of charge for all interested users.

CASE STUDY: THE USE OF SIM4TREE FOR PARTICIPATIVE DECISION MAKING ON THE STRATEGIC MANAGEMENT OF THE FOREST COMPLEX "BOSLAND"

Sim4Tree's potential for supporting multi-stakeholder strategic decision making is illustrated by its use in the Bosland project. The Bosland project aims at sustainable forest management in the ±45-km^2 forested area of Bosland, situated in the northeast of Flanders, to provide a multitude of functions (e.g., timber production, conservation area, recreation) for a variety of stakeholders (e.g., owners, forest industry, visitors). This vision translates into the management plans for the diverse owners. These drafted individual plans, however, consist largely of *ad hoc* management interventions and do not provide an assessment of the future overall forest development. As such, the Bosland association felt the need for an overarching strategic management plan. This is where Sim4Tree came in: based on the state of the forest in 2010, the effect of different management scenarios on the forest development up to 2070 was simulated, and the optimal scenario was identified in a participatory process.

The Bosland Forest and Management Objective

The Bosland project area consists of about 93% of forested area, where the remaining 7% is made up of small patches of non-forest habitat, water bodies and infrastructure. The forest largely consists of homogeneous stands on sandy soils and is dominated by Corsican pine (*Pinus nigra* var. *maritima*) and Scots pine (*Pinus sylvestris*), which constitute 94% of the standing stock. The envisaged management for the forest complex essentially comes down to a targeted conversion of stands to achieve a more heterogeneous forest in terms of tree species, tree ages and habitats, while maintaining reasonable harvests and growing stock. More specifically, this implies striving for a larger fraction of native hardwood species, high-quality species and trees, non-productive forest and other habitats, while allocating management types and conversions at the most optimal places, at the most optimal time. Since the creation of additional non-productive forest and other habitats, which is tightly linked to recreational and conservation functions, will occur in predefined zones, the objective for Sim4Tree was to support the strategic planning of conversions towards native hardwood species and high-quality species and trees.

Simulation and Optimization of Management in Sim4Tree

The defined objective can be achieved in various ways. To structure the plethora of choices, a hierarchy containing three essential management options, each including sub-options, was designed and used to define 36 management scenarios (see Figure 2). These were correspondingly differentiated in Sim4Tree, giving rise to 36 scenarios, which were then simulated (to illustrate this, the essential scenario parameters are listed in the Supplementary Materials, Table S2). To illustrate the influence of the various management options, the evolution in species composition is displayed for three scenarios in the Supplementary Materials, Figure S4.

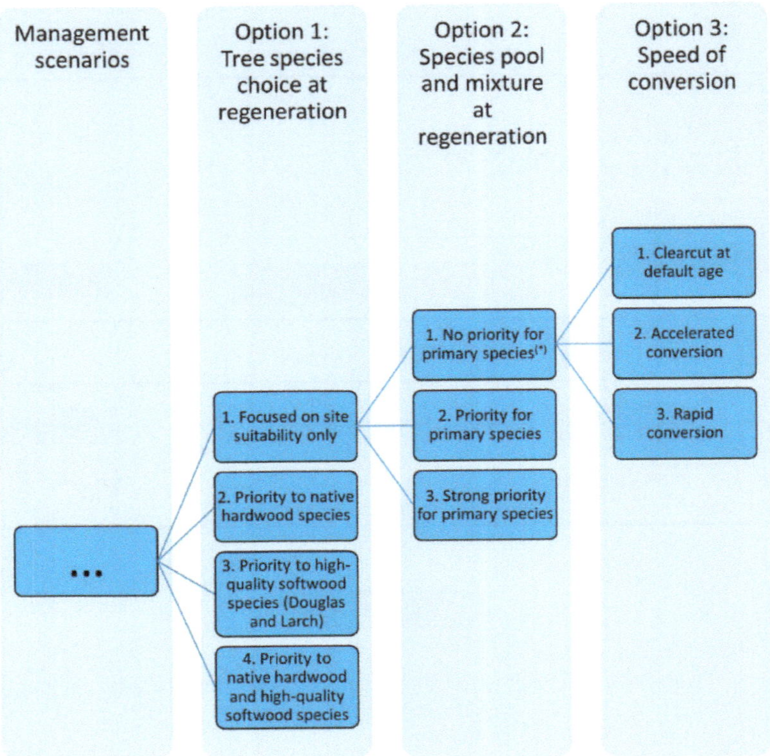

Figure 2. Overview of the hierarchical structure in management options for Bosland. Management scenarios are defined by choosing amongst three hierarchical management options. The name of a scenario (e.g., 312) indicates the chosen sub-options. By considering all possible combinations of sub-options, a total of 36 scenarios is defined. * The differentiation between primary and secondary species depends on the tree species choice at regeneration and the site quality.

Subsequently, the optimal management scenario was identified in four steps. In the first step, each scenario was held against a set of ecosystem service threshold values and excluded in case of non-compliance. More specifically, it was based on compliance of the forest state in 2070 with the current forest legislation, which consists of a set of norms and principles for sustainable forest management composed by the Flemish government. In a second step, the remaining scenarios were compared based on their provision of 14 different ecosystem services (see Supplementary Materials, Table S3), which reflect the targeted management objectives and, by extension, the Bosland project aims. The scenarios were ranked using the IIPT algorithm (see Supplementary Materials, Table S3), using a set of criteria weights defined by the stakeholders (see the next section). In a third step, the data were exported and subjected to an alternative MCDA, using the program, Web-Hipre [41] (see Supplementary Materials, Figure S5), as well as to a sensitivity analysis to account for the effects of weight choices, using the program, JSMAA [42] (see Supplementary Materials, Table S4). In the last step, the three highest ranked scenarios were selected and presented to the stakeholders (see the next section).

The Participatory Process

Although Sim4Tree was vital for providing the technical information, based on which the strategic management plan could be designed, the effective realization of this plan was a result of a broader participatory process. Indeed, Sim4Tree both required input from a stakeholder committee, which included representatives from the forest managers, the municipalities, the wood trading and processing industry and local interest groups, and, in turn, equally only provided input for the eventual decision making. With regard to input for Sim4Tree, the development of the specific set of management scenarios for attaining the envisaged objective, *i.e.*, the definition of the different scenarios, was only possible thanks to the stakeholders' expertise and was based on differentiating expectations with regard to future wood market, policy development and climate change. Likewise, the criteria and weights for evaluating the different management scenarios against the envisaged objective were decided upon by the committee. In turn, Sim4Tree offered the input for the additional MCDA and the sensitivity analysis, which all together offered the stakeholders a scientifically-founded set of optimal scenarios, amongst which the most optimal one, however, was eventually chosen not solely based on its quantified rank, but on a holistic and expertise-guided assessment of its performance. In addition, the committee provided specific suggestions for further refinement and optimization of the three selected management scenarios, whereafter the simulation-optimization-selection cycle could be repeated. Finally, the most optimal scenario only formed a blueprint, subject to the limitations of the software, based on which the stakeholders drafted the final strategic management plan, which was then translated into the individual management plans.

Shortcomings: Accuracy and Uncertainty in Sim4Tree

By using mathematical descriptions and concepts to describe the real-world forest development, Sim4Tree's simulations inevitably imply inaccuracies. Conceptually, we can distinguish between three sources of inaccuracy. Firstly, the database concept requires pixels to be independent, implying that neither interactions nor explicit spatial ecosystem services can be simulated (see earlier). Secondly, there are inaccuracies related to the input data. This comprises, on the one hand, measurement errors of the initial forest status (initialization). For example, for Flanders as a whole, the initialization (N1) is estimated to have a maximum deviation of 10% with regard to forested area and a deviation of respectively up to 5% and 20% for the area of common and uncommon tree species, due to lack of a single, up-to-date map for the distribution of tree species in Flanders. On the other hand, this refers to the uncertainties for the future biotic and abiotic environment, such as climate change (the SRES-scenarios) and the future wood market (cost-benefit structures). Lastly, the model introduces inaccuracies by not or only partly accounting for the effect of certain factors. The yield tables can, for instance, only account for a very specific subset of contemporary management regimes and for time steps of five years, while they cannot account at all for factors such as genetic effects and pest and disease dynamics.

The three sources of inaccuracy add to each other. To assess overall inaccuracy, simulated and measured standing stock were compared for short-term simulations of five forests on level N2. Differences between 1% and 20% were typically observed, with outliers up to 80% for individual species. Nevertheless, this should be put into perspective for three reasons. Firstly, when enlarging the area (N1), these deviations are expected to largely eliminate each other. Secondly, when different scenarios are compared to each other for optimization, the scenario errors are expected to be very similar and, therefore, will only marginally influence the outcome of interest. In a general sense, Sim4Tree does not claim to make exact predictions, rather the results should be interpreted in relative terms by comparing scenarios. Lastly, and most importantly, only the inaccuracies due to the database concept are inherent to the Sim4Tree DSS, while the others stem from the "external" data and tools that it uses. It has shown that the software can flexibly replace and/or integrate these with new improved data and tools (see earlier).

DISCUSSION

Although Sim4Tree has been developed in first instance as a result of the lack of instruments assisting adaptive forest management planning tailored to Flanders, along the way, needs were revealed corresponding to the key targets of DSS development (see the Introduction). Indeed, the forestry sector in Flanders is equally confronted with the complex challenges of sustainable forest management in a dynamic environment, and as such, the inclusion of multiple spatio-temporal scales and ecosystem services, the demand for robust results and the possibility of participatory decision making were pertinent to the design of Sim4Tree all the way through. Accordingly, the software can be considered an effort towards addressing these three key targets of forest management DSS development and was held against these.

With regard to integration, Sim4Tree is an example of how a toolbox approach can be used to support multiple spatio-temporal scales. The pixel concept can be applied to any spatial scale, while the required detail, in terms of temporal resolution and management options, can be correspondingly adjusted by adding the output of a suitable model to the database. The plug-in architecture of the database allows for this and the integration of other data sources (e.g., other geographical datasets) and methods (e.g., other optimization algorithms), although the model flexibility does not extend to real-time models. In this respect, Sim4Tree could be considered a DSS generator, while Sim4Tree v2.5 is then a specific DSS [43]. The ease of plugging-in equally accounts for the GUI, which can be extended with additional components, such as other GIS plug-ins, visualization tools and ecosystem services. This might be mostly necessary for the latter: the currently limited array of ecosystem services, the predominant focus on timber production and the impossibility of including explicit spatial ecosystem services remain main shortcomings. Nevertheless, it was shown in the Bosland case study that this can be solved by calling in Sim4Tree for only a part of the considered ecosystem services, which makes it a part of a bigger toolbox. In this respect, the software cannot yet be automatically linked to other DSS or services, yet its adherence to the Microsoft Windows operating system and PostgreSQL RDBMS implies favorable conditions. From another viewpoint of integration, it should be noted that the fundamental design of Sim4Tree is not geographically specific and that it has been added to the ForestDSS meta-database [20] to promote visibility and cooperation.

The lack of robustness of the DSS is for now the weakest point. Yield tables for Flanders are only available for a very limited set of management regimes and should only be considered as an approximation in light of the recent innovative and diversifying management strategies. Similarly, although the software can account for climate change, the hybrid approach is only an approximation and has several shortcomings (see earlier). Other risks and uncertainties, such as natural disturbances and socio-economic developments, as well as features, such as genetic characteristics and species interactions in mixed and uneven-aged forests, cannot be taken into account. Nevertheless, if appropriate models become available, these factors could be included to a certain extent. For instance, a certain mixture of tree species and ages could be assigned to a pixel, as long as there is a model that can assign a corresponding development path, which restricts it to a predefined set of combinations (no real-time models), while in general, the spatial representation is restricted by not considering within-pixel variation. As a rule, the user should be mindful of the shortcomings and, because of the uncertainties and possible inaccuracies, interpret the results in relative terms. In addition, the uncertainties call for using Sim4Tree for adaptive management, whereby the quantitative optimization is complemented with expert knowledge and the planning is systematically adjusted to deviations and future developments.

Lastly, participatory elements occupy a central role in Sim4Tree. In the first place, the software has been designed to incorporate stakeholder opinions, which is materialized by the MCDA approach. This functionality is vital to the decision making process, although it should be acknowledged that Sim4Tree usually only forms a part of the participatory process and should be dynamically complemented with other methods and stakeholder interactions to realize a full-fledged decision making process. This was clearly illustrated by the Bosland case study. Furthermore, the fact that Sim4Tree does not try to cover the entire participatory process keeps it from being too specific and, as such, bears the advantage that it can be creatively used in various applications. Secondly, stakeholders' and potential users' feedback and suggestions for DSS contents and features have been of prime importance throughout the software development and have been decisive for the software design in all of its aspects. The attempt to satisfy differing user needs and to favor effective widespread use is above all reflected by the three defined decision levels, N1, N2 and N3. Much effort has been put on user-friendliness (e.g., intuitive stand-alone GUI, various visualization options, including GIS, extensive manual and documentation) and on maintaining transparency and credibility for both experts and non-experts through the straightforward methodology (e.g., initialization, yield tables and visualized results are all very intuitive; the MCDA approach leads to a transparent and consistent evaluation). As a result, Sim4Tree might essentially contribute to decision making, as long as its possibilities and limitations within the participatory process are well understood.

In summary, with Sim4Tree v2.5, we have a DSS at our disposal that can aid in the participatory decision making process on the strategic policy and management of forest regions and complexes in Flanders. Despite various shortcomings and uncertainties, its inherent structural flexibility implies great potential for further improvement of the performance and for extension of the applicability, which is all the more important given future prospects of a continued increase in the demand for integration and participation and in computer power. Related to this, its low requirement for input data, at least on decision levels N1 and N2, is an important asset in light of expected increasing labor costs for inventories and field observations. Future software adaptations will primarily focus on shortcomings that impede Sim4Tree's use on a larger scale. In fact, a list of possible future adaptations was compiled, and these were ranked according to their importance and feasibility. This revealed, amongst a total of 75 suggestions, four absolute priorities: incorporating models other than the yield tables, operationalizing decision level N3, accounting for sub-optimal tree densities (no maximal basal area) and implementing various cost-benefit profiles. In addition, future adaptation will be steered by feedback and suggestions from users.

CONCLUSIONS

This paper has presented the Sim4Tree DSS, which supports participatory decision making for forest management and policy. It provides simulations of forest development, ecosystem services potential and economic performance through time, at multiple spatio-temporal scales, under user-defined management and climate regimes, and allows for an optimization by ranking different scenarios. The current Sim4Tree v2.5 is parameterized for the region of Flanders, Belgium, and has achieved substantial merits towards including participatory approaches and integrating knowledge sources within a toolbox. Its lack of robustness remains the weak point, however. Nevertheless, the general plug-in architecture of Sim4Tree allows for further adaptions and, correspondingly, (even) further improvements with regard to these three aspects. Rather than focusing on *ad hoc* needs of particular projects, the software is designed to keep up with newly emerging decision problems and questions at various spatio-temporal scales.

ACKNOWLEDGMENTS

We thank the Flemish Fund for Science and Technology (IWT) and all members of the IWT Strategic Basic Research (SBO) Sim4Tree consortium, including Gaby Deckmyn and Reinhart Ceulemans (University of Antwerp), Dries Vansteenkiste and Joris Van Acker (Ghent University), Marc Devrieze and Joost Malfliet (KU Leuven, University of Leuven), for their contributions to the first version of Sim4Tree. We also thank the Flemish Agency for Nature and Forests (Ondersteunend Centrum van het Agentschap voor Natuur en Bos, OC-ANB) for the financial support and Wim Aertsen, Pieter Moonen, Lieve Borremans and Thomas Van de Peer for their contributions to Sim4Tree v2.5. Lastly, we want to thank all members of the different stakeholder commissions, which delivered input for the development of Sim4Tree along the way.

SUPPLEMENTARY MATERIALS

Table S1. Tree species included in Sim4Tree v2.5, either with specific yield tables or linked to the yield tables of another species

Tree Species with Own Yield Table	Tree Species Linked to Yield Table
Ash (*Fraxinus excelsior*)	Elm (*Ulmus* spp.)
	Field maple (*Acer campestre*)
	Sycamore (*Acer pseudoplatanus*)
Austrian pine (*Pinus nigra* var. *austriaca*)	-
Beech (*Fagus sylvatica*)	-
Birch (*Betula pendula* & *Betula pubescens*)	-
Common alder (*Alnus glutinosa*)	Grey alder (*Alnus incana*)
Corsican pine (*Pinus nigra* var. *maritima*)	-
Douglas fir (*Pseudotsuga menziesii*)	Silver fir (*Abies alba*)
Larch (*Larix kaempferi* and *Larix decidua*)	-
Norway spruce (*Picea abies*)	-
Pedunculate oak (*Quercus robur*)	Sessile oak (*Quercus petraea*)

Poplar, hybrid (*Populus euramericana*)	Willow (*Salix* spp.)
	Aspen (*Populus tremula*)
	Grey poplar (*Populus canescens*)
Red oak (*Quercus rubra*)	-
Scots pine (*Pinus sylvestris*)	-

Table S2. Sim4Tree scenario parameters in the Bosland case study

Scenario parameter	**Selected choice in the Bosland case study**
Decision level	N2
Model	Yield tables
Simulation period	2010–2070
Climate scenario	Baseline (no climate change)
Land use types	High forest (productive + non-productive), forest gaps, water
Silvicultural regime	Thinning from below
Timing of final cut	Fixed at default age and gradual conversion
Regeneration	Regeneration to a set of species
Practical planning	No limitation of the overall annual area of final cuts
Cost-benefit analysis	Not applicable

Table S3. Optimization criteria and weights for the Bosland case study, along with the simulated values for 3 different management scenarios and their rank in the iterative ideal point thresholding (IIPT) optimization. Optimal values are indicated in green. Criteria and weights were chosen by the stakeholder committee

Criterion	**Measurable Ecosystem Service**	**Unit**	**Evaluation Period**	**Weight in MCDA**	**Target**	**Illustrative Scenario Values**		
						Scenario 322	**Scenario 112**	**Scenario 412**
Maintenance of growing stock	Standing volume of hardwood	M^3/ha	Final time step	0.041	Maximize	28	45	27
	Standing volume of softwood	m^3/ha		0.054		104	87	100
Maintenance of annual harvests (thinnings + clearcut)	Harvested volume of hardwood	m^3		0.045		10,214	14,319	9176
	Harvested volume of softwood	m^3		0.051		112,565	100,757	110,133
Fraction of the area containing valuable species	Area native hardwood	Ha		0.057		524	760	604
	Area Douglas fir and larch	Ha		0.064		168	71	168
Fraction of thick trees	Fraction of trees in largest diameter class	%		0.054		17	17	18
Evenness of the area distribution between the different species	Shannon evenness index	-		0.054		0.56	0.52	0.59
Evenness of the age distribution	Deviation from the mean	-		0.070	Minimize	120	131	120

Maximized production	Produced volume of hardwood	10^3 m^3	Cumulative over the simulation period	0.086	Maximize	26	41	23
	Produced volume of softwood	10^3 m^3		0.086		1036	1003	1032
	Produced volume of Douglas fir and larch	10^3 m^3		0.113		13	10	15
Avoiding harvest fluctuations	Deviation from the mean	-		0.113	Minimize	3.3	3.6	3.3
Ensuring presence of clearcut areas	Deviation from the mean	-		0.093		2.5	2.6	2.5
						Rank **64** in IIPT	Rank **77** in IIPT	Rank **100** in IIPT

Table S4. Results of the sensitivity analysis for the 5 highest ranked scenarios. By varying the criteria weights, the frequency by which a scenario is ranked first, second, up to tenth, is calculated. The total score is the sum of these frequencies, each divided by their corresponding rank

Scenario	Ranked as Most Optimal (% of Total)	Ranked as Second Most Optimal (% of Total)	Ranked as Third Most Optimal (% of Total)	Total Score
411	48	12	7	61
412	5	9	9	23
112	1	4	8	21
122	0	2	5	19
221	1	5	6	17

(a)

Time step	Species	Area (km^2)	Average age (years)	Standing stock (m^3)	Harvest volume (m^3)	Amount of trees per hectare
2066 - 2070	Beech	213	43	18 284	2 858	2 894
2066 - 2070	Corsican pine	892	23	108 424	77 453	1 996
2066 - 2070	Douglas fir	74	34	15 535	3 772	1 118
2066 - 2070	Forest gap	94	-	-	-	-
2066 - 2070	Heathland	95	-	-	-	-
2066 - 2070	Larch	99	34	16 529	2 128	959
2066 - 2070	Norway spruce	52	37	8 850	3 163	1 690
2066 - 2070	Pedunculate oak	742	38	65 731	8 390	2 556
2066 - 2070	Scots pine	1 696	27	205 627	158 047	1 705

(b)

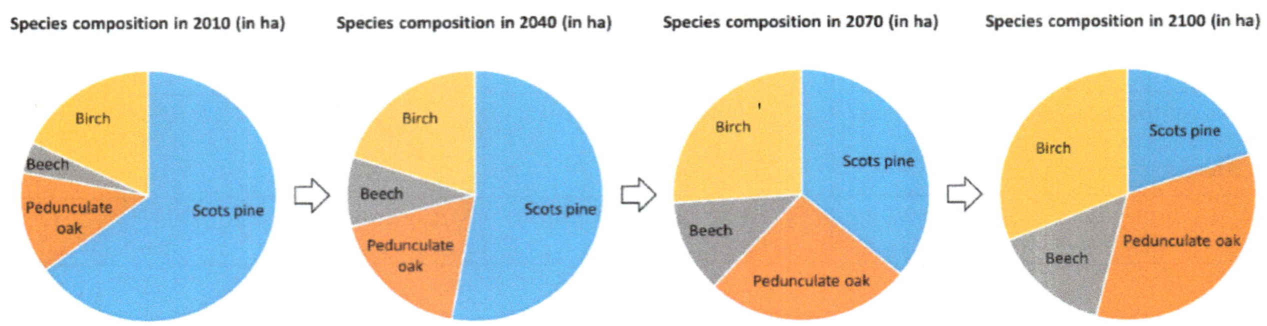

(c)

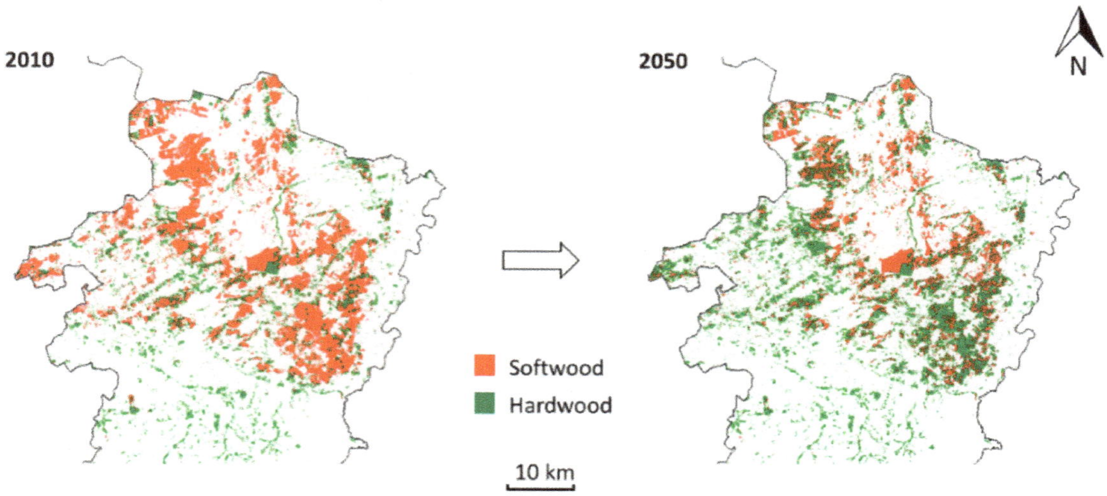

Figure S1. Example of a tabular output (**a**), graphical output (**b**) and map output (**c**).

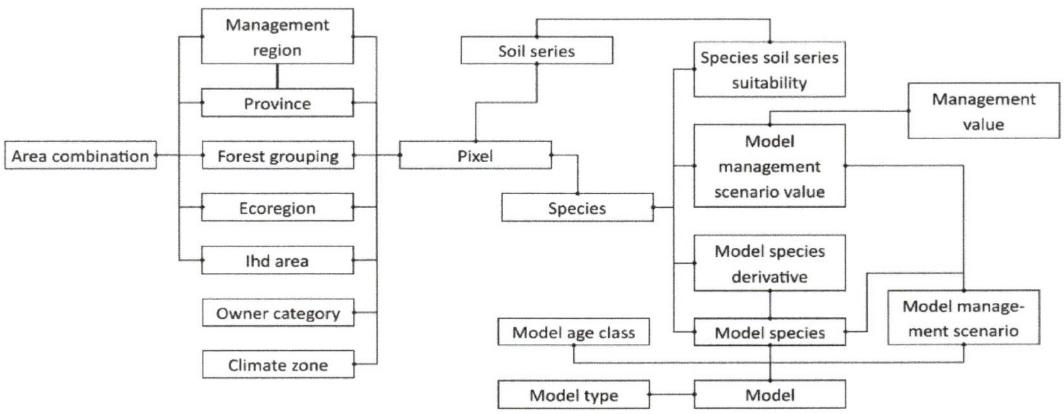

Figure S2. Excerpt from the relational database management system (RDBMS).

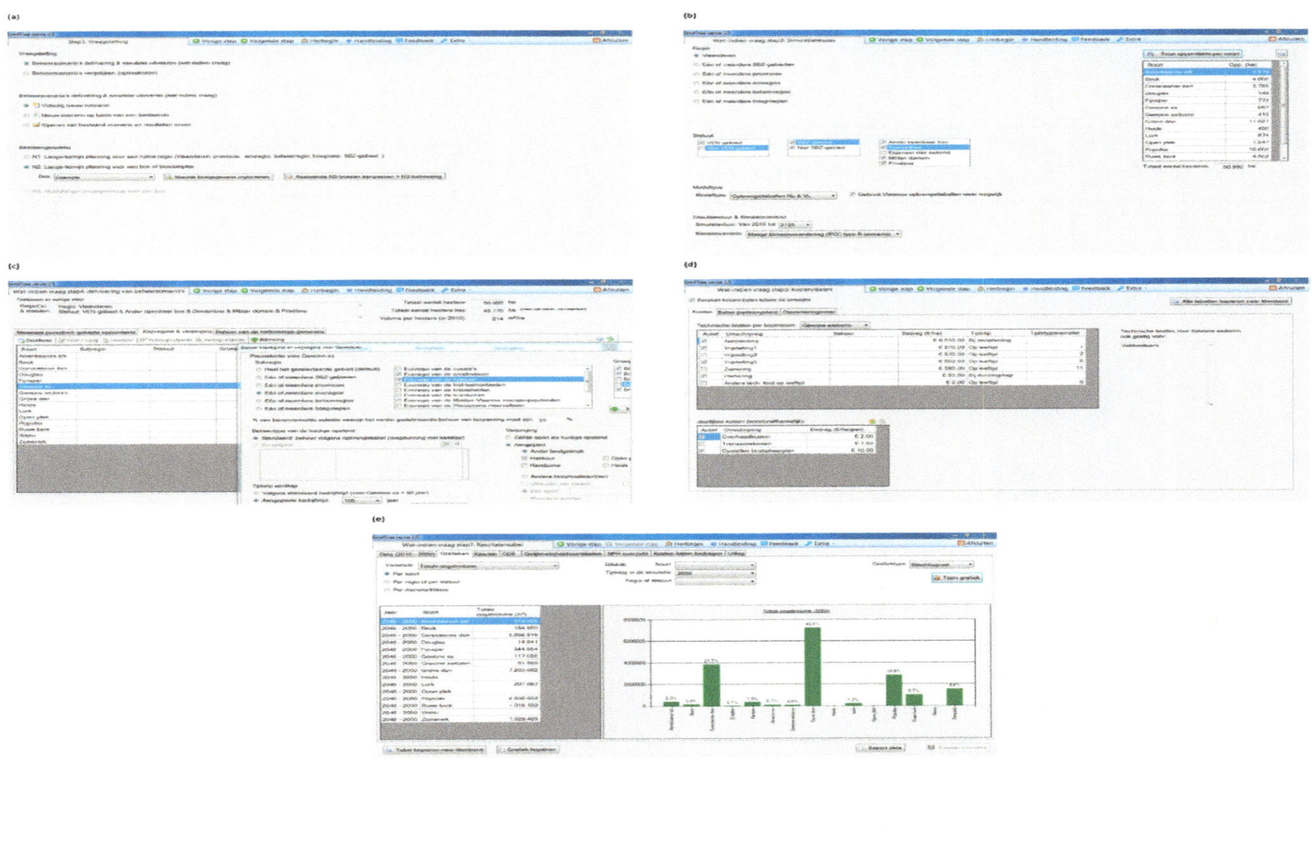

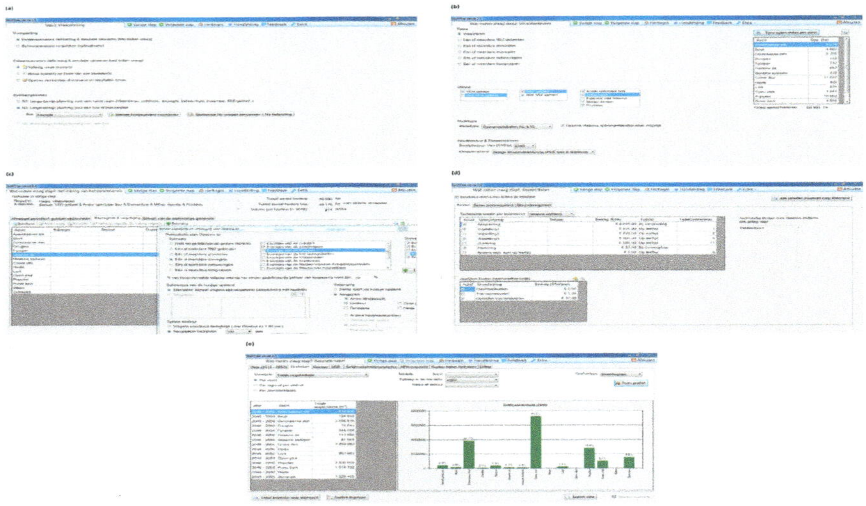

Figure S3. Graphical user interface. (**a**) Step 1: initialization; selection of question type and decision type; (**b**) Step 2: initialization; selection of model, simulation period and climate change scenario; (**c**) Step 3: definition of management scenario; (**d**) Step 4: definition of cost-benefit structure; (**e**) Step 5: results.

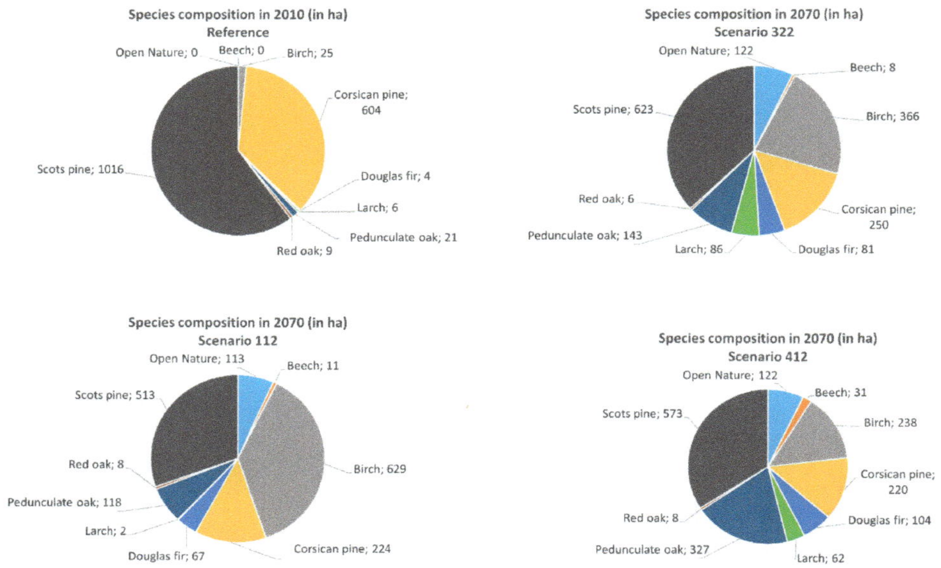

Figure S4. Evolution of tree species composition in Bosland from 2010 (**a**) to 2070 (**b**,**c**,**d**) (total area = 1685 ha), according to three different management scenarios.

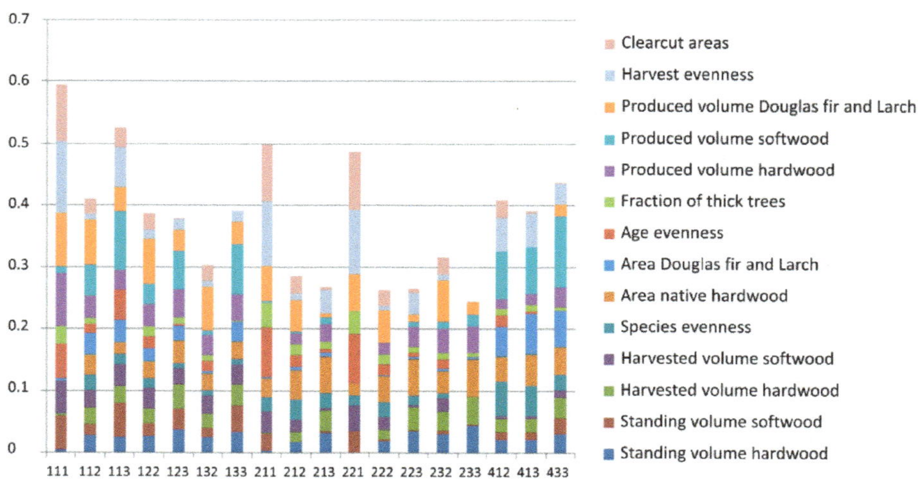

Figure S5. Result of the Web-Hipre multi-criteria decision analysis (MCDA) of the simulated scenarios, complied with the forest legislation. The length of each bar represents the weighted score for the corresponding criterion.

AUTHOR CONTRIBUTIONS

Floris Dalemans contributed to the climate functionalities of the software, to the user manual and other documentation, to the software tests and to the list of future possible adaptations and first-authored the paper. Paul Jacxsens designed and implemented the database in PostgreSQL RDBMS, designed and coded the software functionalities in the C#.NET framework and contributed to the documentation, software tests and future possible adaptations. Pieter Moonen was a principal investigator in the Bosland case study and wrote parts of the corresponding section of the paper. Vincent Kint initiated Sim4Tree development and contributed together with Bart Muys to its first version; together with Jos Van Orshoven, they also coordinated the development of v2.5, contributed to the documentation and proofread and contributed to the drafts of the paper. Moreover, Vincent Kint contributed to the software tests and future possible adaptations and wrote parts of the Sim4Tree decision support system section of the paper.

REFERENCES

1. Farrell, E.P.; Führer, E.; Ryan, D.; Andersson, F.; Hüttl, R.; Piussi, P. European forest ecosystems: Building the future on the legacy of the past. *For. Ecol. Manag.* **2000**, *132*, 5–20.
2. Cubbage, F.; Harou, P.; Sills, E. Policy instruments to enhance multi-functional forest management. *For. Policy Econ.***2007**, *9*, 833–851.
3. Hahn, W.A.; Knoke, T. Sustainable development and sustainable forestry: Analogies, differences, and the role of flexibility. *Eur. J. For. Res.* **2010**, *129*, 787–801.
4. Department of Forest Resource Management, Swedish University of Agricultural Sciences. *Computer-Based Tools for Supporting Forest Management*; Borges, J.G., Nordström, E.M., Garcia Gonzalo, J., Hujala, T., Trasobares, A., Eds.; Department of Forest Resource Management, Swedish University of Agricultural Sciences: Umea, Sweden, 2014; p. 503.
5. De Groot, R.S.; Wilson, M.A.; Boumans, R.M.J. A typology for the classification, description and valuation of ecosystem functions, goods and services. *Ecol. Econ.* **2002**, *41*, 393–408.
6. Spiecker, H. Silvicultural management in maintaining biodiversity and resistance of forests in Europe—Temperate zone. *J. Environ. Manag.* **2003**, *67*, 55–65.
7. Bugmann, H.; Palahí, M.; Bontemps, J.-D.; Tomé, M. Trends in modeling to address forest management and environmental challenges in Europe: Introduction. *For. Syst.* **2010**, *19*, 3–7.
8. Pretzsch, H.; Grote, R.; Reineking, B.; Rötzer, T.H.; Seifert, S.T. Models for forest ecosystem management: A European perspective. *Ann. Bot.* **2008**, *101*, 1065–1087. [PubMed]
9. Fürst, C.; Lorz, C.; Vacik, H.; Potocic, N.; Makeschin, F. How to support forest management in a world of change: Results of some regional studies. *Environ. Manag.* **2010**, *46*, 941–952.
10. Reynolds, K.M.; Twery, M.; Lexer, M.J.; Vacik, H.; Ray, D.; Shao, G.; Borges, J.G. Decision support systems in forest management. In *Handbook on Decision Support Systems 2*; Burstein, F., Holsapple, C.W., Eds.; Springer: Berlin, Germany, 2008; pp. 499–533.
11. Muys, B.; Hynynen, J.; Palahí, M.; Lexer, M.J.; Fabrika, M.; Pretzsch, H.; Gillet, F.; Briceño, E.; Nabuurs, G.J.; Kint, V. Simulation tools for decision support to adaptive forest management in Europe. *For. Syst.* **2010**, *19*, 86–99.
12. Sheppard, S.R.J.; Achiam, C.M. Public participation in forest decision making. In *Encyclopedia of Forest Sciences*; Evans, J., Youngquist, J., Eds.; Academic Press/Elsevier: Oxford, UK, 2004; pp. 1173–1182.
13. Lawrence, A.; Stewart, A. Sustainable forestry decisions: On the interface between technology and participation. *Math. Comput. For. Nat. Sci.* **2011**, *3*, 42–52.
14. Menzel, S.; Nordström, E.-M.; Buchecker, M.; Marques, A.; Saarikoski, H.; Kangas, A. Decision support systems in forest management: Requirements from a participatory planning perspective. *Eur. J. For. Res.* **2012**, *131*, 1367–1379.
15. Sheppard, S.R.J.; Meitner, M. Using multi-criteria analysis and visualisation for sustainable forest management planning with stakeholder groups. *For. Ecol. Manag.* **2005**, *207*, 171–187.
16. Marques, A.F.; Ficko, A.; Kangas, A.; Rosset, C.; Ferreti, F.; Rasinmaki, J.; Packalen, T.; Gordon, S. Empirical guidelines for forest management decision support systems based on the past experiences of the expert's community.*For. Syst.* **2013**, *22*, 320–339.
17. Vacik, H.; Lexer, M.J. Past, current and future drivers for the development of decision support systems in forest management. *Scand. J. For. Res.* **2014**, *29*, 2–19.
18. Packalen, T.; Marques, A.F.; Rasinmäki, J.; Rosset, C.; Mounir, F.; Rodriguez, L.C.E.; Nobre, S.R. A brief overview of forest management decision support systems (FMDSS) listed in the FORSYS wiki. *For. Syst.* **2013**, *22*, 263–269.
19. Segura, M.; Ray, D.; Maroto, C. Decision support systems for forest management: A comparative analysis and assessment. *Comput. Electron. Agric.* **2014**, *101*, 55–67.
20. Marques, A.F.; Rosset, C.; Rasinmaki, J.; Vacik, H.; Gordon, S.; Nobre, S.; Falcao, A.; Weber, D.; Granitzer, M.; Eriksson, L.O. Collaborative development of a semantic wiki on forest management decision support. *Scand. J. For. Res.* **2014**, *29*, 30–43.
21. Calama, R.; Tomé, M.; Sanchez-Gonzalez, M.; Miina, J.; Spanos, K.; Palahí, M. Modelling non-wood forest products in Europe : A review. *For. Syst.* **2010**, *19*, 69–85.
22. Mäkelä, A.; Grace, J.C.; Deckmyn, G.; Kantola, A.; Campioli, M. Simulating wood quality in forest management models. *For. Syst.* **2010**, *19*, 48–68.
23. Reynolds, K.M.; Schmoldt, D.L. Computer-aided decision making. In *Computer Applications in Sustainable Forest Management*; Shao, G., Reynolds, K.M., Eds.; Springer: Netherlands, 2006; pp. 143–169.
24. Rammer, W.; Schauflinger, C.; Vacik, H.; Palma, J.H.N.; Garcia-Gonzalo, J.; Borges, J.G.; Lexer, M.J. A web-

based ToolBox approach to support adaptive forest management under climate change. *Scand. J. For. Res.* **2014**, *29*, 96–107.

25. Reynolds, K.M.; Borges, J.G.; Vacik, H.; Lexer, M.J. ICT in forest management and conservation. In *Information Technology and the Forest Sector. IUFRO World Series Volume 18*; Hetemaki, L., Nilsson, S., Eds.; International Union of Forest Research Organizations: Vienna, Austria, 2005; pp. 150–171.

26. Bhargava, H.K.; Power, D.J.; Sun, D. Progress in web-based decision support technologies. *Decis. Support Syst.* **2007**,*43*, 1083–1095.

27. Lexer, M.J.; Seidl, R. Addressing biodiversity in a stakeholder-driven climate change vulnerability assessment of forest management. *For. Ecol. Manag.* **2009**, *258*, S158–S167.

28. Fontes, L.; Bontemps, J.-D.; Bugmann, H.; van Oijen, M.; Gracia, C.; Kramer, K.; Lindner, M.; Rötzer, T.; Skovsgaard, J.P. Models for supporting forest management in a changing environment. *For. Syst.* **2010**, *19*, 8–29.

29. Hanewinkel, M.; Peltola, H.; Soares, P.; Gonzalez-Olabarria, J.R. Recent approaches to model the risk of storm and fire to European forests and their integration into simulation and decision support tools. *For. Syst.* **2010**, *19*, 30–47.

30. Kramer, K.; Buiteveld, J.; Forstreuter, M.; Geburek, T.; Leonardi, S.; Menozzi, P.; Povillon, F.; Schelhaas, M.J.; Teissier du Cros, E.; Vendramin, G.G.; *et al.* Bridging the gap between ecophysiological and genetic knowledge to assess the adaptive potential of European beech. *Ecol. Modell.* **2008**, *216*, 333–353.

31. Kramer, K.; van der Werf, D.C. Equilibrium and non-equilibrium concepts in forest genetic modelling: Population- and individually-based approaches. *For. Syst.* **2010**, *19*, 100–112.

32. Uran, O.; Janssen, R. Why are spatial decision support systems not used? Some experiences from the Netherlands.*Comput. Environ. Urban Syst.* **2003**, *27*, 511–526.

33. IPCC. *Special Report on Emissions Scenarios*; Nakicenovic, N., Swart, R., Eds.; Cambridge University Press: Cambridge, UK, 2000; p. 570.

34. Van Orshoven, J.; Gilliams, S.; Muys, B.; Stendahl, J.; Skov-Petersen, H.; van Deursen, W. Support of decisions on afforestation in north-western Europe with the Afforest-sDSS. In *Environmental Effects of Afforestation in North-Western Europe*; Heil, G.W., Muys, B., Hansen, K., Eds.; Springer: Dordrecht, The Netherlands, 2007; pp. 227–247.

35. Jansen, J.J.; Sevenster, J.; Faber, P.J. *Opbrengsttabellen Voor Belangrijke Boomsoorten in Nederland*; IBN-DLO: Wageningen, The Netherlands, 1996; p. 240.

36. Meuleman, B.; Dufrane, F.; Carletti, G. *Productietabel Voor Corsicaanse den in Vlaanderen*; LISEC: Genk, Belgium, 1991; p. 91.

37. Malczewski, J. *GIS and Multicriteria Decision Analysis*; Wiley: New York, NY, USA, 1999; p. 408.

38. Mendoza, G.A.; Martins, H. Multi-criteria decision analysis in natural resource management: A critical review of methods and new modelling paradigms. *For. Ecol. Manag.* **2006**, *230*, 1–22.

39. De Meyer, A.; Estrella, R.; Jacxsens, P.; Deckers, J.; Van Rompaey, A.; Van Orshoven, J. A conceptual framework and its software implementation to generate spatial decision support systems for land use planning. *Land Use Policy* **2013**,*35*, 271–282.

40. Estrella, R.; Cattrysse, D.; van Orshoven, J. Comparison of three ideal point-based multi-criteria decision methods for afforestation planning. *Forests* **2014**, *5*, 3222–3240.

41. Mustajoki, J.; Hämäläinen, R.P. Web-HIPRE: Global decision support by value tree and AHP analysis. *INFOR J.* **2000**,*38*, 208–220.

42. Tervonen, T. JSMAA: Open source software for SMAA computations. *Int. J. Syst. Sci.* **2014**, *45*, 69–81.

43. Gilliams, S.; Raymaekers, D.; Muys, B.; van Orshoven, J. Comparing multiple criteria decision methods to extend a geographical information system on afforestation. *Comput. Electron. Agric.* **2005**, *49*, 142–158.

CHAPTER 3

THEORIZING THE IMPLICATIONS OF GENDER ORDER FOR SUSTAINABLE FOREST MANAGEMENT

Jeji Varghese[1] and Maureen G. Reed[2]

[1]Sociology and Anthropology, University of Guelph, Guelph, ON, Canada N1G 2W1

[2]School of Environment and Sustainability and Department of Geography and Planning, University of Saskatchewan, Saskatoon, SK, Canada S7N 5A6

ABSTRACT

Sustainable forest management is intended to draw attention to social, economic, and ecological dimensions. The social dimension, in particular, is intended to advance the effectiveness of institutions in accurately reflecting social values. Research demonstrates that while women bring distinctive interests and values to forest management issues, their nominal and effective participation is restricted by a gender order that marginalizes their interests and potential contributions. The purpose of this paper is to explain how gender order affects the attainment of sustainable forest management. We develop a theoretical discussion to explain how women's involvement in three different models for engagement—expert-based, stakeholder-based, and civic engagement—might be advanced or constrained. By conducting a meta-analysis of previous research conducted in Canada and internationally, we show how, in all three models, both nominal and effective participation of women is constrained by several factors including rules of entry, divisions of labor, social norms and perceptions and rules of practice, personal endowments and attributes, as well as organizational cultures. Regardless of the model for engagement, these factors are part of a masculine gender order that prevails in forestry and restricts opportunities for inclusive and sustainable forest management.

INTRODUCTION

In Canada, it can no longer be assumed that timber is the sole product of forestry. The broadening of interests in forests has been characterized by increased pressures from diverse interest groups to be involved in sustainable management of public resources. For example, forestry advisory committees across the country have been established to contribute directly into management decisions about forestry and thereby contribute local knowledge to the social, economic, and ecological dimensions of forest sustainability. Social sustainability includes consideration of society's responsibility, which is one of six criterion of sustainable forest management (SFM) in Canada. This criteria "addresses the effectiveness of institutions in managing resources in ways that accurately reflect social values, the responsiveness of institutions to change as social values change, how we deal with the special and unique needs of particular cultural and/or socio-economic communities, and the extent to which the allocation of our scarce resources can be

Citation: Jeji Varghese and Maureen G. Reed, "Theorizing the Implications of Gender Order for Sustainable Forest Management," International Journal of Forestry Research, vol. 2012, Article ID 257280, 11 pages, 2012. doi:10.1155/2012/257280

considered to be fair and balanced (page 17 [1])". This social sustainability criterion includes fair and effective decision making (Criteria 6.4) and raises questions about inclusiveness, that is, how to ensure that a broad range of interests and values are included in these processes, given the heterogeneity of the population, particularly in terms of differences in terms of influence, dependency, and, cross-cutting variables such as gender and ethnicity.

In this conceptual article, we identify three models of engagement in SFM—expert-based, stakeholder-based, and civic engagement—and consider the ways in which each model includes issues and concerns of one social group historically excluded from forestry management—women. This focus does not presuppose that women are the only or most important consideration for sustainable forest management. Nor does it suggest that all forestry issues are segregated by gender. Nevertheless, empirical research conducted in Canada, India, Sweden, Kenya, Nepal, and Thailand has suggested that the interests of women and men in forestry are significantly different and that these differences are typically not reflected in forest policy, see [2]. We suggest that the gap between interests and outcomes may arise because the forestry sector is subject to a "gender order" that privileges men's contributions to forestry, constrains women's participation in forestry management, and ultimately reduces the capacity of the forestry sector to achieve inclusive forestry management as a key component of social sustainability.

We begin our discussion by drawing on previous research that demonstrates that women bring distinctive interests and values to forestry issues. Next, we provide a brief discussion of gender order and inclusive engagement. The consideration of inclusive engagement hinges on conceptualizations of participation and representation. We consider different definitions and how they apply. We then define the different models for involvement—expert-based, stakeholder-based, and civic engagement. We consider, in theory, how women's involvement in each model might be advanced or constrained. We then analyze experiences of women under each model according to five criteria. Finally, we draw some conclusions on how the gender order of forestry affects opportunities for inclusive forest management.

Method/Approach

We approach this paper by conducting a meta-analysis of a range of empirical studies to illustrate key factors that limit women's involvement across a range of engagement options. We draw on research we have participated in across the country from the mid 1990's to the present as well as research conducted by other academics at Canadian universities and researchers at the Canadian Forest Service to examine expert and stakeholder models. For example, to examine expert models we focus on studies of women foresters in British Columbia, Alberta and Saskatchewan see [3–5]. To examine stakeholder models, we focus on research of Burns Lake and Revelstoke Community Forests in British Columbia see [6, 7], the Prince Albert Model Forest in Saskatchewan see [8], in-depth case studies of three public advisory committees in Alberta see [9] and two committees in Manitoba and Nova Scotia see [10] and a national survey of both members and chairpersons of forest advisory committees that was conducted by a network of social scientists in 2004 see [11, 12]. To examine civic engagement models we draw mainly on published studies from outside of Canada, as this model is less prevalent within Canada.

RATIONALE: WHY WOMEN'S VOICES MATTER IN FORESTRY DEBATES

Despite the fact that most Canadian women do not use forest resources for subsistence purposes, they still hold interests and perspectives in forest management that are distinctive from men's and hence, their active involvement in giving advice about how forests should be managed could influence decisions about sustainability. Furthermore, forestry communities are experiencing rapid economic, social, and ecological changes that affect both women and men. At the community level, for example, climate change may have significant effects including reduced health status of residents during extreme events, altered paid and unpaid work patterns within communities, changes in livelihood and household relations, and long term health concerns for indigenous residents who have traditionally relied on country foods of the boreal forest. Because there remain marked differences in the roles, activities and expectations of men and women living in forestry-based communities, women and men will have different capacities to adapt to changing conditions [13].

Research has long documented that women express greater concern about the environment than men and express greater support for the protection of forest ecosystems [12–18]. They also typically express greater risk aversion to and concern for climate change—an environmental concern that will be significant for those living in boreal forests of Canada [19, 20]. Additionally, in Sweden, researchers have observed that women perceive social impacts of forest management differently than men [21], while women employed in the USDA Forest Service are more likely than men to hold positive expectations and higher levels of trustworthiness regarding environmentalists, range users, citizen activists, and tribal representatives [22].

Yet, women have not mobilized these perspectives, and concerns into collective social or political actions affecting forestry directly. With respect to the environment, women have been found working in grassroots organizations [23], engaging more in environmentally friendly behaviors that can be integrated into their everyday life [24], and working actively with forest certification programs [25]. With respect to forestry, typically women are not leaders of forestry companies or national or international environmental organizations [23, 26, 27] and they do not participate as actively in other decision-making positions to advance these interests [14].

Thus, while there is ample reason to believe that women bring different values, perspectives, expectations, and concerns to forestry issues, women have yet to advance these concerns directly. We suggest that this discrepancy can be explained by understanding how a particular gender order shapes ideas about what interests become represented in forest management forums. In the next section, we define gender order, describe how it operates within forestry communities, and illustrate how it shapes ideas and processes associated with inclusive forest management.

DEFINING KEY CONCEPTS

Our theoretical approach is taken from feminist scholarship on gender order, public involvement in environmental management, and inclusivity as a social indicator of sustainable forest management.

Theory on Gender Order

Theoretical work on gender order has focused on organizations and work sites. It is also applicable to "communities" in the sense of territorial and occupational communities related to forestry. A gender order refers to the "dichotomous order of gender whereby maleness and femaleness are perceived as opposites and attributed different forms of behavior, different roles, and different places (page 246 [28])". A gender order is not a natural order, but one that is created and recreated in everyday work interactions through organizational cultures that establish the normative rules, values, and meanings in the workplace. According to Gherardi and Poggio [28] these organizational cultures both establish and institutionalize gender positionings, frequently with the effect of "keeping women in "their place" (page 246 [28])".

That forestry work is highly gendered has long been documented. "Canadian forestry continues to be dominated by a masculine gender order that separates men and women and favours male workers in general… [whereby] potential contributions women might make to management and planning for the sustainability of forestry and forestry communities are overlooked (page 78 [29])". In North America and Scandinavia, forestry occupations have created and elevated the importance of a working man's culture or community [30–34]. This bias has been generated by the mystique of logging that created an image of forestry as hard, outdoor, physical labor that is dirty, dangerous, and ultimately masculine. Despite an increase in the range of forestry occupations in the late 20th century that includes planning, regulation, as well as the introduction of labor saving devices, the idea of forestry as men's work still prevails. In forestry communities, this image is married to a traditional division of labor where men have "rightfully" enjoyed forestry jobs and high incomes and have served their families as primary breadwinners, while women have stayed at home to raise children or have only worked for "pin money" in forestry households. While this "traditional" (Traditional is set within quotations because of the debate over whether traditional refers to long-standing and prevalent gender roles or ones that are relatively recently instituted or locally naturalized [35].) division of labour has crumbled in urban settings, it has remained remarkably resilient in Canadian rural places [3]. For example, even where they are currently employed in "nontraditional" forestry occupations, women in forestry communities still carry the disproportionate share of juggling work-home schedules to meet childcare and other domestic duties [5, 36].

Furthermore, organizations, including advisory committees, model forests and other participatory venues, may be shaped by local practices, activities, norms, and attitudes that are taken for granted in everyday life. For example, in a study of Australian executive culture, Sinclair [37] described the organizational culture as a masculine domain "dominated by values, norms, symbols, and ways of operating that are oriented to men (page 6)" and that these characteristics were essential to the way men constructed their identity as leaders. The problem is "not that women's interests and men's interests are opposed—frequently they coincide—but rather it is to conduct their business as though they are always in synch (page 6)" [37]. Forest management operates within an organizational and occupational culture that is highly gendered. It has historically been biased toward male norms and thus can reproduce gender inequality even today—if unintentionally [22, 34, 38, 39]. So even though women serve on advisory tables, as minority representatives, women are faced with male occupational and organizational cultures that shape the extent to which women's participation is effective. One might suppose that this is a reflection of a deeper structural gender order that permeates society, but other previously male-dominated fields (e.g., business, medicine, engineering) have made positive strides to being more inclusive of women [40] in a manner that is not currently reflected to the same degree within forestry.

This ordering of gender in the structure of forestry and forestry communities has become institutionalized as data used to describe the industry and forestry communities are limited, researchers typically do not segregate their findings on work-labor studies by gender, and government policy makers assume that forestry workers are male [29]. Together, these practices give the impression that women are not numerically, economically, or socially important to forestry, and consequently, women's potential contributions to management and planning for the sustainability of forestry and forestry communities are frequently overlooked. This pattern also affects decisions about when, how, and who to involve in sustainable forest management.

Inclusive Engagement

An enormous literature across a range of topic areas has dealt with public involvement in political and social life. Environmental management in general and forestry in particular are no exception. "Natural resource scholars and practitioners who are interested in public engagement often refer to concepts such as representation, participation, or involvement, whereas democratic theorists often refer to the term inclusion (page 533)" [41]. Our conception of inclusion is that it involves both "nominal" and "effective" participation.

Nominal participation refers to the simple demographic representation of particular groups in society. Representatives are assumed to share the values, attitudes, and socioeconomic characteristics of those they represent; in a sense, they are mirrors of the larger population [42, 43]. Wellstead et al. [42] and Pitken [43] refer to this form of representation as descriptive or mirror representation. Participants may be seen as "standing for" the larger group. Applying this to forest management, if women are present in advisory committees, they are assumed to represent women's interests. This is a fairly limited and often inaccurate portrayal of how people actually operate. The limitation of nominal participation is that it tends to assume that individuals embody and represent a series of static characteristics rather than act on their beliefs. In the case of gender and forestry, a focus solely on nominal participation would assume that only women will bring forward concerns related to their gender and/or that women may restrict their contributions to such concerns. Despite their small numbers, we know that women come to decision making forums with knowledge and perspectives that are not necessarily relevant to gender (e.g., knowledge of the biophysical system), while they may have concerns about the topics of discussion or the procedures by which decisions are made that are different from those of their male counterparts.

Effective participation requires that participants are active and engaged in forest management. An assessment of effective participation would consider the rules of decision making, such as whether a process is viewed as accessible, transparent, fair, and consistent with existing laws, regulations, or policies [8, 44]. It might also consider the attributes of individuals such as their knowledge of particular issues, their ability to mobilize resources, and their comfort in speaking out on particular issues. Thus, effective participation involves assessing the ability of members to enhance the equity and efficiency of decisions and to advance their goals.

Effective participation also gives consideration to how power operates within SFM processes and the influence of power on the capacity of individuals and groups to advance their interests [3, 6]. Biases can be subtle, almost invisible, and yet, they can influence effective participation. They can emerge in procedures that determine selection criteria and processes, identify sources and relevance of data, value alternative experiences and knowledge, set the location and timing of meetings and payments for attendance, and provide level and type of agency support from government for the process [45, 46].

We suggest that nominal participation is linked to effective participation, particularly where participating groups are minorities. That is, the number of people from particular social groups will influence the "rules of entry" and "rules of practice"—both formal and informal—and thereby alter the conditions for effective participation. The absence of particular social groups may give rise to the assumption that the rules and procedures adopted by the more restricted group represents the "norm" against which any variation is considered deviant. Here, the notion of critical mass is relevant. For instance, Dahlerup [47] observed that among Scandinavian women politicians, once women became a significant minority (passing a threshold of some 30% seats in Parliament or local councils), there was less stereotyping and openly exclusionary practices by men, a less aggressive tone in discussions, a greater accommodation of family obligations in setting meeting times, and a greater weight given to women's concerns in policy formulation. The necessity of critical mass within forest advisory committees has also been documented more recently by Richardson et al. [10].

Effective participation may also be achieved through two forms of representation (subjective attached and objective unattached) noted by Pitken [43] and Wellstead et al. [42] whereby participants are "standing for" or viewed as acting on behalf of other groups. Subjective attached representation refers to acting for those with a subjective interest [43]. In the context of forestry, most advisory committees operate on this model; that is, participants are acting on behalf of others who share a stake in forestry matters. In the context of issues with a gender dimension, those with a subjective attached form of representation would be sensitive to gender issues and would acknowledge the differential stake that women may have in SFM. Hence, those perspectives would still be taken into account.

Objective unattached representation refers to acting for those who are uninterested [43]. This form of representation would involve individuals who might not be directly interested in forestry, but who might represent interests on behalf of all Canadians or even on behalf of the forest itself. Participants who speak for intrinsic values or future generations would fall into this category. With respect to gender, this form of representation would also imply that even if women were not present, there would be sensitivity to differences in perspectives by gender by those who participate. There is not much experience with this form of objective unattached representation in the Canadian forest sector. Value-based committees (with which we have little experience in Canada), citizen juries documented in the United States (again, little experience in Canada) or broader-based venues that are less engaged (e.g., open houses, questionnaire surveys, voting), and social science research (a few limited examples in Canada) may be seen as examples of attempts to achieve objective unattached representation.

Each of these forms of inclusion implies different ways in which gender-based issues are understood and brought to the table. Nominal participation/descriptive representation would require that women be present to ensure their interests are met, while subjective attached and objective unattached representation suggest that women need not necessarily be present. They do, however, require that on the part of participants there is an awareness of and an ability to understand and speak for the different perspectives that women and men may have.

INCLUSIVENESS OF EXPERT-BASED, STAKEHOLDER-BASED, AND CIVIC ENGAGEMENT MODELS

Within this subsection, we examine three models of engagement for their extent of inclusiveness in SFM. To do this, we compare how the role of women across expert-based, stakeholder-based, and civic engagement models differ and examine factors that constrain women's access and effectiveness in each model.

The pyramid in Figure 1 illustrates fewer numbers of people involved in more elite models at the top and greater number of people involved as we move toward the bottom. Hence, by definition, the models at the top of the pyramid offer smaller degrees of inclusion of the range of views than do the models towards the bottom of the pyramid. Yet, the level of input is less detailed and less sustained as the scope of involvement widens. Nevertheless, in all three models, both nominal and effective participation of women may be constrained (or enhanced) by several factors including rules of entry, divisions of labor, social norms and perceptions and rules of practice, personal endowments and attributes, as well as organizational cultures.

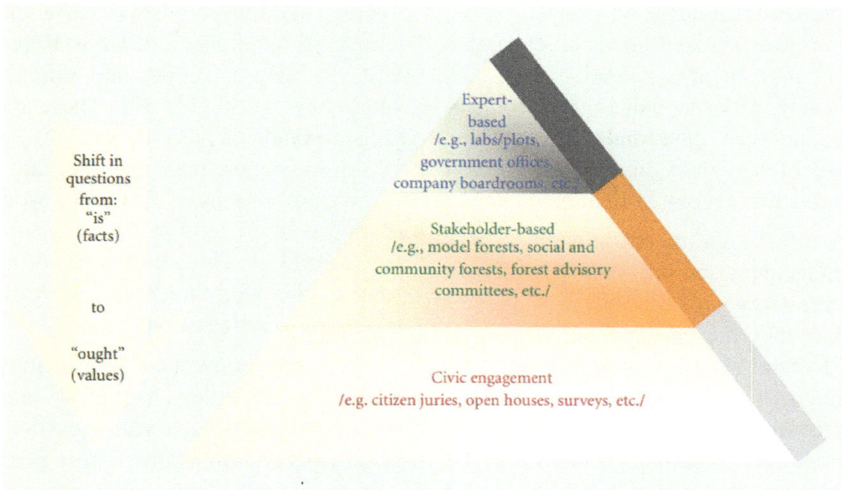

Figure 1: Three Models of SFM Engagement.

These factors provide points of analysis or assessment of each of the three models described in more detail below. It is important to note that these factors act on different scales and can interact with each other. Rules of entry refer to how access is gained or constrained for nominal participation. Effective participation is enabled or constrained by the other factors. Division of labor considers how paid and unpaid work as well as specialization within paid work may be divided along gender lines. Social norms and perceptions and rules of practice include broader societal expectations about acceptable individual behavior within social settings as well as informal guidelines about how to behave collectively. Personal endowments and attributes include education level, property status, marital status, age, and so forth. Organizational culture refers to the shared values and norms that guide behavior within an organization and between the organization and others. We argue that regardless of the model for engagement, these factors are part of a masculine gender order that prevails in forestry.

Expert-Based Models of Engagement

Expert-based models provide representation and voice to experts (whether foresters, scientists, policy makers, etc.); thus privileging their knowledge in SFM. The underlying assumption within this model is that expert knowledge produces "facts," and that decision making needs to rely on these "objective facts" or "truths." In order to be involved in expert-based models of forest management, expertise is typically restricted to knowledge about forestry science, regulation or policy, and business. While decisions made about forestry affect forestry communities more broadly, it is rare that expertise is sought about the social dimensions of forestry communities, including public health, social welfare, culture, and so on that may be affected by changes in forestry or land use policy.

Rules of entry into expert panels require that participants have experience in positions of specific forestry expertise and/or authority. Very few experts in forestry science and business are women. For example, between 4 and 16 percent of registered professional foresters are female in each province [48]. Furthermore, Teske and Beedle [48] noted that in that year, there were no women Chief Executive Officers in any of the forest companies, although there were several women vice presidents. Despite increases in the number of women with formal education in Canada, women continue to be significantly underrepresented in management positions in the public and private sectors [5]. Thus, within expert-based models, the gendered division of labor and rules of entry typically restrict women's participation and reduce the range of issues deemed relevant to community sustainability. The lack of women in management positions within the industry and regulation of forestry translates directly into their absence in expert panels.

Gendered social norms and perceptions and associated rules of practice also influence expert-based models. There remains a social construction of appropriate female behavior (e.g., soft speech, deference to males, making the coffee) that operates even in the absence of overt structures. Even when entry rules are favorable, women may not participate as effectively due to a range of preexisting social norms that reinforce stereotypes about acceptable female behavior, appropriate ways of communicating, and social interaction between men and women. For example, female registered professional foresters (RPFs) working in BC in the 1990s reported being allowed to attend policy meetings to serve the coffee rather than to provide expertise [3].

Stakeholder-Based Models of Engagement

Stakeholder-based models give representation and voice to stakeholders, organizations/groups, and so forth that have a stake in forest management. In theory, stakeholder-based models are viewed as being more inclusive than expert-based models, in that they include local people and their knowledge. But of course this raises questions about who determines who has a stake and how groups may be represented? In terms of stakeholder-based models of engagement in Canada, such as within-community forests, model forests, and forest advisory committees, women have had a limited nominal and effective participation. In terms of nominal participation, for example, while 5 of 15 managers (staff) of model forests are female, only 2 of 15 model forest board chairs/presidents (decision makers) are female. In addition, few partner representatives within model forests and stakeholders are female. A case study of a community forest in British Columbia revealed no woman representative [6]. At a much broader scale, an analysis of 102 forest sector advisory committees across Canada revealed that women comprised only 17 percent of the membership [12]. These women held significantly different values about forestry and environmental protection than their male counterparts, they had lower levels of trust for industry, and they rated their experiences on the committee less favorably than men [12].

Formal rules of entry affect nominal participation within stakeholder models of engagement and may reinforce gender bias. Stakeholder representatives are typically selected as part of their employment or they may be selected if they are representatives of formal domains such as unions, management representatives, and chambers of commerce. Women are more likely to be part of organizations that address public health, education, children's sports, and community events that are not as hierarchically organized and/or are less likely to be considered part of "forestry." Men are more likely to be part of unions or business associations that are organized toward forestry employment are thus more likely to be considered stakeholders to represent local interests [5, 10, 29]. Women that do attend are more likely to attend because of their work as ENGO representatives [12]. In addition, Richardson et al. [10] quoted one female member of a forest management advisory committee as saying "because we do not have a lot of women in the industry we do not naturally gravitate towards thinking about or suggesting women to be on the committee (page 527)." This quotation demonstrates that it becomes normalized to think that women are not interested or knowledgeable, and therefore, to overlook opportunities to engage women in these processes. Thus, while nominal participation may be constrained by a lack of women in professional networks of the forestry/resource sector, this gap gains positive reinforcement by prevailing assumptions about women's desire or ability to participate in stakeholder advisory groups.

Nominal and effective participation within stakeholder models of engagement are also affected by a gendered division of labor. Forestry as a profession is subdivided into a number of different area, some involves hard physical labor, in rough inaccessible terrain and remote areas and others in increasingly technical and office-related functions. This diversity of function, in combination with the socialization of men and women, has led to stereotypes about what type of work men and women in forestry prefer. As a result, certain aspects of forestry are seen as more suitable to males or females. Fullerton [49] and Martz et al. [5] found that even though on average women have higher levels of education than men, within the forest industry, women are overrepresented in clerical and administrative occupations and the ranks of unemployed, and that women are underrepresented in operations, scientific, and management categories. In a gender analysis of Alberta foresters, Varghese et al. [4] found that there was a perception of male domination in operations, whereas silviculture (which is considered to have less arduous fieldwork since it takes place in harvested areas) was perceived to be equally accessible to both men and women. Even within government positions related to forestry, women are overrepresented in clerical and administrative occupations and underrepresented in operations, scientific, and management categories [49]. In addition, the gendered division of labor in society continues to shape women's involvement in community and related service work. Despite stronger environmental sentiments, women are less able to be involved in direct activism or extra-curricular voluntary efforts because of the division of labor. Reproductive responsibilities limit the nature, scope, and efficacy of their activism or community work [24, 50]. These responsibilities "not only structure one's disposable time, they also shape the costs and risks of movement participation (page 913)" [24]. Even where women are employed full-time, women, particularly young married women, tend to be more constrained by their family responsibilities. Thus, the timing of meetings and the need to balance other household/childcare work can be serious barriers to women's participation

Personal endowments and attributes can also shape expert women's contributions to forestry management. These factors include education levels, role of the family and in the community, and socioeconomic status. These personal attributes combine with social norms to provide entry points or barriers for some groups over others. For example, women living in rural areas typically have higher rates of formal education than men, but they have less formal employment experience in forestry. In a local culture where education-by-apprenticeship is more highly valued, women's educational status may not grant them greater access to or effectiveness in forestry management processes. Richardson et al. [10] demonstrated how female members of forest management advisory committees were conscientious of their minority positions and felt uncomfortable offering their views on issues unless they were confident in their views.

Social norms and perceptions and rules of practice reinforce the division of labor and rules of entry. Related to the division of labor is a perception that women do not have a direct stake in forestry management or are not qualified to represent community interests. For example, women on Vancouver Island were shut out of some forestry activities and associated discussions because they were told that they "could not handle the language" [3]. If women were to participate, they may try to raise issues that are deemed irrelevant. For example, when women on Vancouver Island tried to engage men in discussions about local economic development, they were told that men in the community were not interested in participating in women's knitting circles. Richardson et al. [10] found masculine norms were taken for granted within forest management advisory committees. Many participants within her study felt that women on these advisory committees displayed behaviors typically attributed to men (e.g., being outspoken and confident) rather than behaviors typically attributed to women (e.g., shyness or drawing on emotional arguments).

These perceptions and practices reinforce and reproduce a masculine organizational culture that limits women's nominal and effective participation and reproduces gender inequality—if unintentionally. Gherardi [51] notes that even where masculine organizational cultures appear to include women, participants typically express their distrust of equality and implicitly threaten those who seek to change the rules. All of these challenges reduce the number and efficacy of women participants in stakeholder-based models that operate within highly masculinized cultures.

Civic Engagement Models of Engagement

Civic engagement models reflect more of an increased acknowledgement of the public nature of the bulk of forest resources and hence are based more on representing the diverse set of values that are present among the general public, not only the local communities. Civic engagement models however are not well developed in the Canadian forest sector, so we turn to examples in other jurisdictions. Citizen juries, which were originally developed in Europe [52, 53], have been used in the United States to incorporate the concerns of "ordinary citizens" in contested public debates [54].

Rules of entry. Generally, citizen juries are randomly selected with some consideration of balancing across selected characteristics such as age, education, and gender [53], however there are also examples where marginalized populations, such as women, are targeted [55]. However, in the latter case, this "process of selecting jurors may be contested—for example, those who do not feel it is important for women to participate directly (page 79 [56])" reflecting gendered social norms/perceptions and rules of practice impacting nominal participation.

Social norms/perceptions and rules of practice and organizational cultures can also diminish effective participation. As citizens' juries are modelled on an idealization of the legal process [57] with an emphasis on expert testimony and rationalism [58], Ward et al. argue that it is "easy to be patronizing, tolerating "non-rational" forms of argumentation from "others" (e.g., from women or indigenous groups), or as a second-best when rational argumentation fails (page 286 [59])" Pickard's [60] criticisms of citizen juries related to their lack of accountability, authority, legitimacy, and representativeness raises broader questions about the role of civic engagement in forestry management.

Other mechanisms that would involve a wide range of participants who speak for others include open houses, surveys, and workshops. Although people may speak on their own behalf, the commitment by each individual is less sustained and the input is less detailed than that of the other mechanisms. In addition, although women may secure nominal participation, effective participation is marred by the lack of influence of most civic engagement models [60].

Table 1 provides an overview of some types of measures that would enable us to assess women's nominal and effective participation for each form of representation within each model of engagement. Citizen engagement models also seem to share many of the drawbacks of stakeholder models. As we move from expert to civic models, we improve the chances for nominal participation of women because the criteria for selection become less specialized, however, none of the models addresses effective participation because the barriers are systemic, insomuch that the barriers are often invisible.

Table 1: Measures to assess female participation and representation in SFM engagement models

Participation	Forms of Representation	SFM engagement models		
		Expert	Stakeholder	Civic
Nominal	Descriptive	For example, numbers of female "experts"	For example, numbers of female stakeholder representatives and/or gender issue focused stakeholders	For example, numbers of females in forest advisory committees
Effective	Descriptive	For example, female experts are able to voice their concerns in a meaningful way and be heard	For example, female stakeholders reps are able to voice their concerns in a meaningful way and be heard	For example, females are able to voice their concerns in citizen juries in a meaningful way and be heard
	Subjective attached	For example, experts act on behalf of women forestry experts whether present or not	For example, stakeholder representatives act on behalf of women/women's groups who have a stake in forestry whether present or not	For example, citizen juries act on behalf of women's/women's groups who have an interest in forestry whether present or not
	Objective unattached	For example, experts acting on behalf of women who do not have a direct stake in forestry even when they are not present	For example, stakeholder representatives act on behalf of women's groups who do not have a direct stake in forestry even when they are not present	For example, citizen juries act on behalf of women who do not have a direct stake in forestry even when they are not present

Table 2 highlights some examples of how the gender order pervades each of the three models of engagement. Women historically have been absent (lack of nominal participation) or present but effectively silenced (limited effective participation) in forestry management. In comparing engagement models, we see that there has been limited access and limited effective participation for women in all three models. We have shown that a masculine gender pervades regardless of the engagement model.

Table 2: Examples of influence of gender order on inclusivity of engagement models

Processes of gender order	**Engagement Models**		
	Expert (e.g., expert panels)	**Stakeholder (e.g., advisory committees)**	**Civic (e.g., citizen juries)**
Rules of entry	The few female experts in forestry science and business limit nominal participation	Stakeholders often represent formal domains limiting nominal participation as women often involved in informal domains considered external to forestry.	Process of selecting citizen juries contested based on social norms.
Division of labor	Few females in senior management positions limit nominal participation.	Female foresters are often assigned to office-related forestry work. Women often involved in community and related service.	
Social norms & perceptions and rules of practice	Pre-existing social norms that reinforce gender stereotypes limits effective participation	Perception that women do not have a direct stake in forestry decisions or are not qualified to represent community interests.	Citizen juries can be patronizing of nonrational argument.
Personal endowments and attributes	Stereotype that effective managers require masculine skills.	Lack of critical mass may influence confidence of women.	
Organizational cultures	Gender order within forestry as a profession	Gender order within industry and society.	Emphasis on rationalism in citizen juries excludes other forms of arguments

DISCUSSION AND CONCLUSION: HOW THE GENDER ORDER OF FORESTRY AFFECTS THE OPPORTUNITIES FOR INCLUSIVE SUSTAINABLE FOREST MANAGEMENT

Figure 2 is a concept map that provides an overview of the links among the concepts we have discussed in this paper. By starting at the bottom of the concept map and working our way to the top, we examined how gendered rules of entry, divisions of labor, social norms and perceptions, personal endowments and attributes, and organizational cultures result in and reinforce a particular gender order that constrains both nominal and effective participation within three engagement models for forestry management. We did this by examining the links among the different types of participation and representation and the implications of these differences for the engagement models. Despite different models or strategies, these factors remain as persistent and often invisible barriers. We argued that engagement models are embedded within a masculine gender order that severely restricts women's nominal and effective participation.

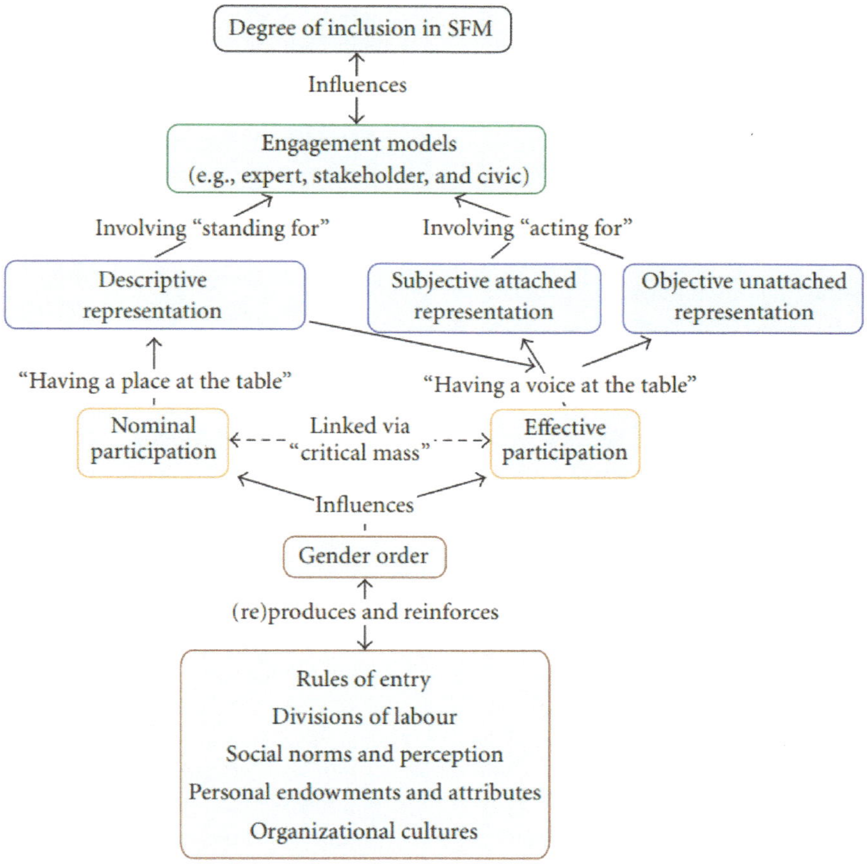

Figure 2: Concept map highlighting framework for considering gender order in SFM.

Like in forestry itself, women are distinctly under-represented in SFM. One reviewer raised an interesting question about possible causes for this underrepresentation: "is it possible that there exists a deliberate resistance to women's involvement on the part of some actors on the basis of the likelihood that women's perceptions and concerns will challenge dominant narratives in the industry? In other words, perhaps it is not women per se who are being excluded but rather the set of values and concerns they have a propensity to hold?" Inclusiveness across a range of social groups (representing diverse values and concerns) is necessary to work toward achieving social sustainability of forests. In order to enhance inclusiveness in SFM, we need to pay attention to how women and other marginalized groups may be excluded from SFM and/or effectively participating in SFM. This requires a mix of strategies that would increase the nominal participation of women and also improve the opportunities for women to be effective. Moving beyond a simple add women and stir approach, we need to address those constraints that limit effective participation by identifying and addressing the underlying masculine gender order in forestry. Because of the link between nominal and effective participation, we need strategies that address both.

Simple practical strategies to address social constraints to SFM recognize the competing time interests of women. For example, providing an honorarium, childcare/eldercare (or costs), considering the timing of meetings recognizes time constraints. In addition, ensuring a "critical mass," providing gender sensitivity training may address some of the constraints women face in actively engaging in these processes. Capturing gender differences in statistics provides essential information to acknowledge and address possible gender issues. But we recognize that these solutions focus on simplistic ideas about representation and inclusiveness. Consequently, we also advocate for more difficult changes, such as raising the profile of social sustainability alongside ecological and economic sustainability; addressing how systemic biases within forestry communities and the forest industry consider production, consumption, and appreciation of forest resources as gender-neutral; giving greater attention to how power relations affect the definition and use of local community knowledge, with particular attention to how that knowledge is produced and how participation is constrained; greater attention to strategies for incorporating the resulting diversity into SFM (e.g., gender differences across different models, such as consensus, conflict resolution, and majority wins). Greater participation in this case by women may not yield more democratic processes if we do not also have the right mix of participants, if we have not addressed the ground rules for debate, or if we ignore the way in which power operates at the negotiating table including how power enters speech itself. For those who are motivated to increase women's participation in forestry governance, increased awareness of the gender order and a realization that forestry does not mirror society but rather appears to be an outlier generally would also enable consideration of how other sectors have addressed some of the structural issues. One of the benefits of our meta-analysis is that we can develop "new processes" with our eyes open in advance and try to address some of the "structural" challenges rather than simply "tinker" with the engagement models hoping it will fix things.

Before closing, we acknowledge that not all women want to participate in forestry governance, just as not all men do. But those women who do want to participate may be constrained by a number of factors, including the masculine nature of direct venues as well as the organization preferences of women. As we strive to improve processes for sustainable forest management, it is important to continually increase our understanding of how our policies and daily practices construct gender. Particularly as gender positioning affects the formation and implementation of policy and has enormous influence in the differential capacity of women and men to contribute to ecological, economic, and social sustainability. A gender focus will help to sensitize researchers and policy makers to multiple inequities and help create opportunities for more inclusive concepts, analyses, and ultimately more inclusive policies and practices that place equal value on the contributions of women and men, and particular social groups without privileging any particular one. In this way, we can gain a better understanding of what constitutes inclusive forest management for all residents of forestry communities and try to identify strategies to deliver it.

While we discuss gender as a point of exclusion, we realize that the category of women is a heterogeneous one. Some groups of women are not marginalized, some groups may face forms of marginalization that are similar to other groups, while others may face a double jeopardy. For example, indigenous people and those associated with nontimber forest products have also been marginalized and may share similar forms of exclusion. Women within these social groups may experience multiple forms of exclusion from forestry management due to their gender and to the social status of their social or cultural group [61]. The exclusions can perpetuate overlapping and/or cumulative invisible losses, whereas inclusionary forest management can increase transparency and make these losses visible [62]. It is possible that the framework we have developed may apply to other marginalized or excluded groups, however, only empirical research would reveal its suitability.

ACKNOWLEDGMENTS

The authors would like to acknowledge and thank their research participants over the years. Funds for research reported here were gratefully received from the Sustainable Forest Management Network. Students and colleagues in Canada and Sweden have helped hone their ideas. They also thank the anonymous reviewer for their insightful comments.

REFERENCES

1. Canadian Council of Forest Ministers (CCFM), Defining Sustainable Forest Management in Canada: criteria and Indicators 2003, Natural Resources Canada, Ottawa, Canada, 2003.
2. G. Lidestav and M. Reed, "Preface: gender and forestry," Scandinavian Journal of Forest Research, vol. 25, no. 9, pp. 1–5, 2010. · ·
3. M. G. Reed, Taking Stands: gender and the Sustainability of Rural Communities, UBC Press, Vancouver, Canada, 2003.
4. J. Varghese, P. Nelson, and R. Boehm, Towards Equal Participation: Women Foresters in Alberta, Gender and Policy Action Committee, Alberta, Canada, 1999.
5. D. Martz, M. G. Reed, I. Brueckner, and S. Mills, Hidden Actors, Muted Voices: The Employment of Rural Women in Canadian Forestry and Agri-food Industries, Final Report to the Status of Women, Policy Research Fund, Ottawa, Ontario, 2006.
6. M. G. Reed and K. McIlveen, "Toward a pluralistic civic science?: assessing community forestry," Society and Natural Resources, vol. 19, no. 7, pp. 591–607, 2006. · ·
7. J. Varghese, Impacts of, and challenges for, local ownership in the forest sector, Ph.D. Dissertation, University of Alberta, 2005.
8. J. Varghese, Prince Albert Model Forest Public Participation Review. A report prepared for the PAMF Local Level Indicators Working Group, Saskatchewan, Canada, 2000.
9. J. R. Parkins and D. J. Davidson, "Constructing the public sphere in compromised settings: environmental Governance in the Alberta Forest Sector," Canadian Review of Sociology, vol. 45, no. 2, pp. 177–196, 2008.
10. K. Richardson, A. J. Sinclair, M. G. Reed, and J. R. Parkins, "Constraints to participation in Canadian forestry advisory committees: a gendered perspective," Canadian Journal of Forest Research, vol. 41, no. 3, pp. 524–532, 2011. ·
11. J. R. Parkins, L. Hunt, S. Nadeau, M. Reed, J. Sinclair, and S. Wallac, "Public participation in forest management: results from a national survey of advisory committees," Information Report NOR-X-409, Northern Forestry Centre Information Report Series, 2006.
12. M. G. Reed and J. Varghese, "Gender representation on Canadian forest sector advisory committees,"Forestry Chronicle, vol. 83, no. 4, pp. 515–525, 2007. ·
13. M. G. Reed, "Environmental governance and gender in canadian resource industries and communities," in Resource and Environmental Management in Canada: Addressing Conflict and Uncertainty, B. Mitchell, Ed., pp. 526–553, Oxford University Press, New York, NY, USA, 4th edition, 2010.
14. K. D. V. Liere and R. E. Dunlap, "The social bases of environmental concern: a review of hypotheses, explanations and empirical evidence," Public Opinion Quarterly, vol. 44, no. 2, pp. 181–197, 1980. · ·
15. P. Mohai, "Men, women, and the environment: an examination of the gender gap in environmental concern and activism," Society & Natural Resources, vol. 5, no. 1, pp. 1–19, 1992. ·
16. D. J. Davidson and W. R. Freudenburg, "Gender and environmental risk concerns: a review and analysis of available research," Environment and Behavior, vol. 28, no. 3, pp. 302–339, 1996. ·
17. D. B. Tindall, "Social values and the contingent nature of public opinion and attitudes about forests,"Forestry Chronicle, vol. 79, no. 3, pp. 692–705, 2003. ·
18. H. Uliczka, P. Angelstam, G. Jansson, and A. Bro, "Non-industrial private forest owners› knowledqe of and

attitudes towards nature conservation," Scandinavian Journal of Forest Research, vol. 19, no. 3, pp. 274–288, 2004. · ·

19. D. J. Davidson, T. Williamson, and J. R. Parkins, "Understanding climate change risk and vulnerability in northern forest-based communities," Canadian Journal of Forest Research, vol. 33, no. 11, pp. 2252–2261, 2003. · ·

20. G. Johnsson-Latham, A Study on Gender Equality as a Prerequisite for Sustainable Development, Environment Advisory Council, Stockholm, Sweden, 2007.

21. S. Arora-Jonsson, "Relational dynamics and strategies: men and women in a forest community in Sweden," Agriculture and Human Values, vol. 21, no. 4, pp. 355–365, 2004. · ·

22. K. E. Halvorsen, "Relationships between national forest system employee diversity and beliefs regarding external interest groups," Forest Science, vol. 47, no. 2, pp. 258–269, 2001. ·

23. J. Seager, Earth Follies: Coming to Terms with the Global Environmental Crisis, Routledge, New York, NY, USA, 1993.

24. D. B. Tindall, S. Davies, and C. Mauboules, "Activism and conservation behavior in an environmental movement: the contradictory effects of gender," Society and Natural Resources, vol. 16, no. 10, pp. 909–932, 2003. ·

25. L. K. Ozanne, C. R. Humphrey, and P. M. Smith, "Gender, environmentalism, and interest in forest certification: Mohai›s paradox revisited," Society and Natural Resources, vol. 12, no. 6, pp. 613–622, 1999.

26. B. Livesey, "The politics of Greenpeace," Canadian Dimensions, vol. 28, pp. 7–12, 1994.

27. S. Müller, "Report on the professional and family situation of women working in the environmental field," in Proceedings of the 2nd European Feminist Research Conference, Feminist Perspectives on Technology, Work and Ecology, T. Eberhart and C. Wachter, Eds., IFF IFZ Interdisciplinary Research Center for Technology, Work and Culture, Graz, Austria, July 1994.

28. S. Gherardi and B. Poggio, "Creating and recreating gender order in organizations," Journal of World Business, vol. 36, no. 3, pp. 245–259, 2001. · ·

29. M. G. Reed, "Reproducing the gender order in Canadian forestry: the role of statistical representation,"Scandinavian Journal of Forest Research, vol. 23, no. 1, pp. 78–91, 2008. · ·

30. T. Dunk, It's a Working Man's Town: Male Working-class Culture in Northwestern Ontario, McGill-Queen's University Press, Quebec, Canada, 1991.

31. M. S. Carroll, Community and the Northwestern Logger: Continuities and Changes in the Era of the Spotted Owl, University of Colorado Press, Boulder, Colo, USA, 1995.

32. B. Brandth and M. S. Haugen, "Breaking into a masculine discourse women and farm forestry,"Sociologia Ruralis, vol. 38, no. 3, pp. 427–442, 1998. ·

33. B. Brandth and M. S. Haugen, "From lumberjack to business manager: masculinity in the Norwegian forestry press," Journal of Rural Studies, vol. 16, no. 3, pp. 343–355, 2000. · ·

34. UNECE/FAO Team of Specialists on Gender and Forestry, Ed., Time for Action: Changing the Gender Situation in Forestry Report, Food and Agriculture Organization of the United Nations, Rome, Italy, 2006.

35. J. K. Gibson-Graham, "Stuffed if I know!: reflections on post-modern feminist social research, Gender,"Place and Culture, vol. 1, no. 2, pp. 205–224, 1994.

36. V. Preston, D. Rose, G. Norcliffe, and J. Holmes, "Shift work, childcare and domestic work: divisions of labour in Canadian paper mill communities," Gender, Place and Culture, vol. 7, no. 1, pp. 5–29, 2000. · ·

37. A. Sinclair, Trials at the Top, The Australian Centre, University of Melbourne, Melbourne, Australia, 1994.

38. P. Davidson and R. Black, "Women in natural resources management: finding a more balanced perspective," Society and Natural Resources, vol. 14, no. 8, pp. 645–656, 2001. ·

39. B. Brandth, G. Follo, and M. S. Haugen, "Women in forestry: dilemmas of a separate women›s organization," Scandinavian Journal of Forest Research, vol. 19, no. 5, pp. 466–472, 2004. · ·

40. WinSETT Centre, "QuickStats # 11: Women in Paid Work," 2011, http://www.ccwestt.org/Portals/0/publications/QuickStatsNo11%20Jul%202011.pdf.

41. J. R. Parkins and R. E. Mitchell, "Public participation as public debate: a deliberative turn in natural resource management," Society and Natural Resources, vol. 18, no. 6, pp. 529–540, 2005. · ·

42. A. M. Wellstead, R. C. Stedman, and J. R. Parkins, "Understanding the concept of representation within the context of local forest management decision making," Forest Policy and Economics, vol. 5, no. 1, pp. 1–11, 2003. · ·

43. H. Pitken, The Concept of Representation, University of California Press, Berkeley, Calif, USA, 1967.

44. A. Conley and M. A. Moote, "Evaluating collaborative natural resource management," Society and Natural Resources, vol. 16, no. 5, pp. 371–386, 2003. · ·

45. M. Alston and J. Wilkinson, "Australian farm women—shut out or fenced in? The lack of women in agricultural leadership," Sociologia Ruralis, vol. 38, no. 3, pp. 391–408, 1998. ·

46. B. Agarwal, "Participatory exclusions, community forestry, and gender: an analysis for South Asia and a conceptual framework," World Development, vol. 29, no. 10, pp. 1623–1648, 2001. · ·

47. D. Dahlerup, "From a small to a large majority: women in Scandinavian Politics," Scandinavian Political Studies, vol. 11, pp. 275–298, 1988.

48. E. Teske and B. Beedle, "Journey to the top—breaking through the canopy: Canadian experiences,"Forestry Chronicle, vol. 77, no. 5, pp. 846–853, 2001. ·

49. M. Fullerton, "Gender structures in forestry organizations: Canada," in Time for Action: Changing

the Gender Situation in Forestry, UNECE/FAO Team of Specialists on Gender and Forestry, Ed., p. 2026, Food and Agriculture Organization of the United Nations, Rome, Italy, 2006.

50. S. McGregor, Beyond Mothering Earth: Ecological Citizenship and the Politics of Care, UBC Press, Vancouver, Canada, 2006.

51. S. Gherardi, Gender, Symbolism and Organizational Cultures, Sage, London, Uk, 1995.

52. University of Newcastle upon Tyne. "Citizens" Jury calls for moratorium on commercialisation of GM to continue University of Newcastle upon Tyne, 2005, http://www.ncl.ac.uk/press.office/press.release/content.phtml?ref=1063031063.

53. G. Smith and C. Wales, "The theory and practice of citizens› juries," Policy and Politics, vol. 27, no. 3, pp. 295–308, 1999. ·

54. Jefferson Center, "Citizen Jury Handbook," 2005, http://www.jefferson-center.org/.

55. M. P. Pimbert and W. T. Prajateerpu, A Citizens Jury/Scenario Workshop on Food and Farming Futures for Andhra Pradesh, India, International Institute for Environment and Development, London, UK, 2002.

56. N. Kanji and S. F. Tan, "Understanding local difference: gender (plus) matters for NGOs," Participatory Learning and Action, vol. 58, no. 1, pp. 74–81, 2008.

57. J. Gobert, Justice Democracy and the Jury, Ashgate Publishing Company, Aldershot, UK, 1997.

58. A. Coote and J. Lenaghan, Citizens' Juries: Theory into Practice, IPPR, London, UK, 1997.

59. H. Ward, A. Norval, T. Landman, and J. Pretty, "Open citizens› juries and the politics of sustainability,"Political Studies, vol. 51, no. 2, pp. 282–299, 2003. · ·

60. S. Pickard, "Citizenship and consumerism in health care: a critique of citizens› juries," Social Policy and Administration, vol. 32, no. 3, pp. 226–244, 1998. ·

61. M. G. Reed and D. Davidson, "Terms of Engagement: the involvement of Canadian rural communities in sustainable forest management," in Reshaping Gender and Class in Rural Spaces, B. Pini and B. Leach, Eds., pp. 199–220, Ashgate Publishing Company, Aldershot, UK, 2011.

62. N. J. Turner, R. Gregory, C. Brooks, L. Failing, and T. Satterfield, "From invisibility to transparency: identifying the implications," Ecology and Society, vol. 13, no. 2, article 7, 2008.

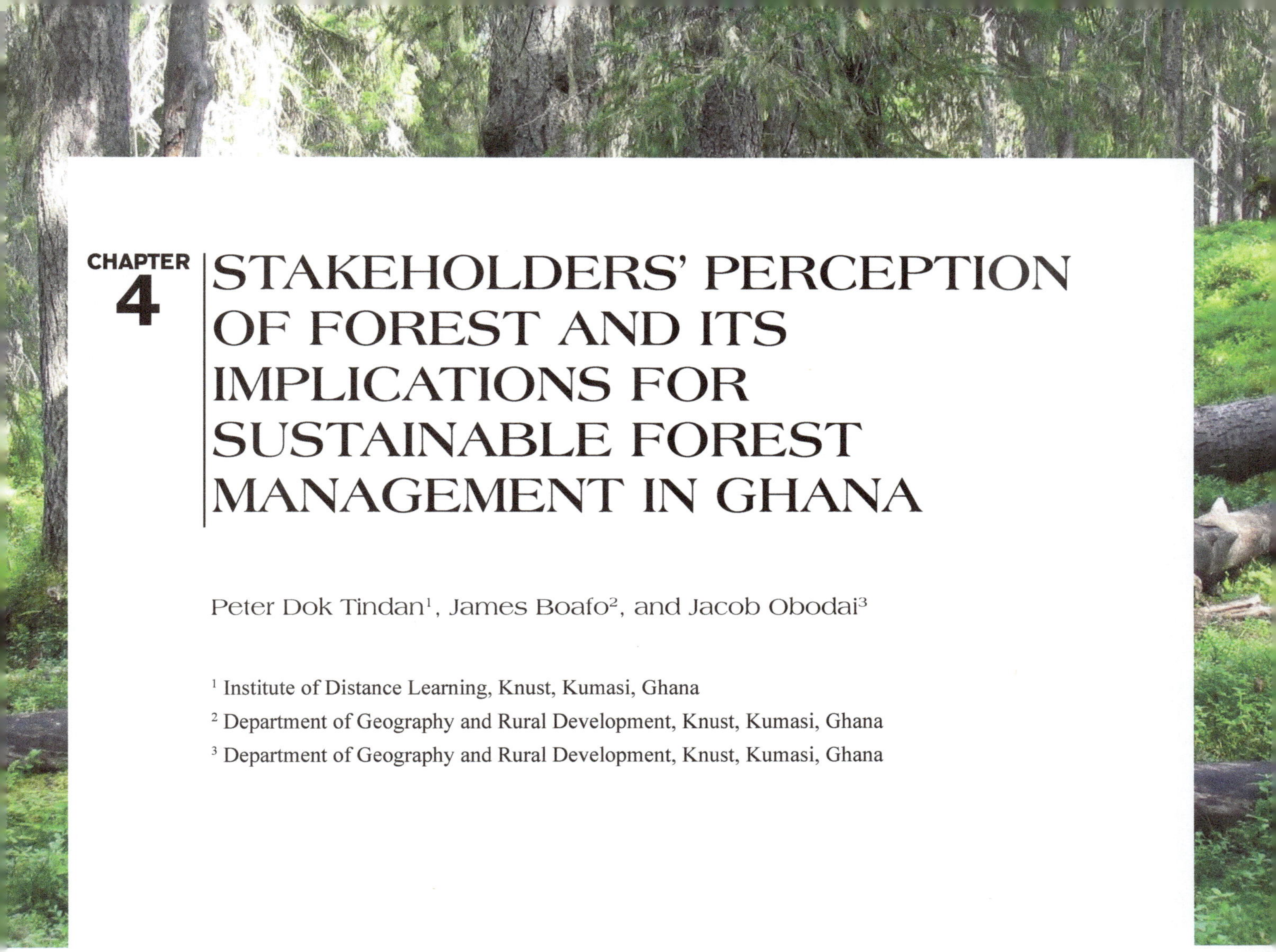

CHAPTER 4

STAKEHOLDERS' PERCEPTION OF FOREST AND ITS IMPLICATIONS FOR SUSTAINABLE FOREST MANAGEMENT IN GHANA

Peter Dok Tindan[1], James Boafo[2], and Jacob Obodai[3]

[1] Institute of Distance Learning, Knust, Kumasi, Ghana

[2] Department of Geography and Rural Development, Knust, Kumasi, Ghana

[3] Department of Geography and Rural Development, Knust, Kumasi, Ghana

ABSTRACT

The objective of the study is to examine stakeholders' perception of forest and its implication for sustainable forest management in Ghana. This has become necessary in the light of the fact that operational application of forest management remains troublesome leading to the development of inconsistent forest management practices in the country. The study was approached using a cross-sectional design imploring semi-structured interviews and questionnaires as data collection method and instrument respectively in Juaso Forest District in the Asante Akim Central Municipality of Ghana. The research findings indicate that stakeholders' participation in sustainable forest management in the Juaso Forest District has been met with a myriad of challenges. These challenges are deeply rooted in the stakeholders' perceptions of what constitute forest in the communities. These perceptions have defined the conditions required for sustainable forest management at the local level. It is noted that sustainable forest management is a collaborative efforts, hence at the local level efforts should be made to ensure participation of all stakeholders in the decision making, policy design and implementation.

Keywords

Stakeholders, Forest, Access, Nature, Right, Management, Collaborative, Ghana.

INTRODUCTION

Stakeholders as those who are likely to be affected by a policy or project and who have the power to help or obstruct its acceptance at the decision making level [1]. These can range generally from entities of national and local government, civil society organizations, private organizations, citizens and international donors. Stakeholders are found in every sector of the economy.

The situation is similar in the forestry sector where there are many stakeholders involved in various forestry activities. The stakeholders in the forestry sector come from different background including foresters, rangers, policy makers, farmers, and local communities (traditional leaders, assembly members, community-based organizations among others). The

Citation: Peter Dok Tindan, James Boafo, and Jacob Obodai, "Stakeholders' Perception of Forest and its Implications for Sustainable Forest Management in Ghana," International Journal of Innovation and Applied Studies, vol. 9, no. 3, pp. 1021–1031, November 2014.

roles and position of these stakeholders are defined based on their distinct expertise. In this connection, the diverse activities in the forestry sector, for instance management, are carried out by individuals who have expertise in management. The local communities however involved in any forest active also have expertise in forest management based on local knowledge and adaptation since their livelihoods are predicated on forest resources.

Forests in general are complex and contested spaces, not fixed entities whose nature can be stated in an absolute way This is because the way forests are understood and valued is inextricably linked with the ideas of these diverse stakeholders who view or lay claim on them [2].

According to [3], a forest is defined as an ecosystem characterized by more or less dense and extensive tree cover, often consisting of stands that vary in species composition, structure, age, class, and associated processes. The United Nations Environmental Programme (UNEP) explains that to understand what constitute forests, two criteria can be used and these include the crown cover threshold and land use criteria. The UNEP highlights a 10% minimum threshold of crown cover for defining both opened and closed forests. In using the land use criterion forests are defined in line with the "main purpose or use" of the forest resources. And this perspective is similarly shared by [4] who claim forests as tree covered areas not predominantly used for purposes other than forestry. Based on this opinion, they have distinguished forests into two types; natural forests and planted secondary or plantation forests.

It is also important to emphasize that the various perceptions of the stakeholders within the forestry sector influence their contribution towards sustainable forest management. In Ghana, forest ownership and management is fraught with the imbalances of trust. Governmental institutions such as the Forest Services Division (FDS), in collaboration with the local communities, present options for collaborative forest management. However, lack of absolute trust and positive expectations, particularly from the local people have invariably derailed the motive of such drives and efforts. Again, the fact of the ad hoc programmes by forestry institutions in curbing rapid deforestation from over-exploitation have usually marginalized the local people in the planning processes. Since the lack of trust in forest ownership and management is a major drive of deforestation, the next section of the paper reviews the trajectory of forest tenureship and management in Ghana.

THE TRAJECTORY OF FOREST TENURESHIP AND MANAGEMENT IN GHANA

Forest power including tenureship in Ghana has a long history dating back to the pre-colonial era, where forests were owned in common by families, clans, and stools (communities) [5], [6]. However, subsequent forest reforms such as the Forest Ordinance of 1927 authorized the colonial government to put part of Ghana's forest into reserves [6]. After the reservation was completed in 1939, a forest policy was adopted in 1948 (Forestry Commission [7]. This policy was criticized by the World Bank preparatory missions for the Forest Resource Management Project (FRMP) as being inadequate in dealing with the problem of deforestation in Ghana [8].

The ownership and management of forest reserves were solely the preserve of the Government of Ghana in trust for the local people. These functions were entrusted into the hands of Forest Services Division at the local level. The 1927 Forest Ordinance also provided an alternative mode of forest management by the owner(s) under the direction of the then Forestry Department (now known as Forest Services Division). Thus, the local communities saw the creation of the reserves as a complete removal of their rights and benefits, and an expropriation of their land and forests [8]. Subsequently, the Concessions Act of 1962 vested the right to grant timber concessions and the management of all timber resources, both on and off reserves, in the central government. In addition, through the Forest Protection Decree of 1974, the management of trees within forest reserves and the rights to own, plant, use, and dispose of trees are controlled by the State [9].

The subsequent forest reforms and policies in Ghana aimed at protecting the forest resources, particularly timber both on and off reserves with no or limited involvement of the local communities. In view of the over-centralization of forest management without any decentralized role for communities and industries, the non-integration of rural production systems into forest management, and the demise of off-reserve forests, the government in the 1980s and early 1990s, called for "social forestry" or "rural forestry" (was promoted by the then Forestry Department) as a way of encouraging communities to establish village, communal, individual or family woodlots [10].

Successive postcolonial governments in Ghana continued the colonial legacy of forest management, which excluded the local people and focused on commercial interests [11]. Currently in Ghana, the chieftaincy institutions exercise resource ownership functions on behalf of the local people whereas the government exercises administrative and management functions of forested lands by holding the lands and trees on them in trust [12]. In this case, people do not have any control over forest resources, particularly the economic species, on their own parcels of lands, which may form part of a concession leased out by the government to timber companies. Victims of such a situation may feel excluded from sharing the benefits of forest resources and may cause damage to these resources.

However, forest communities in Ghana have their livelihoods predicated on the availability, access, and utilization of products of the forest [13], [14]. The exclusion of the local communities in forest management in Ghana denies them the right and ability to access forest resources, including NTFPs, upon which their livelihoods subsist [15]. Consequently, concerns raised by major stakeholders along with increasing international attention on deforestation stirred the revision of the 1948 Forest Policy, which culminated in the adoption of the new Forest and Wildlife Policy (FWP) in 1994 (1994-FWP) [16].

The overall aim of the 1994-FWP was conservation and sustainable development of the nation's forest and wildlife for maintenance of environmental quality and perpetual flow of benefits to all parts of society. It provided a framework for some important legislative reforms including the Forestry Commission Act (Act 5), which consolidates sector agencies [17]. The 1994-FWP also emphasized collaborative forest management (CFM) by enhancing the role of communities in forestry at policy, managerial and implementation levels [18], where the government, led by the Forest Services Division, partnered with forest communities to manage the forest [19]. Thus, the policy emphasized the decentralization of forest management responsibilities to communities. The FC therefore created a Collaborative Natural Resources Management Unit (CNRMU) and, at the community level, the Community Forest Committees (CFCs) and Community Biodiversity Advisory Groups (CBAGs) [17]. Despite the fact that the 1994-FWP recognized the importance of Collaborative Natural Resource Management (CNRM) and Community-based Natural Resource Management (CBNRM), these had no legal backing [18], [17]. Thus, communities do not participate in forest management as of right but as licensees, and also CFCs do not participate in planning or policy formulation about important issues like access rights or benefits sharing [17].

More recently it has been noted that one key factor perpetuating deforestation is the changing perception of stakeholders on forest delineation or reserve. Thus, the perception of forest delineation particularly by the local people varies based on the nominal attention given to trees on off-reserves as compared trees in on-reserves by the FSD. This has become a disincentive towards sustainable forest management. Consequently, the various stakeholders namely, the local government entities in charge of forests and the local communities perceived forest from a different angle, which has influenced their roles in sustainable forest management. It is therefore necessary to identify the position(s) and role(s) of every stakeholder(s) but more importantly how effectively they participate in the pursuit of the objectives of forest management. The understanding of participation per this study is a process through which stakeholders have influenced and share control over the development initiative, decisions and resources, which affect them [20]. Theoretically the research opines that, stakeholders' perceptions of forest influence the position and/or value they place on the resources. This in turn also influences their role in forest management and usage.

Figure 1 is a flow chart showing the interrelationship between stakeholders' perception and the implication for sustainable forest management in Ghana.

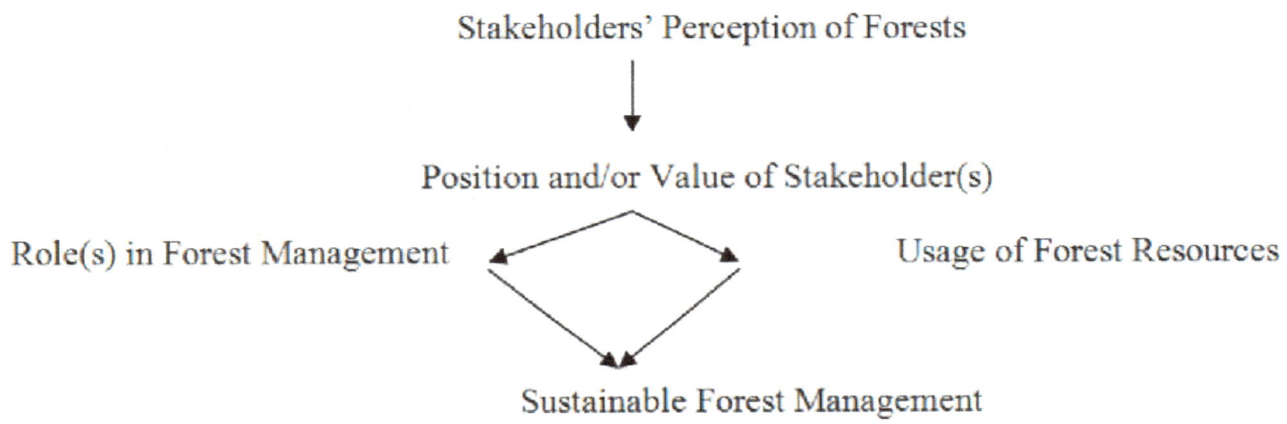

Figure 1: Stakeholder perception and sustainable forest management.

STUDY AREA

The study was conducted in two communities: Dwease and Praaso in the Juaso forest district in the Asante Akim Central Municipal Assembly (AACMA) of Ghana. The Municipality lies between latitudes 6^0 30' North and 7^0 30' North and longitudes 0^0 15' West and 1^0 20' West. It shares boundaries to the north with Asante Akim North district, to the east and to south with Asante Akim South and to the west with Ejisu-Juaben municipality. The populations of the two communities are estimated at 5160 and 5148 for Dwease and Praaso respectively [21].

The vegetation type of the Municipality can be categorized under the moist semi-deciduous forest belt. The vegetation pattern is distributed in three dimensions; Open Forest which covers 576 km^2 and spreads over the highlands of the municipality. The other two include the Closed Forest covering 230 km^2 on the range and the Wooded Savannah covering 246 km^2. The different tree species produce tropical woods such as Wawa, Ofram, Sapele, Sanfina, Okyere (Kofo), Onyina, Kyenkyen, Otie and Yaya, which have high economic value. Apart from wood products, foodstuffs (delicacies such as snails and mushroom) and other raw materials are obtained from the forest for industrial and domestic use [21].

About 82% of the population in the municipality is engaged in private informal occupations like agriculture, trading, vocational services (hair dressing, sewing and driving), carpentry and masonry. Agriculture stands out as the major occupation employing 53.9% of the people especially those aged 15 years and above. The major staple food crops produced include maize, cassava, plantain, cocoyam and yam while cocoa is the major cash crop. Agricultural production is however largely on subsistence basis (72%) of the farmers cultivating less than 3 acres of land) while about 6% engaged in large-scale farming [21]. Figure 2 is a map of the Asante Akim Central Municipality with the studied highlighted in the shades.

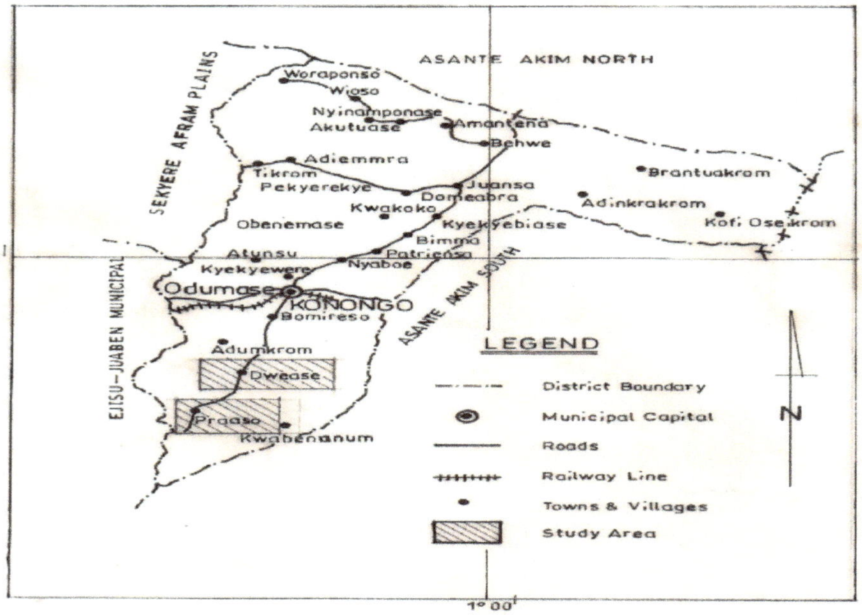

Figure 2: Map of Asante Akim Central Municipal Assembly indicating sampled communities in shades.

RESEARCH METHODS

The cross-sectional design was used for this study because of its suitability. The focus was to indicate the pattern of association between and among the key variables, which were sought, and the cross-sectional design is a good choice in that regard. A cross-sectional design entails the collection of data on more than one case and at a single point in time in order to collect a body of data in connection to two or more variables, which are examined to detect a pattern of association [22].

With regard to the sampling technique, both stratified random and purposive sampling techniques were used because of the mixed research strategy adopted. The stratified random sampling was used to select household heads for the questionnaire administration. The stratification was based on gender (male and female strata), and respondents were then randomly sampled from these strata depending on who expressed interest of being involved in the research. The purposive sampling was used to select subjects based on their relationship with the research questions [22] and this included key informants like Forest Services Division, traditional rulers and assemblymen to conduct the interviews. The total sampling frame consisted of two staff from the Forest Services Division (Assistant manager and a forest guard based in Dwease), Chief of Dwease and the Assemblyman of Praaso. In addition, a total of One Hundred and Ten (110) households respondents were sampled and this consisted of Sixty (60) from Dwease and Fifty (50) from Praaso.

Both primary data and secondary information were collected for this study, with the primary data resulting from a two month field survey in the two communities while the secondary information resulted mainly from peer-reviewed articles, books, magazines and other sources with relevance to the study. Semi-structured interviewing was used for the key informants (traditional rulers, assembly and the Forest Services Division). Though different interview guides were used for the various informants, the content of each basically covered the demands of the key research question. Self-administered questionnaire was also used to solicit the individual views of heads of households regarding the key research question and this was augmented by personal observations, in which the researchers adopted the overt role of being "observers-as-participants".

The pre-coded questions from the questionnaire administration were analyzed using the Statistical Package for Social Sciences (SPSS) and the result presented in frequency tables. The qualitative data was analyzed thematically, by identifying key themes in the interviews and these were outlined through indexing or coding the data and reflecting and interpreting the data. The major ethical consideration observed was with regards to informed consent; however, the essence of the research was explained to each participant before engaging him or her either in an interview or questionnaire administration. But this does not eliminate the fact that, they did not have detailed knowledge about the outline of questions (the way the questions were to be asked and how). The need to preserve such information became necessary in order not influence the perception of respondents and to reduce biases, which could jeopardize the reliability of the research findings.

RESULTS AND DISCUSSIONS

Stakeholders' Perception of Forest and Forest Types in Ghana

In due recognition of the fact that forest ownership in Ghana is considered to be dual (legal estate in the government and the beneficial estate in the local communities), the key stakeholders identified were broadly delineated into state and local partners. In terms of the state partners, forest governance and management is officially vested in the Forestry Commission of Ghana (FC) who carries out this responsibility through its decentralized body; the Forest Services Division (FSD) in addition to other public bodies and agencies like the Wildlife Division, Timber Industry Development Division, Wood Industries Training Centre; and the Resource Management

Support Centre. It was however discovered that the FSD is the only a functional body within the state apparatus with recognized role(s) towards forest management.

From Table 1, it is identified that 77 of the total household respondents (70.6%) stated that the Forest Services Division (FSD) is the major institution of forest governance and management while the remaining 33 respondents (26.6%) indicated that local partners including the Chiefs, Assemblymen and local residents are responsible for forest governance and management.

Table 1: Respondents' views on the stakeholders of forest governance and management

Stakeholders	Frequency	Percent (%)
Forest Services Division	77	70.6
Local partners	33	29.4
Total	110	100

The understanding of forest as perceived by the various stakeholders involved in this study unanimously connotes tree-dominated environment. The study identified the individual responses on the perception of forest to be directly in line with the two most critical factors; *crown cover threshold* and *land use criteria* used in distinguishing forest types as highlighted by the UNEP. Based on these factors four categories of viewpoints were identified in relation to the understanding of forests in the studied communities. These include "*Crown cover threshold only*"; "*Land use criterion only*"; "*Both Crown cover threshold and Land use criteria*"; and "*Other, please specify*" as shown in Table 2.

Table 2: Household respondents' criteria for defining forest

Criteria	Frequency	Percent
Crown cover threshold only	39	35.5
Land use criterion only	24	21.8
Crown cover threshold and Land use criteria	37	33.6
Other, please specify	10	9.1
Total	110	100.0

It is identified from Table 2 that, 39 respondents (35.5%) defined forest using the crown cover threshold criterion only. Based on this, their perception of forest is expressed as *"a land with large and closed association of tree covering and interaction of other biodiversity"* (Personal interviews, 2013). Though the local people do not necessarily understand the technicalities of what constitute the crown cover threshold to distinguish between opened and closed forest types as indicated by the UNEP, their descriptions is congruent with the perception of forest by [23] as the associations of large, woody, perennial tree species, generally several times the height of a human, and with a more or less closed canopy of leaves overhead.

A total of 24 respondents (21.8%) also explained their understanding of forest following the land use criterion only. By this criterion, forest is *"a land that has never or is partially being cultivated"*. The focus of this perspective is informed by the role of human intervention in the regeneration or degradation of the forest environment and this clearly relates with the definition of forests by [24] as tree covered areas not predominantly used for purposes other than forestry. Majority of the household respondents identified that a greater proportion of the forests can be classified as plantation or secondary forests because they have their farmlands located within the forested area. Even respondents argued that the Dome River Forest reserve located in Dwease can be classified as secondary forest because parts of it had been used for agro-forestry under the Modified Taungya System (MTS).

For the 37 respondents (33.6%) who defined forest using both the crown cover threshold and land use criteria, a forest "*is any land with closed covering of trees and other biodiversity that has never or is partially being cultivated*". It is observed and also indicated by respondents that because much of the forests have been cultivated or are still being cultivated and this has essentially affected the crown cover development of trees and caused the loss of indigenous trees and other vegetative species.

While the crown cover and land use criteria seem to dominated in the expressions on what constitute forests among the local residents, as pointed out in Table 2, it was also observed that, 10 respondents (9.1%) defined forests based on different factors. One of such factors highlighted was the *"maturity of the trees"*. By this the respondents meant the *"size"* of the trees measured in terms of the thickness of the trunk, which they use to determine whether a tree is ready to be harvested. Based on this view, it was common to hear some respondents argue that there is no forest in the area. This is because logging especially the operation of chainsaw has destroyed if not all, majority of the trees needed to classify the tree-dominated ecosystem as forest(s).

Again, some indicated the institutional policy framework that defines the tree ownership and management. Under this, forests can be a land owned and managed solely by "the government" (herein defined as the Forest Services Division) or jointly owned and managed "the FSD and traditional rulers" for a given community. This is confirmed in the definition of forest by the FSD. According to the FSD;

"A forest is a piece of land earmarked and set aside by prescribed policy with the minimum height of the trees being 5meters that involves a complex interaction of biodiversity, water bodies, and other physical land characteristics" (Personal interview 2013).

The *"5meters"* height of trees within a forest can be used to define the crown cover threshold as indicated by the household respondents. The definition of height measured in "meters" partially relates with the understanding of forest as advanced by [25]. According to FAO, forest refers to as a land spanning more than 0.5 hectares with trees higher than five (5) meters and a canopy cover of more than ten (10) percent, or trees able to reach these thresholds in situ; it does not include land that is predominantly under agricultural or urban land use. The use of height is a suitable denominator to delineate forests from other terrestrial biome and more specifically, other plant communities. Forests for that matter are easily recognized by the height and tree crown cover.

Even though the FSD perceives the height of trees as a necessary factor in defining forest, the emphasis is placed on whether the land has been demarcated by policy to be protected. In addition, the responsibility of governance and management must be vested with the Forest Service Division and this could be in trust for one or more communities. Based on this explanation a forest is connoted same as a forest reserve. In relation to the two communities, Dwease has a forest because it shares in the ownership and management of the Dome River Forest Reserve while Praaso does not because there is no forest reserve located in this community. In addition, it also has no share in the ownership and management of the reserve at Dwease. The import is that, even though Praaso has extensive trees dominated vegetative species above the minimum threshold of 5 meters, such areas are not regarded as forests according to the understanding of forests as inquired at the local FSD office.

It is realized generally that the perception of forest as an ecosystem characterized by more or less dense and extensive tree cover, often consisting of stands that vary in species composition, structure, age, class, and associated processes clearly captured the fragmented ideas expressed by the research respondents. The distinction of forest types into primary and secondary forest types seeks to examine whether the forest has undergone succession and to what extent the indigenous forest resources have been maintained in terms of species composition, structure, age, and class. It is observed and also indicated by respondents that much of the forests have been cultivated or are still being cultivated and it has essentially affected the crown cover development of trees and has caused loss of indigenous trees and other vegetative species.

Based on this analysis, a permutation of the varied perceptions of forest by the household respondents revealed that the forest types in the study area can be categorized into primary-opened forest; primary-closed forest; secondary-opened forest; and secondary-closed forest as indicated in table 3.

Table 3: Forest types (delineated using both crown cover and land use criteria)

Forest type	Frequency	Percent
Primary-closed forest	15	13.6
Primary-opened forest	18	16.4
Secondary-closed forest	4	3.6
Secondary-opened forest	73	66.4
Total	110	100.0

From Table 3, majority of the respondents perceived the predominant forest type in both communities to be secondary-opened and this implies that the forest is being cultivated and has opened canopy. Majority of the respondents noted that, most of the forested areas in the communities is being used to cultivate both tree and food crops however, the local people did not necessarily understand the technicalities of what constitute the crown cover threshold to distinguish between opened and closed forest types.

This was explained by staff of the FSD to be determined using the conditional score, which refers to the degree of visibility under the trees canopies for a distance of 800 meters long. The scores are ranged from 1 to 5 with a score of 1 being the highest; this is equivalent to 100% and 5 being the lowest score. Forests with highest condition score often lack undergrowth which aids clear visibility under the canopies and such category are classified as closed forests. Those with low scores have more undergrowth, which produces visibility under the canopies, and these are defined as opened forests. Based on this, the Dome River forest reserve located in Dwease is argued to have a conditional score between 3 and 4, which is low and thus is classified as an opened forest. Also, since it has been used for agro-forestry, it is obviously classified as secondary-opened forest type.

Implications for Sustainable Forest Management in Ghana

It is necessary to highlight that the various understandings of the forest as expressed by respondents reveal some critical issues which are pertinent for inquiring into the conditions for local level sustainable forest management. It is realized that the interpretation of the concept of sustainable forest management by different actors and institutions vary. Notwithstanding, a common consensus is the need to recognize the participation of all stakeholders in forest management activities. The views expressed particularly by the local people in both communities are congruent with [26] who stressed that for forest management to be participatory there is the need to involve local people in decision-making and in the control and management of the forests.

It is pertinent to state that the local stakeholders duly recognize their positions and roles towards sustainable forest management nonetheless they argued that in practice, forest governance and management in their communities has been solely handed by the state apparatus. For instance the chief of Dwease remarked that;

"Though the Forest Services Division is supposed to foster collaboration with us for forest management (FM), I have realized our efforts are not appreciated. We suggested a voluntary forestry team from the community to help monitor

the chainsaw operations but it has been refuted by the FSD. Yet most of the forest guards used by the FSD cannot read and write and this has promoted ineffective data collection, which is a great challenge towards sustainable forest management in this community" (Personal Interview 2013).

This observation confirms [27] who notes that, the Forest Ordinance No. 13 of 1927 (Cape 157) along with subsequent enactments in the Forestry sector have operated to expropriate all forest lands from the communities (both the legal estate and management powers) to the government. This has often created conflicts regarding the sharing of forests benefits as well as issues relating to ownership and management.

From the above analysis, it is realized that, management of forests in Ghana is characterized by disputes and disoriented ideas. This orientation has critically affected the conditions and practices required for sustainable forest management especially at the local level. As a result, if Ghana is to achieve the objectives of sustainability in the forestry sector, then fundamental issues relating to the positions and roles of the various stakeholders on forest management and governance ought to be addressed. In the case of the studied communities, the critical issues, which should be looked, include *"access to", "nature of"* and *"right to"*.

On the issue of *access to*, those who explained their understanding of forest based on the ownership criterion indicated that the Dome River Forest Reserve is *less accessible*. This is because by definition, such a land has been earmarked by the Forest Services Division (FSD) as a reserve and access to such forest type is highly regulated. This contrasts greatly with access to the off-reserves forest types. It is explained that, because these are generally being cultivated, they are more of de facto open-access resources widely open to any entrepreneur and individuals in the communities who are able to negotiate unofficial access to the forests independently of an official legal regime in place.

In order to ascertain this claim, the heads of households were questioned on the kind of forest resources they mostly extract. The respondents opined that, they either extract Non-Timber Forest Products (NTFPs) only or Timber Forest Products (TFPs) only or a combination of both NTFPs and TFPs as shown in Table 4.

Table 4: Household responses on the type of forest products extracted

Type of forest products extracted	Frequency	Percent
NTFPs only	85	77
TFPs only	3	3
Both NTFPs and TFPs	22	20
Total	110	100

The findings in Table 4 indicate that about 77 % of household respondents stated that they extract NTFPs only, 3% extract TFPs only while the remaining 20% extract both TFPs and NTFPs. It was noted that the extraction of NTFPs is more likely to contribute more importantly to household livelihoods than TFPs. Most respondent explained that they engage more in the extraction of NTFPs because it is difficult to obtain a logging permit in order to have a concession for TFPs. It is confirmed at the Forest Services Division, that there are stringent measures to regulate the extraction of timber products. The management explained that, before giving concession to any individual or company or groups of companies, some criteria are sought. These include that prospective person or company should be formally registered; possess a letter of tax clearance from the Ghana Revenue Authority; letter from the Ghana Timber Association (GTA) indicating registration as a member; and should also poses a track record of good conducts and not owing the Forestry Commission in anyway.

Moreover, because most of the local people cannot meet the terms of these criteria, they tend to extract more of the NTFPs because these are easily accessible within the fringes of the forests as compared to TFPs. Access of TFPs by the local people is done mainly through chainsaw operation which is considered illegal. Even though some of the local people indicated that they have taken measures to manage the off-reserve forests against fire and illegal logging, these measures are however dependent on the economic value they place on the particular tree species located on their lands. It is argued that Mahogany (*Khaya senegalenis*) and Odum (*Milicia excels*) are more likely to be preserved because these are considered to be economic species.

Therefore individuals with these trees on their farmlands assumed that they could receive some share of the monetary benefit when the tree is being logged for sale and for that matter they take good care of the trees. The challenge however is that, such individuals could lose the supposed benefits because they may not know when these trees may be harvested especially with the increased illegal logging. As a result, it is highlighted by some farmers that in some cases they would rather prefer to cut down all timber species, (even when these trees are not matured) because apart from the fact that they do not get the monetary benefits from logging, the process also sometimes destroys their crops and there is also often less or no compensation for the losses.

Another issue to guarantee effective conditions for local level sustainable forest management in Ghana is the definition of the *nature of* forest types. This is quite necessary because it has implications for the kind of strategies adopted for sustainable management of the forest within an area. The study discovered that there are a number of strategies which have been adopted to ensure sustainable management of forests in the two communities: Dwease and Praaso. These strategies, which are being adopted, varied in the context of the *nature of* the forest types. For instance, it is identified that the strategies adopted by the FSD are mainly used in the management of forests within reserves. The outcome of these strategies directly connects with the institution's perception of what constitute forest(s).

It is discovered at the FSD that there are three working cycles; protection, production and research and regulation which defines the strategies adopted to promote sustainable forest management. However in practice, it is identified the real and workable strategies currently being employed by the FSD within the two communities which include the "Enrichment planting; and the conventional administrative function of regulating illegal logging".

The Enrichment planting involves the process of introducing some trees species (both indigenous and exotic) into a degraded forest and these are inter-planted amidst the already existing trees. The activity was being carried out in the Reserve located in Dwease and this involved the planting of eight indigenous tree species namely, Wawa (*Triplochiton scleroxylon*), Mahogany (*Khaya senegalensis*), Mazonia, Kokrodua (*Pericopsis elata*), Kusia (*Nauclea diderrichii*), Edinam (*Entandrophragma angolense*), Amire (*Terminalia ivoresnsis*) and Framo (*Terminialia superb*). The choice is influenced by the belief that these species are being threatened to extinct.

A key concern is that this strategy is practiced only within the forest reserves and not the off-reserve zones. Also, since this activity is solely championed by the FSD through its Plantation department, there is therefore no cooperation with the local communities and no benefits through income generation or losses for smallholders in the either of the communities. Based on this, it is evident that currently there is no real strategy in terms of tree planting which is being adopted by the FSD to ensure the sustainability of off-reserve forests which are even more extensive in the two communities.

In terms of its regulatory responsibility, it was ascertained the FSD has established research plots to monitor tree growth and development. In addition, there is also regulation of timber extraction, and the Division carries out this responsibility by providing the legal framework or guidelines to determine which trees are harvested, where they can be obtained and by whom. These regulatory policies particularly with regard to timber forest products are also now keenly being emphasized because Ghana is a signatory to the Voluntary Partnership Agreement (VPA). Moreover, the FSD in collaboration with Trophenbos International Ghana (TBI) is helping to register chainsaw operators within the Jauso forest district, which is currently operational in Juaso and yet to be extended to other communities including Dwease and Praaso. The specific aims include helping them to understand the consequences of their activities on the sustainability of forests, enhance their knowledge on branding and marketing their wood products and capacity building in business and financial management.

It is also conclusive that the *nature of* forest types as understood by the FSD is linked with its management strategies that are directly geared towards the management of Timber Forest Products (TFPs) rather than Non-timber Forest Products (NTFPs). Such approach compromises local participation in sustainable management efforts because the local people often have less access to the TFPs and/or the resulting benefits accruing from their development. This makes them feel cheated and excluded from benefit sharing of forest resource and influences their attitude and perception on policies towards sustainable forest management.

Whiles the issue is far beyond the mere refusal on the part of the FSD to implement policies of forest management in off-reserves, the challenge is rather linked with the *"right to"* management. This defines the benefits and responsibilities of the appropriate stakeholders, especially on tree tenure. Though sustainable forest management should be collaborative according to the 1994 Forest and Wildlife Policy (FWP), the research findings reveal that policies on forests' governance have expropriated the *right to* management solely to the state apparatus (FSD) and this has narrowed the effective role that could be played by the leaders and individuals at the community level. According to the Convention on Biological Diversity (CBD), local empowerment of resource users is a potential key strategy to achieve long-term sustainability [28]. This transfer of right however, must also include transfer of responsibilities for the protection of the resources.

According to [29], they view the security of tenure of natural resources as an important issue if local communities are to use the natural resources in their localities sustainably. In their study in three forest fringe communities; Nkawie, Juaso and Goaso, they identified that tenure determines in large part, whether the local people are willing to participate in the management and protection of forest and tree resources. It must not be forgotten however that throughout the world communities use and manage their forest in diverse ways, and forest use rights and management responsibilities vary widely, depending on historical factors, social and political contexts and national policies [30].

Therefore the failure of the state apparatus to effectively manage national forests is evidenced in Dwease and Praaso because chainsaw operators and agriculturalists continue to negotiate illegal access to the secondary opened forest types and causing tremendous loss of tree cover and other biodiversity. Thus, to foster sustainable management of forests at the local level, it is recommended that the tree tenure in off-reserves should be reviewed in order to critically consider the stakes of the local leaders and individual especially land owners. This could redefine how local perception on the *"nature of"* and *"right to"* have influenced the condition towards sustainable management of forests.

CONCLUSION AND RECOMMENDATIONS

The study has discovered that the call to sustainable management of natural resources particularly in the forestry sector is highly needed in Ghana especially in the Juaso forest district. However, the research findings have also confirmed that stakeholders' participation in sustainable forest management in both communities has been met with a myriad of challenges. These challenges are deeply rooted in the stakeholders' perceptions of what constitute forest and the classification of forest, which are present in the communities. These perceptions have defined the conditions required for local level sustainable forest management. The conditions for the ineffective sustainable forest management in these communities have been determined by three factors; "access to"; *"nature of"* and *"right to"*.

The discussion has pointed out that, the FSD, which is the major stakeholder, has a limited number of workable management strategies or policies which are currently been used to promote the sustainability of forests in the Juaso forest district. In addition, the adopted strategies have not really promoted collaboration with local partners that have challenged effective stakeholder analysis for policies on forest management. Also, because these strategies and regulatory polices of FSD are directly geared towards the sustainable management of TFPs to the neglect of NTFPs, it has therefore hindered sustainable management of forests in off-reserves regions in the district. On the part of individuals in the communities, it is recognized that, they practice the sustained-yield ecological principle as their way of promoting sustainable forest management. Even though this strategy has not led to the regeneration of forests; especially in off-reserves, it has bridged the gap in the form of compensation for environmental benefits or advance payments for future sales of products.

It is therefore recommended that, with the re-launch of the National Forest Plantation Development Programme (NFPDP) and its redirected focus to cover private lands located outside forest reserves [31], the FSD should adopt formidable yet realistic strategies to sustainably managed forests in both on-and-off-reserves. In this regard, the concept of community forestry, which seeks to give greater control and management of forest to communities, is a formidable strategy to ensure sustainable forest management in the Juaso forest district. In this connection, revitalizing the Collaborative Forest Management strategy at the community level will be a step in the right direction since it promotes effective participation of all the stakeholders especially the traditional leaders and local people in decision-making processes. Even though in some areas, the institutional framework and regulations may be weak, the key concern of the Forestry Commission and its diverse bodies should be to strengthen these because in many cases they have still proven to be effective in managing forests sustainably at local level.

REFERENCES

1. Brinkerhoff, D. and Crosby, B, "Citizen Participation in the Policy Process" Chapter 3 in *Managing Policy Reform: Concepts and Tools for Decision-Makers in Developing and Transitioning Countries*. Kumarian Press, Connecticut, USA, pp 51-56, 2002. Available: http://site.ebrary.com/lib/agder/Doc?id=10069004 (12 September 2013).
2. Adams, W. M., *Green Development: Environment and Sustainability in the Third World,* 3rd Edition, Routledge, London and New York, 2009.
3. Purnomo, H. A, *Modeling Approach to Collaborative Forest Management*, 2003 [Online] available at: http://repository.ipb.ac.id/handle/123456789/42524 (10 September 2013).
4. J. Carle and P. Holmgren, *Definitions Related to Planted Forests in Forest Resources Development Service,* FAO, Rome, pp 2-9, 2000.
5. C. E. Owubah, D. C. Le Master, J.M. Bowker, and J.G. Lee, Forest tenure system and sustainable forest management: the case of Ghana, *Forest Ecology and Management*, 149 (2001) 253-256, 2001.
6. E. Acheampong and E. Marfo, The Impact of Tree Tenure and Access on Chainsaw Milling in Ghana, *Ghana Journal of Forestry*, Vol. 27, pp. 68-86, 2011.
7. Forestry Commission (FC), *Forest and Wildlife Policy*, 1994, Forestry Commission, November 1994, Ministry of Lands and Forestry, Accra – Ghana.
8. E.N.A Kotey, J. Francois, J.G.K. Owusu, R. Yeboah,, K.S. Amanor, and L. Antwi, *Falling into Place: Policy that works for forests and people series No. 4,* International Institute for Environment and Development, London, 1998.
9. Y. Agidee, Forest Carbon in Ghana: Spotlight on Community Resource Management Areas, Katoomba Group's Legal Initiative Country Study Series, Forest Trends, Washington, DC, 2011.
10. Food and Agricultural Organization (FAO) *Multi stakeholder forest management: a case from the humid zone in Ghana,* Based on the work by Dominic Blay, Forestry Research Institute of Ghana, Forest Management Working Paper FM/32. Forest Resources Development Service, Forest Management Division. FAO, Rome (unpublished) 2007. (Online) Available: http://usi.earth.ac.cr/glas/sp/FAO/FAO9953e00.pdf (30 January, 2013).
11. Kidane-Mariam, T. 2003. Environmental and habitat management: The Case of Ethiopia and Ghana. Environmental Management, 31(3), 313-327
12. Forestry Commission, Ghana`s Forests: Reporting Progress. Timber Industry Development Division, Ghana Forestry Services, Accra, 2006.
13. Appiah, D. O. 2009. Personifying sustainable rural livelihoods in forest fringe communities in Ghana: A historic rhetoric? Journal of Food, Agriculture & Environment Vol.7.
14. A.K. Asamoah, A.M. Pinard, R.J. Cobbinah, L. Damnyag, E. Nutakor, S.K. Nketiah, B. Kyere, and C. Nyarko, Chainsaw milling and trade in West Africa: Socio-economic impact of chainsaw milling and the lumber trade in Ghana, *DFID, Renewable Natural Resource Research Strategy (RNRRS) (FRP) PMP 05-08*, Natural Resource International, UK, 2007.
15. K. Oteng-Kufuor, New institutional economics and the failure of sustainable forestry in Ghana, *Natural Resources Journal*, 44, 743-760, 2004.
16. Ministry of Lands and Forestry (MLF), *Forest and wildlife policy, Republic of Ghana*, 1994.
17. Care International, *Ghana Land and Forestry Policy Support Facility Sub-project No 1*, 2005.
18. K. Opoku, *Forest Governance in Ghana: An NGO perspective*, a report produced for FERN by Forest Watch Ghana, March 2006 (Online) Available: http://www.fern.org/sites/fern.org/files/media/documents/document_3643_3644.pdf
19. W. Lindsay, Community-Based natural resource management: Case studies from community forest management projects in Ghana, Mexico, and United States of America, *NRES 523- International Resource Management*, 2008. Available: http://www.uwsp.edu/forestry/StuJournals/Documents/IRM/wood.pdf (30 August 2012).
20. United Nations University (2007) Open Educational Resource on Environmental Impact Assessment based on the UNEP EIA Training Manual 1-4, available: http://eia.unu.edu (5 July 2013).

21. Asante Akim Central Municipal Assembly, Draft Strategic Environmental Report for the District Medium Term Development Plan under the Ghana Shared Growth and Development Agenda 2010-2013, pp 9-68, 2010.

22. Bryman, A., *Social Research Methods,* 3rd edition, Oxford University Press Inc, New York, 2008.

23. Gabler, R.E, Petersen, J.F. and Trapasso, L.M, *Essentials of Physical Geography,* eighth edition, Thomson Brooks, Belmont, pp. 240, 2007.

24. Carle, J. and Holmgren, P. (2000) "Definitions Related to Planted Forests", in Forest Resources Development Service, FAO of United Nations, Viale Delle Terme di Caracalla, I-00100 Rome, Italy, pp 2-3 and 9.

25. Food and Agriculture Organization, "Criteria and Indicators for Sustainable wood fuels", in FAO Forestry, Paper 160, Electronic Publishing Policy and Support Branch, Viale Delle Terme di Caracalla, Rome, pp. 5-11, 2010.

26. Y. Agidee, Forest Carbon in Ghana: Spotlight on Community Resource Management Areas, Katoomba Group's Legal Initiative Country Study Series, Forest Trends, Washington, DC, 2011.

27. K. Kasanga, Current Land Policy Issues in Ghana, *Land Reform 2003 (3) Special Edition*, pp 41-54, 2002 (Online) available: http://www.fao.org/docrep/006/y5026e/y5026e0a.htm Accessed 9 August 2013.

28. Convention on Biological Diversity, *Sustainable Forest Management, Biodiversity and Livelihood: A Good Practice Guide*, Montreal, 2009.

29. J. E. M, Arnold (2001) *Forests and People: 25 years of Community Forestry*. FAO, Rome.

30. M. Poffenberger, *Communities and Forest Management*, IUCN, Washington, 1996.

31. Forestry Commission (FC) *National Forest Plantation Development Programme Annual Report*, 2011

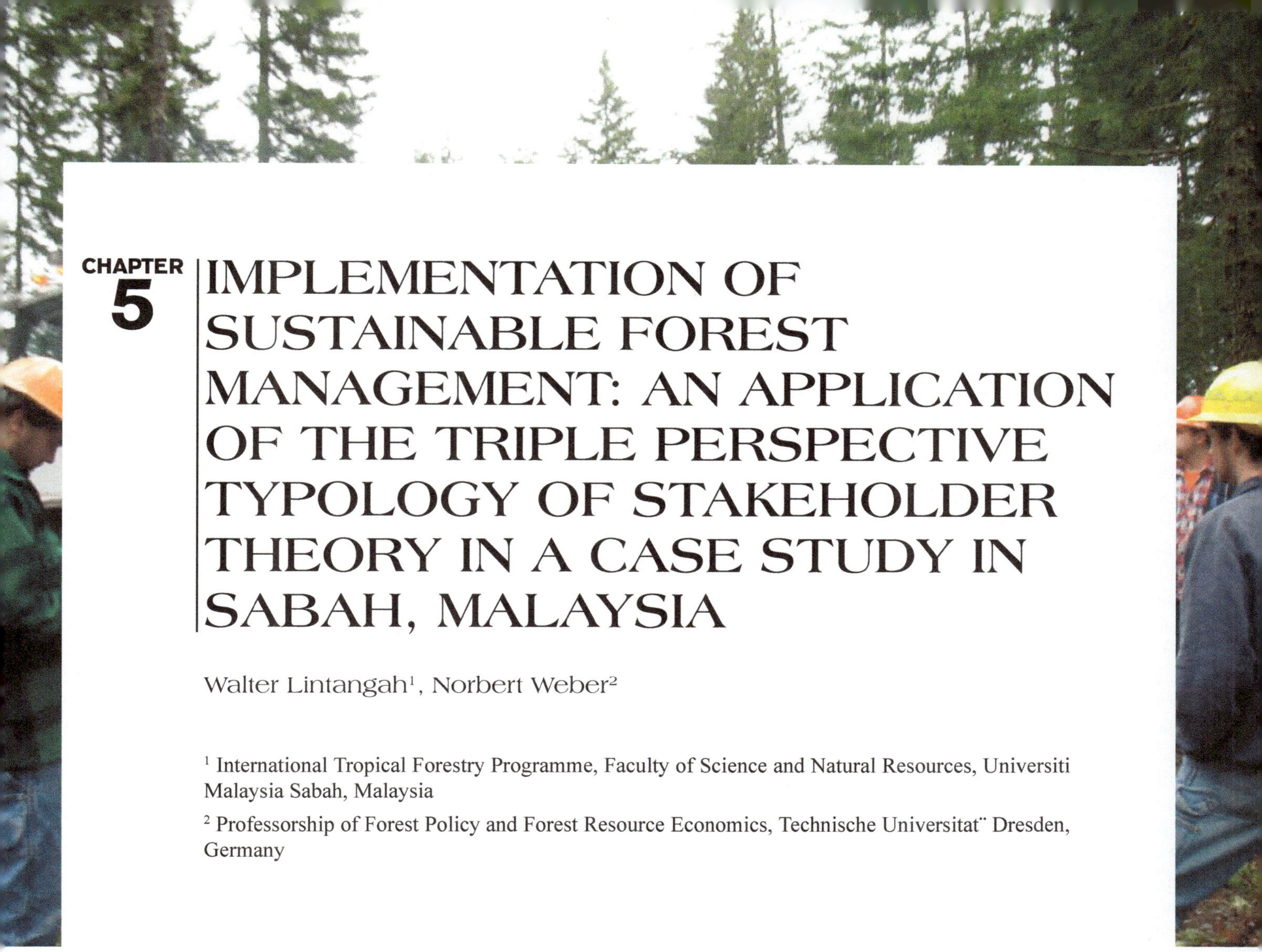

CHAPTER 5

IMPLEMENTATION OF SUSTAINABLE FOREST MANAGEMENT: AN APPLICATION OF THE TRIPLE PERSPECTIVE TYPOLOGY OF STAKEHOLDER THEORY IN A CASE STUDY IN SABAH, MALAYSIA

Walter Lintangah[1], Norbert Weber[2]

[1] International Tropical Forestry Programme, Faculty of Science and Natural Resources, Universiti Malaysia Sabah, Malaysia

[2] Professorship of Forest Policy and Forest Resource Economics, Technische Universitat¨ Dresden, Germany

ABSTRACT

The progress of the concept of Sustainable Forest Management (SFM) is dynamic and its success or failure during implementation can be evaluated in different ways. In a detailed survey in four Forest Management Units (FMUs) in Sabah, the current implementation of SFM at the FMU level was assessed based on the triple perspective typology of stakeholder theory. This approach encompasses conceptual, corporate and stakeholder centric point of view. The conceptual perspective explores the SFM concept and how it relates to the FMU holder – stakeholder interaction at the FMU level, the corporate perspective examined how the FMU holders put the concept into practice and address their stakeholders, while the stakeholder perspective analyses how the stakeholders attempt to accomplish their claims and interests through the corporate-centric (FMU holders) at the management level. The stakeholder analysis provides the platform for stakeholder identification, categorization and their general perception and behavior towards the overall performance of SFM objectives. The Stakeholder Relation Management (SRM) that integrates the FMU holders and stakeholder participation under the SFM concept were also identified. Different FMU holders are engaged with distinct objectives to be achieved, which determine their direct relationship with the stakeholders. The stakeholders were attributed to static and dynamic groups, which are determined by their existence, claims and interests, and involvement in various SFM programs and activities at the FMU level. They provided distinguished preference and agreement toward various issues and characteristics related to SFM objectives, implementation and stakeholder participation at the management level. Most of the respondents of the multi-interest stakeholder group agreed with SFM main contributions towards the elements of environmental objectives, followed by economic objectives and elements of social objectives. The approach for SFM assessment based on the different lenses of conceptual, corporate and stakeholder centric provides complementary evidence on the pragmatic implementation of SFM at the forest management level.

Citation: Walter Lintangah, Norbert Weber, "Implementation of sustainable forest management: an application of the triple perspective typology of stakeholder theory in a case study in Sabah, Malaysia," Journal of Forest and Landscape Research 1 (2015): 1–11; DOI: 10.13141/jflr.v1i1.251.

Keywords

Sustainable Forest Management (SFM); Stakeholder Analysis; Triple Perspective Typology; Stakeholder Relation Management (SRM); Forest Management Unit (FMU); participation

INTRODUCTION

Sustainable Forest Management (SFM) has evolved as a concept to manage forests according to the forest principles with the main goal "to contribute to the management, conservation and sustainable development of forests and to provide for their multiple and complementary functions and uses" (IUFRO, 2005). Under the forest principles, it is proclaimed that "forest resources and forest lands should be sustainably managed to meet the social, economic, ecological, cultural and spiritual needs of present and future generations" (UNCED, 1992). However, the SFM concept is dynamic, and the success of its implementation is governed by many factors. The principal and technical aspect of the concept has been developed over time with the involvement of many parties including local and international experts and organizations. People are key actors in the social-ecological system and disregard for local claims and needs has resulted in the failure of many forestry projects (Putz, 1994; Vanhanen et al., 2010). According to Nilsson (2001) to secure sustainable forest management, it must be comprehended that sustainable forestry is more an issue of human behavior than an issue of trees and forests. How to build consensus among all stakeholders surrounding the issues of sustainability is one of the primary challenges facing the resource-based projects (Appiah, 2013, p. 37). The fundamental differences in visions about the forest, for example, have resulted in the failure of international negotiations for an international legally binding forest treaty (Arts, 2002; Lebel et al., 2004). Vanhanen et al. (2010, pp. 201-222) stipulated the forest stakeholders as people who depend directly on forests or participate in their management such as forest communities, forest managers and companies, conservationists, forest policy makers, development organizations, and scientist. They are facing the challenges related to understanding vulnerability, identifying adaptation options, and implementing adaptation with the changing of economic, social, global political environments and adaptation to climate change. Forest stakeholders are concerned and continuously have to deal with the questions of how the forests should look, what kind of products, services and experience it should be able to provide, and what functions the forest should perform (Gamborg and Larsen, 2003, p. 559). The conventional approach to describe and monitor status and trend of forests and forest management is through the application of established tools such as forest certification and the criteria and indicators (C&I) (Rametsteiner and Simula, 2003).

Sabah is the second largest of the thirteen states in the Malaysian Federation with a total land mass of about 7.49 million hectares. About 4.4 million ha or 58.8% of the land area is remaining under forest cover (MTC, 2010) with approximately 48.17% or 3.6 million ha of the land has been gazetted as Permanent Forest Reserve (PFR) to accommodate different purposes towards economic, social and environmental functions of the forest (SFD, 2011b). The SFM policy was launched in Sabah in 1997 after the successful implementation of SFM concept in Deramakot Forest Reserve (DFR), which was managed with the assistance and collaboration of the Malaysian Federal Government and international support of the German Agency for Technical Cooperation (Deutsche Gesellschaft fur Technische Zusammenarbeit ¨ GTZ, now GIZ). Currently, there are 32 Sustainable Forest Management License Agreement (SFMLA)/Long Term License Agreement (LTL) signed by the Forest Management Unit (FMU) holders, with the task of managing most of the Commercial Forest Reserves (Class II) in the state. Apart from the SFMLA/LTL holders, the state government is also directly involved in managing nine FMUs that are under the jurisdiction of the Sabah Forestry Department (SFD). The experiences of SFM implementation in Sabah illustrate the importance of the role of stakeholders in the progress of the SFM concept at the forest management level. The stakeholder may influence the success or failure of the SFM in practice. Among the identifiable success factors in the progress of SFM implementation within the state are commitment, experience and resources of a broad community of interest that are elements of the partnership approach (SFD, 2011a, p. 211). The stipulated sources of challenge for implementation include " i) the complexity of problems, ii) understanding the dynamics of nature and iii) developing practical management planning system that respects both the complexity of the problem and the dynamics of nature" (SFD, 2011a, p. 235). The identified impediment for implementation is the concern of financial support required for SFM funding, the presence of local communities in SFM license areas and the demand from wood processing industries (SFD, 2010, p. 201). Further issues are related to the lack of relations and mechanisms for corporate-stakeholder communication as a means of solving or discussing matters pertaining to SFM implementation involving both the FMU holders and their stakeholders (SFD, 2010, p. 209). These problems signify that research on the corporate stakeholder relationship under the SFM implementation is still of imperative necessity. This study was carried out so as to understand the contemporary implementation of SFM in Sabah with the focus on FMU holder-stakeholder relations at the forest management level.

THEORETICAL FRAMEWORK AND METHODS

The stakeholder theory is "a theory about how business actually does and can work" (Freeman et al., 2010, p. 3). The primary purpose of the concept is to address three main problems of "understanding and managing a business, integration of ethics, responsibility, and sustainability with the usual economic view of capitalism, and understanding what it takes to be successful in current business world" (Freeman et al., 2010, p. 3). This theory provides the foundation for stakeholder identification, classification, and categorization and to understand their behavior (Aaltonen, 2011). The stakeholder analysis as a tool under this concept is a process through which project managers try to apprehend and read the project's stakeholder environment in order to be able to define the right type of action concerning different stakeholders (Aaltonen, 2011). The three perspectives of stakeholder theory identified as Triple Typology Perspective of Stakeholder Theory [TPT] was introduced by Steurer (2006) to approach stakeholder management from three thematic perspectives of conceptual, corporate and stakeholder centric. The conceptual perspective explores how particular concepts (such as SFM) relate to business–stakeholder interactions. The corporate perspective focuses on how corporations address their stakeholders, and the stakeholder perspective analyses how stakeholders try to influence corporations (Steurer, 2006). The SRM un-der this concept is delineated as the transmission mechanism that connects the conceptual centric from societal groups to the business world (Steurer et al., 2005, p. 265). The Triple Perspectives Typology is incorporated into the second-order stakeholder theory by Donaldson and Preston (1995), which differentiates the descriptive, normative and instrumental aspect of stakeholder theory. The interrelations of the three concepts of corporate, conceptual and stakeholder centric and their relationship with the approaches for stakeholder relation management under the framework are shown in Figure 1.

The conceptual centric under this study is related to the overall SFM policy. The emphasis on this concept was to comprehend the general organization and implementation of SFM concept and its interrelation with the corporate and stakeholders at the FMU level, respectively. SFM as the conceptual centric deliberates the normative human intervention towards the sustainability of the economic, social and environmental aspects of conservation, and the use of forests. It also covers administrative, legal and technical topics (Food and Agriculture Organization of the United Nations, 2013). The corporate perspective denotes to the FMU holders that manage the forest under SFM principles. Under the business case for sustain-able development, the corporate perspectives seek to find how the firms can further their economic sustainability by giving consideration to social and environmental issues (Dyllick and Hockerts, 2002). The criterion for corporate sustainability constitutes a firm's eco-efficiency and socio-efficiency for the business case; eco-effectiveness and sufficiency for the natural case; and socio-efficiency and ecological equity for the societal case (Dyllick and Hockerts, 2002). The stakeholder centric is confined to many stakeholders who are directly or indirectly affected or can affect SFM in practice. This perspective analyses how stakeholders attempt to influence corporations (Steurer, 2006). It is focused on stakeholder behavior and understanding towards the current implementation of SFM at the FMU level. Stakeholder relation management is related to the question how the FMU holders and stakeholders manage their relationship to accomplish their interests and claims under the SFM implementation. It can advance an organization's performance in many ways including through change in "the organization's externally imposed mandates, funding sources, decision-making protocols, or accountability mechanisms" (Steurer et al., 2005, p. 7). Table 1 outlines the different approaches applied for the assessment.

Methods and Material

Various assessments of the SFM implementation at the FMU level were carried out under the different perspectives of conceptual, corporate and stakeholder. The method of the study is comprehensively referred to Lintangah (2014). Stakeholder analyses, questionnaire surveys, discussion via face-to-face interviews and literature reviews on reports and records associated with the SFM implementation ('lessons learned reports') were methods employed to collect information. The respondents include personnel of SFD and FMU holders, local communities and various stakeholders identified through expert interviews and snowball technique (Reed et al., 2009). Household samplings were used to investigate the stakeholders under the local community situated nearby or within the FMU area. The Q methodology (Watts and Stenner, 2012) and factorial analysis were used to identify the different stakeholder groups based on their behavior and perception towards the current implementation of SFM. The approaches to assess SFM implementation under the different perspectives of conceptual, corporate and stakeholders are shown in Table 1. The 'conceptual' perspective evaluated the various elements of SFM policy under implementation, which constituted from sources comprising literature reviews, expert interviews, technical manuals, and managerial guidelines for SFM implementation at the FMU level. Furthermore, individual perception on the capacity levels of SFM implementation at the FMU level was identified through questionnaires survey and Q methodology. The questionnaire surveys apprehended the stakeholders' level of agreement, based on Likert Scale, towards the scenario of SFM implementation under the distinguished social, economic and environmental issues. The Q methodology was carried out to determine the stakeholders' attitude towards various statements (concourse) related to the SFM implementation at the FMU level.

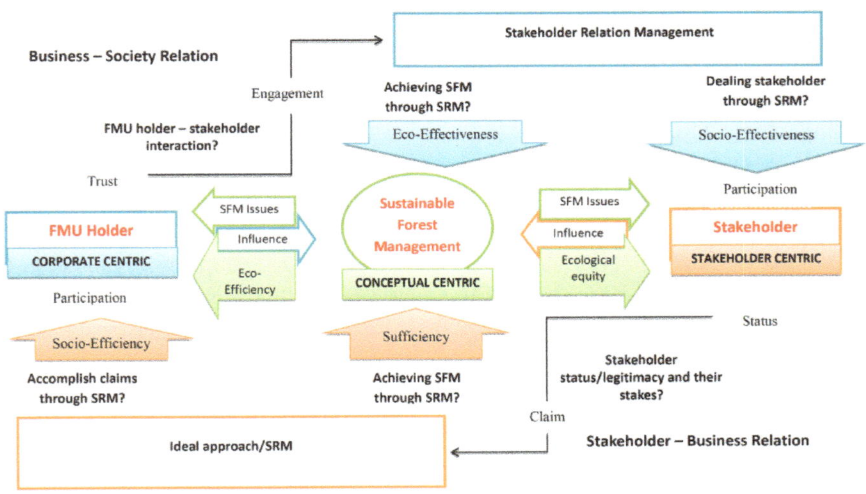

Figure 1. Conceptual framework of the study based on Dyllick and Hockerts (2002); Lintangah (2014); Steurer (2006)

Under the corporate perspective, direct interviews and questionnaire surveys were employed to comprehend the organization and state of SFM implementation at the FMU level. FMU exist in various locations within the forest reserves throughout the state of Sabah. Each FMU was assigned to various FMU holders (corporates) that engage in managing the forests, i.e. state government, private FMU holders and state enterprise. In this study, four FMU areas were selected (two private FMUs, one under the state government, and one state enterprise). The FMUs concerned were FMU 10 (Sabah Forestry Department), FMU 16 (State Enterprise – Sabah Foundation joint venture managed by Rakyat Berjaya (RBJ) with Maxland Sdn. Bhd.), FMU7 (Private – Sabah Forest Industries Sdn. Bhd.) and FMU3 (Private - Timberwell Berhad). The corporate centric assessed the progress and capacity level of SFM implementation, approaches for Stakeholder Relation Management and the identification of Stakeholder Ranking based on their involvement in various programs and activities carried out in the respective FMU area.

Table 1. Triple Perspective Typology approaches for SFM assessment at the FMU level

Perspectives	Conceptual	Corporate	Stakeholders
Target	Overall SFM concept	FMU holders	Stakeholders
Data collection	Expert interviews (policy maker)	Interviews (e. g. forest manager)	Interviews (e. g. community representatives)
	Questionnaire survey	Questionnaire survey	Questionnaire survey
	Card sorting (Q methodology)		Card sorting (Q methodology)
	Literature review		meetings
Type of information	SFM policy and implementation	SFM operations and implementations at the FMU level (objectives, programs and activities)	Stakeholder involvement and engagement under SFM implementation and operation
	Stakeholder identification (potential and static)	Stakeholder identification (dynamic)	Stakeholder identification (dynamic)
SFM assessment	SFM objectives	Progress and capacity level of SFM	Level of stakeholder participation
	Issues pertain to SFM implementation	Approaches to stakeholder relation management	Approaches to stakeholder relation management
		Stakeholder involvement in SFM operations	Level of cooperation and conflicts
			Stakeholder dependency on forest and SFM implementation
Data analysis	Descriptive	Descriptive	Descriptive
	Factor interpretation (Q methodology)	Factor interpretation (Likert scale)	Factor interpretation (Likert scale)

The stakeholders in the FMU areas can be divided into two main categories of internal and external stakeholders. Internal stakeholders comprise the FMU workers, shareholders and forest managers while external stakeholders were referred to stakeholders outside the FMU holder's organization. In this study, the stakeholders were categorized under two main groups, i.e. local community and multi-interest stakeholder group. The multi-interest stakeholder group was further differentiated according to different power relation (hard or soft power) that they possessed (Mitchell et al., 1997; Nye, 2004). This differentiation provides the identification of governmental or intergovernmental agencies, industrial and commercial groups, and other interest groups. Questionnaires were used to obtain responses from the multi-interest stakeholder group on their experiences and level of agreement toward various issues concerning the current SFM implementation in Sabah. The surveys were sent to 200 listed respondents of various institutions through online and printed copy. There were 104 returned by respondents, which represent a response rate of 43.6%. The respondents from the local communities were residents of the selected villages within or on the fringe of the four FMU areas involved in this study. A total number of 332 respondents participated in the study. They comprised respondents from eight villages namely Kampung (Kg.) Gana (FMU 3), Kg. Lupakon, Kaingaran and Kg. Sinua (FMU10), Kg.. Sumambu, Kg. Alutok and Kg. Tilis (FMU 7), and Kg. Dewara and Kg. Mangkuwagu (FMU 16). The assessment under the stakeholder centric included the stakeholder participation and involvement in SFM implementation, approaches for stakeholder relation management and the stakeholders' perceptions of the current SFM implementation at the FMU level.

RESULTS

Conceptual Centric

The descriptive function of the SFM approach served as the normative basis on how the people should manage the forest in a responsible way in order to ensure the maximization of social, economic and environmental benefits for current and future generations. It demonstrates the matters and objectives of the SFM to be adhered to by both the FMU holders and stakeholders. Furthermore, it provides an extensive means of stakeholder identification and the categorization of static and potential stakeholders that grounded from SFM principles, forest legislation, and SFM programmes and activities.

There is a wide range of SFM documentation on international, national, and management levels either functioning specifically or as a combination of descriptive, normative, and instrumental aspects. These include the Forest Principles, Agenda 21, criteria and indicators for the ITTO guidelines on Sustainable Forest Management (ITTO, 2005),the Convention on Biological Diversity (CBD) and various forest-related laws and regulations at both federal and state level, which support the State Forest Policy on SFM. The Sustainable Forest Management License Agreement that is based on the Section 15 of the Sabah Forest Enactment 1968 provides the legal basis for the SFM implementation in the state. At the FMU level, the comprehensive Forest Management Plan attributes all the management prescriptions and activities for SFM implementation on the ground. The FMP is supported by other technical guidelines, operational plans and standards such as the Annual Work Plan, the Comprehensive Harvesting Plan, the Forest Management Standard and the Environmental Impact Assessment.

The various potential stakeholders were identified according to their interests and claims towards the different classes of forest reserve or particular established forest zonations inside the FMU area. The identification of static stakeholders is exclusively based on their existence in the written documents. The identified stakeholders can be grouped and differentiated along a distinct categorization (e.g. potential or actual, active or static, and internal or external). The interaction of the stakeholders under the SFM concept can be explained by the attribute of their power (hard and soft power), and legitimacy and urgency (Mitchell et al., 1997; Nye, 2004). The details of stakeholder's information at different levels (international, federal, state, and FMU level), distinct types of power (hard and soft power) and their different interests and claims can be observed from the SFM documentation, relevant policy and legislation, and also extracted from records and reports at the organizational and operational levels.

The assessment under the conceptual centric results from the level of agreement by the multi-interest stakeholder groups on the SFM achievements towards the environmental, social and economic objectives. Most of the respondents agreed with SFM main contributions towards the elements of environmental objectives, followed by elements of economic and social objectives. Whereas the highest level of agreement concerning the environmental objectives was found regarding the issues of preservation of the natural habitat of wildlife followed by conservation of water catchment areas, and conservation of the diversity of wildlife species. The highest level of agreement with regard to economic objectives included sustaining the source of timber and non-timber forest products, diversified job opportunities and income, and the continuation of the contribution of timber harvesting to state revenue. In terms of social objectives, the most agreed to statement was related to opportunities given to stakeholders to express their opinions and concerns about forest management, opportunities and experiences for recreation, safety of workers and communities, and the conservation of special features of the forest for local communities. The statements least agreed to were related to the sharing of information with local communities, respecting local community rights, accessibility of forest resources to the local communities, and improving the standard of living of the local communities.

The results under the Q Methodology indicated some significant consensus statements among the identified groups towards the various matters and issues related to the current SFM implementation. These comprised the influential factors towards effective SFM progress that include the administration and leadership of SFD, and good collaboration between SFD and FMU holders. Other consensus statements were related to an agreement on the roles of SFM concept in encouraging investment in the forestry sector and its explicit influence towards the rural development in the state. It was agreed that stakeholder participation under the SFM concept could promote learning on the synergy of various forest uses and also can influence the decision-making process.

Corporate Centric

The four SFMLA holders selected in this study represent different arrangements in terms of SFM objectives and institutional backgrounds. There are many programmes and activities of SFM operations conducted at the FMU level, which embrace numerous stakeholders. The operations include plan preparation, social baseline surveys, implementation and supervision of SFM operations, as well as programmes of forest protection, forest conservation, and forest restoration. Different SFMLA holders indicated distinct levels of priority on every SFM element implemented in their respective FMU area. In FMU3, the objectives to be achieved are connected with the three elements of economic production, environmental services and the socio-economic function of the forest. The FMU holder is engaged with most of the SFM projects, but less involved with activities related to wood manufacturing, non-timber forest products and activities associated with recreation and ecotourism. The management of FMU7 is mainly associated with activities related to the economic production function of the forest, which includes timber harvesting, timber marketing, wood manufacturing and industrial forest plantations. This FMU was also highly involved in all activities related to environmental services that associated with the protection of water catchment areas, forest rehabilitation and the protection of biodiversity and wildlife habitats. FMU7 is relatively associated with activities connected to the socio-economic function of the forest, which includes culture, education, research, recreation, and harvesting of forest products for the local community, but recorded a response of 'not agreed' with hunting as an activity in their FMU area. The FMU10 is mainly oriented towards the protection of environmental services of the forest, which includes the conservation of water catchment areas, providing wildlife habitat, forest rehabilitation and biodiversity. It did not engage in activities related to the direct economic function of timber harvesting, except for non-timber forest production for the local communities, and activities related to recreation and tourism. The FMU entangled with activities related to the socio-economic function of the forest, including the element of hunting activity but 'not so much agreed' upon cultural activity and timber production for the local community. In FMU16, the main concern is activities related to the economic production of the forest with the exclusion of activities associated with non-timber forest products, recreation and tourism. The FMU is also highly concerned with activities related to environmental services and the socio-economic function of the forest with the exception of the element of hunting. The overall performance and current capacity level of forest management planning and implementation are differing amongst the FMU holders. However, they generally agreed on the increased level of their experience and planning for forest management and operation in their respective FMU area.

The SFM operations facilitate the identification of various stakeholders with distinctive claims and interests related to the operation undertaken on the ground. They are referred to as dynamic stakeholders. These active stakeholders can move into different positions of salience (latent, expectant or potential) "depending on their attributed possession of power, legitimacy, and/or urgency" (Mitchell et al., 1997). The attribute of urgency determines the dynamic stakeholders, which depends on their contemporary existence, the important or critical relation and the claim of the stakeholder (ibid). The combination of analytical and reconstructive approaches was used to identify and categorize the stakeholders based on their legitimacy and power. Stakeholders' participation in the various elements of SFM programmes and activities were determined by the FMU holders based on pre-listing SFM operations relevant to their respective FMU. The programmes and activities identified as having a greater involvement of stakeholders are those associated with community forests, followed by forest conservation, preparation of development and management plans, forest protection, human-resource development, ecotourism programmes and the administration of the FMU holders. Stakeholders with more connections in various programmes and activities in the FMU area were internal stakeholders (FMU workers), followed by SFD and contractors. Other stakeholders with higher ranking based on the weighted rank included the local community, consultants, business and trade, researchers and scientists, and other state government agencies.

Stakeholder Centric

Multi-Interest Stakeholder Group

The multi-interest stakeholder group, as conceptualized in this study, consists of personnel from SFD and various government agencies, FMU employees, NGOs, consultants, educational and research institutions. Stakeholders under the multiinterest stakeholder group were mainly identified according to their organization and job descriptions. Different organizations were engaged in SFM implementation with distinct roles and purposes, or claims and interests. Most of the stakeholders were connected with SFM implementation owing to their relationship with governmental services (70%). This was followed by stakeholders with occupations in fields connected with environmental conservation (64%), forest management and conservation (64%), wildlife and biodiversity conservation (45%) and forest plantation and restoration (45%). Others had professional relationships with employees or FMU contractors, forest communities, suppliers, forestry-based associations, consultants, manufacturers, educations, researchers and tourism. There were also stakeholders connected with SFM with regard to activities such as forest harvesting, forest plantations, community forestry, trade and industry, recreation, culture and tourism, agriculture, agroforestry and protection of environmental services. The multi-interest stakeholder group was categorized into three main sub-groups that include the SFD group, other government agencies and non-government organizations from the private sector. A further option to categorize the multi-interest group is given by their different response to Likert Scale questions concerning the current implementation of SFM at the management level. They bestowed their different background, interests, claims, and commitments to SFM implementation that also determined their engagement, cooperation, conflicts and level of participation with the FMU holder. A variety of indicators was used to determine the influential factors towards stakeholder commitment to SFM implementation. The main factors proved to be their adherence to sound management of the forest, protection and conservation of the environment, conservation of biodiversity, protection of water catchment areas, forestry education, research, and livelihood of the local communities and their job security.

Local Communities

The study of local communities as stakeholders in SFM implementation was conducted in villages located within or in the vicinity of the selected FMU areas. The communities of the villages within the selected FMUs have a different background in terms of sociodemography, surrounding landscape characteristics and infrastructures development. The indicators of claim and interest of the local communities towards SFM implementation have been derived from their relation and dependency towards the forest. Their dependency on the forest is mainly related to the importance of the forest for their livelihood, including source for fruit and food, traditional medicines, wood and materials for buildings and boat making, and gathering of other non-timber forest products. The important forest products collected by the communities are resin, wood and material for house and boat making, wild honey, rattan and bamboo, firewood, game animals and raw material for handicrafts. Other reasons for entering the forest were associated with hunting, farming and recreational activities. Different villages displayed different levels of interest and claims towards the forest managed under the respective FMU holders. These claims and interests were mainly related to native customary rights on land including water catchment areas, old burial sites, farmland inside the forest, historical sites, places to collect forest resources such as wood for housing, game animals, foods, medicinal plants and other non-timber forest products as well as places for activities such as hunting, recreation, and religious activities. Community participation in SFM implementation differs based on contribution of SFM to their livelihood, cooperation and conflict level with FMU holders, issues and problems, and overall perception of their livelihood affected by the SFM implementation at the FMU level.

The contributions of SFM implementation were seen in elevating the standard of living and creation of job opportunities, development and maintenance of physical infrastructure, increased opportunities in terms of human resource development which include training and attending courses, changes towards a positive lifestyle, provision of health services and improving the quality of the surrounding environment. The respondents rated their overall agreement on the contribution of SFM implementation to the communities at a very low level (40.07%), medium (20.24%), low (18.86%), high (12.12%) and very high level (9.73%). Only a few respondents indicated the level of contribution under 'very high', which was related to improving environmental quality (17.28%), uplifting the living standard (13.11%) and increased job opportunities for the communities (11.11%). Other contributions of SFM rated at a very high level involved less than 10% of the total number of respondents. The contribution of SFM implementation rated at 'very low' by most of the respondents was confined to the provision of opportunities for human-resource development and attending courses (52.62% of all respondents). This was followed by contribution on provision for health facilities (52.17%), contribution to improving the welfare of the communities (46.01%), contribution towards positive lifestyle changes (43.61%), and contribution towards the maintenance of infrastructure (40.92%). Other contributions of SFM implementation at the FMU level, as indicated by respondents, are related to the provision of housing and woodlots for rubber and paddy field cultivation, participation in the forestry community projects implemented by the State Forestry Department, provision for recreation facilities, and development of infrastructure such as roads, bridges, schools and clinics.

There were various issues and problems related to SFM implementation encountered by the different villages at a distinct degree in their respective area. These include dispute of land boundary with the forest reserve under FMU, claim for native land (NCR inside FR): unclear land status and uncertainty of forestry community project sustainability, water catchment areas inside FMU areas, lack of consultations from the FMU holder, lack of interaction and cooperation with the FMU holder, respecting local practice and tradition, restriction on customary practices inside forest areas like agricultural activities and collecting forest produce, and lack of facility provision with regard to road and infrastructure maintenance. Further issues were the lack of understanding of the SFM concept and implementation, the lack of opportunities for job provision by the FMU holders and lack of involvement and cooperation with the relevant government or others agencies.

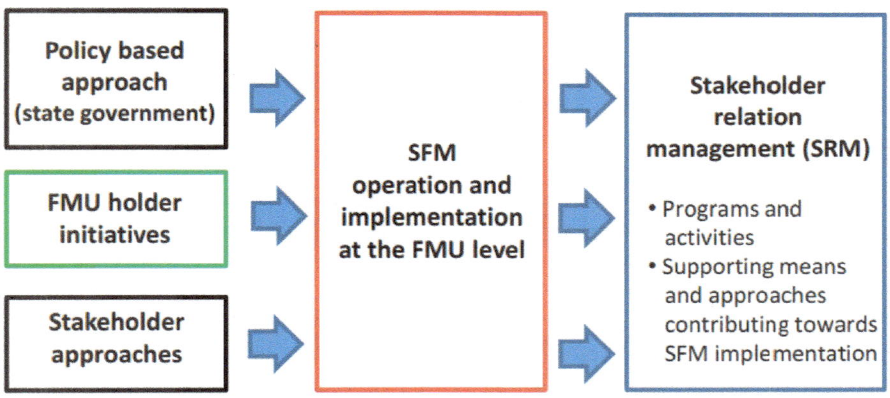

Figure 2. Identification of stakeholder relation management.

Stakeholder Relation Management (SRM)

The stakeholder relation management within SFM implementation is associated with the management of internal and external stakeholder interaction so as to promote and support the attainment of SFM objectives at the FMU level. There are different approaches for stakeholder relation management under the SFM implementation, both by FMU holders and stakeholders, which depend on the different stakeholders' perspectives (Figure 2). The approach of SRM under the FMU holders may consist of Quality Management System, Human Resource Development and Standard Operating Procedures. The SRMs concerning the multi-interest stakeholder group include the so-called interagency planning team, stakeholder workshops and co-operation under the project steering committee. Most of the communities solve their problems by conveying the issues through meetings and representation by the heads of the villages or other relevant agencies or authorities which can deal directly with their problems, such as the SFD and other government agencies. The collaboration and co-operation between FMU holders and international institutions at the FMU levels include development projects, consultancy, funding, research, education and training programmes. The programmes are implemented through Memorandum of Understanding (MOU), contract agreement and various joint venture management committees. Among the highly agreed supportive elements for SRM are technical and management guidelines, such as the Forest Management Plan, the Annual Work Plan, and various federal and state laws. Table 2 summarizes the results of stakeholder relation management under the Triple Typology of Stakeholder Theory.

DISCUSSION AND CONCLUSIONS

In this case study, detailed insight was gained into attitudes and perceptions of individuals and stakeholder groups towards the implementation of SFM at the operational level in Sabah, Malaysia. However, it has to be mentioned here that these findings have not been cross-checked with the state of the forests in the respective areas. Most of the respondents of the multi-interest stakeholder group agreed with SFM main contributions towards environmental objectives, followed by economic objectives and social objectives. Similar to other regions in the world, an important question was whether or not community values were adequately represented under current policies, land management and tenure systems, certification schemes, management planning, and current research priorities (Kozak et al., 2008). A case study on evaluation of SFM implementation in two forest management models in Vietnam and Malaysia by Le et al. (2012) suggested that greater involvement from private sector and other stakeholders, including the local people, can advance the performance of forest management practice at the FMU level.

Table 2. Triple Perspective Typology approaches for SFM assessment at the FMU level

Perspectives	Conceptual	Corporate	Stakeholders
Focus	SFM concept	FMU holders	Stakeholders
Stakeholder identification and categorisation	Potential stakeholder • Static • Legitimate	Dynamic stakeholder • Static • Legitimate	Multi-interest stakeholder Local communities
Stakeholder interests, characteristics and circumstances	Interests and claims • International • National • Management level	Participation and involvement in SFM operation • Chain of operations and responsibilities • Programs and activities	Interests /stake/claims • Can affect or affected by SFM (e.g. forest dependency) • SFM contribution • Issues and problems
Patterns and contexts of interaction between stakeholders	Task and function	Conflict, or cooperation Patterns of communication; trust and influence Flow of information between the actors	Conflict, or cooperation Patterns of communication; trust and influence
Define options for Management (Supporting means and approaches contribute towards SFM)	SRM: Conceptual based approach – SFM Policy, Forest Law and Legislation, International and Regional Cooperation and Agreement on Forest	SRM: Corporate / FMU Holders based Initiatives – e.g. SFMLA/LTL, Forest Management Plan, Forest Certification, SOP	SRM: Corporate – Stakeholder-based initiatives – e.g. Community Forestry Project, Joint Forest Management, Community Based Forest Management

Furthermore, it became obvious that the SFM contributions at the management level are influenced by various factors. These mainly depend on the SFM objectives laid out by the different FMU holders, which were stipulated in the Sustainable Forest Management License Agreement or Long Term License Agreement. The management objectives could be delineated under the Malaysian concepts of Natural Forest Management (NFM), Industrial Tree Plantation (ITP), Conservation Area, and Social Forestry. Moreover, the type of forest reserve classification and the existence of local communities and other stakeholders in the FMU area determine the programs and activities undertaken by the FMU holders in their respective area. Contributions to SFM are stipulated as improving environmental quality, providing job opportunities, facilitating infrastructure development and maintenance, and uplifting the living standard of the communities. The pertinent issues and problems faced by the local communities are related to their claim and dependency on the forest land. Balancing of complex interests between people and forest resources necessitates partnership and negotiating approaches and outcomes, in which "local knowledge is recognized, valued and used" (Nasi and Frost, 2009, p. 4). There are manifold approaches for Stakeholder Relation Management under SFM at the Forest Management Unit level, which include the concepts of partnership, participation, cooperation and collaboration under programs such as community forestry, Community-based resource management (CBRM), Joint forest management (JFM), Corporate Social Responsibility (CSR) programs, inter-agencies involvement, and contract forestry. Adapted to the specific situation, these will promote and support the SFM implementation at the FMU level. SRM creates opportunities for participation by the stakeholders, which in turn promotes effective and efficient implementation of SFM at the FMU level. Especially social aspects can be enhanced through community forestry initiatives. Rebugio et al. (2010) described successful ways to improve community forestry. In their study, characteristic elements were a legally stipulated community forestry policy, and the reinvention of forestry agencies. They claim "a more supportive and facilitative role to assist communities to improve their livelihood and the condition of the forests. As such, the forestry agency has to reinvent itself to be able to cope with this new role, and maintain relevance" (Rebugio et al., 2010, p. 366). Further aspects refer to the promotion of sustainable livelihoods of the people, comprehensive and continuing capacity building of the community, availability of funding support, as well as the consideration of varying interests of local community and other stakeholders to participate in decision-making on forest management and benefit sharing from the forests. These findings could be confirmed by our study as well.

Summing up, the combination of Triple Typology of Stakeholder Theory with Stakeholder Analysis provides an analytical tool for business-stakeholder relationship under SFM implementation at the management level. It is suitable for deepening knowledge and providing critical information for SFM enhancement. The approach can be considered as a well-suited instrument for the assessment of interrelations between corporations and stakeholders. Thus, it can crystallize the focus on three different aspects, based on a theoretical or pragmatic background, respectively. It integrates the human perspective under the stakeholder theory to achieve social objectives for sustainability under SFM policy, within a wider scope of the ecosystem approach and the concept of sustainable development.

ACKNOWLEDGMENTS

This study was financially supported by the Ministry of Malaysian Higher Education (MOHE) and Universiti Malaysia Sabah (UMS). We also would like to thank two anonymous reviewers for their valuable suggestions for improvement.

REFERENCES

1. Aaltonen, K. (2011). Project stakeholder analysis as an environmental interpretation process. International Journal of Project Management, 29(2):165–183, doi:10.1016/j.ijproman.2010.02.001.
2. Appiah, M. (2013). Understanding the complexities and re-source sustainability in Afram Plains, Ghana. Environmental Management and Sustainable Development, 2(1):34–50, doi:10.5296/emsd.v2i1.3040.
3. Arts, B. (2002). 'Green alliances' of business and NGOs. New styles of self-regulation or 'dead-end roads'? Corpo-rate Social Responsibility and Environmental Management, 9(1):26–36, doi:10.1002/csr.3.
4. Donaldson, T. and Preston, L. E. (1995). The stakeholder theory of the corporation: concepts, evidence, and implications. The Academy of Management Review, 20(1):65–91, doi:10.2307/258887.
5. Dyllick, T. and Hockerts, K. (2002). Beyond the business case for corporate sustainability. Business Strategy and the Environment, 11(2):130–141, doi:10.1002/bse.323.
6. Food and Agriculture Organization of the United Nations (2013). Towards sustainable forest management, promoting sustainable management of forests and woodlands.
7. Freeman, R. E., Harrison, J. S., Wicks, A. S., Parma, B. L., and Colle, S. D. (2010). Stakeholder theory: The state of the art. Cambridge University Press, Cambridge.
8. Gamborg, C. and Larsen, J. B. (2003). 'Back to nature' a sustainable future for forestry? Forest Ecology and Management, 179(1-3):559–571, doi:10.1016/S0378-1127(02)00553-4.
9. International Tropical Timber Organization (2005). Revised ITTO criteria and indicators for the sustainable management of tropical forests including reporting format. ITTO Policy Development Series No. 15.
10. IUFRO (2005). Multilingual Pocket Glossary of Forest Terms and Definitions. IUFRO, Wien, Austria.
11. Kozak, R. A., Spetic, W. C., Harshaw, H. W., Maness, T. C., and Sheppard, S. (2008). Public priorities for sustainable forest management in six forest dependent communities of British Columbia. Canadian Journal of Forest Research, 38(12):3071–3084, doi:10.1139/X08-146.
12. Le, D., Lintangah, W., Pretzsch, J., Weber, N., and Bao, H. (2012). Implementation of sustainable forest management in two different forest management unit models in Vietnam and Malaysia. In Tropentag 2012, Gottingen,¨ Germany, September 19-21.
13. Lebel, L., Contreras, A., Pasong, S., and Garden, P. (2004). Nobody knows best: alternative perspectives on forest management and governance in Southeast Asia.
14. International Environmental Agreements, 4(2):111–127, doi:10.1023/B:INEA.0000040415.96194.b5.
15. Lintangah, W. (2014). Stakeholder Analysis in Sustainable Forest Management in Sabah, Malaysia. PhD thesis, Technische Universitat,¨ Dresden.
16. Malaysian Timber Council (2012). Malaysia Forestry & Environment (Facts and Figures).
17. Mitchell, R. K., Agle, B. R., and Wood, D. J. (1997). Toward a Theory of Stakeholder Identification and Salience: Defining the Principle of Who and What Really Counts.
18. The Academy of Management Review, 22(4):853–886, doi:10.2307/259247.
19. Nasi, R. and Frost, P. G. H. (2009). Sustainable forest management in the tropics: Is everything in order but the patient still dying? Ecology and Society, 14(2):40.
20. Nilsson, S. (2001). Future challenges to ensure sustainable forest management.
21. Nye, J. S. (2004). Soft power: The means to success in world politics. Public Affairs, New York, 1st ed edition.
22. Putz, F. E. (1994). Approaches to Sustainable Forest Manage-ment. CIFOR Working Paper No. 4.
23. Rametsteiner, E. and Simula, M. (2003). Forest certification - an instrument to promote sustainable forest management? Journal of Environmental Management, 67(1):87– 98, doi:10.1016/S0301-4797(02)00191-3.
24. Rebugio, L. L., Carandang, A. P., and Dizon, J. T. (2010). Promoting Sustainable Forest Management through Com-munity Forestry in the Philippines. In Mery, G., Katila, P., Galloway, G., Alfaro, R. I., Kanninen, M., Lobovikov, M., and Varjo, J., editors, Forests and society, volume v. 25 of IUFRO world series, pages 355–368. International Union of Forest Research Organizations and Available from IUFRO-WFSE/Metla, Vienna and Vantaa, Finland.

25. Reed, M. S., Graves, A., Dandy, N., Posthumus, H., Hubacek, K., Morris, J., Prell, C., Quinn, C. H., and Stringer, L. C. (2009). Who's in and why? A typology of stake-holder analysis methods for natural resource management.

26. Journal of Environmental Management, 90(5):1933–1949, doi:10.1016/j.jenvman.2009.01.001.

27. Sabah Forestry Department (2010). Annual Report 2010. Sabah Forestry Department (2011a). Annual Report 2011. Sabah Forestry Department (2011b). Fact Sheets of Forest

28. Reserves in Sabah.

29. Steurer, R. (2006). Mapping stakeholder theory anew: from the 'stakeholder theory of the firm' to three perspectives on business-society relations. Business Strategy and the Environment, 15(1):55–69, doi:10.1002/bse.467.

30. Steurer, R., Langer, M., Konrad, A., and Martinuzzi, A. (2005). Corporations, stakeholders and sustainable development I: a theoretical exploration of business–society relations. Journal of Business Ethics, 61(3):263–281, doi:10.1007/s10551-005-7054-0.

31. United Nations Conference on Environment and Develop-ment (1992). Non-legally binding authoritative statement of principles for a global consensus on the management, conservation and sustainable development of all types of forests.

32. Vanhanen, H., Rayner, J., Yasmi, Y., Enters, T., Fabra-Crespo, M., Kanowski, P., Karppinen, H., Mainusch, J., and Valkeapa¨a,¨ A. (2010). Forestry in Changing Social Landscapes. In Mery, G., Katila, P., Galloway, G., I. Alfaro, R., Kanninen, M., Lobovikov, M., and Varjo, J., editors, Re-sponding to global drivers of change, volume 25 of IUFRO world series, pages 201–222. International Union of Forest Research Organizations, Vantaa, Finland.

33. Watts, S. and Stenner, P. (2012). Doing Q methodological research: Theory, method and interpretation. SAGE, Lon-don.

CHAPTER 6

SIGNIFICANT INCREASE IN ECOSYSTEM C CAN BE ACHIEVED WITH SUSTAINABLE FOREST MANAGEMENT IN SUBTROPICAL PLANTATION FORESTS

Xiaohua Wei[1] , Juan A. Blanco[2]

[1] Department of Earth and Environmental Sciences, University of British Columbia, Kelowna, British Columbia, Canada

[2] Departmento de Ciencias del Medio Natural, Universidad Pu´blica de Navarra, Pamplona, Navarra, Spain

ABSTRACT

Subtropical planted forests are rapidly expanding. They are traditionally managed for intensive, short-term goals that often lead to long-term yield decline and reduced carbon sequestration capacity. Here we show how it is possible to increase and sustain carbon stored in subtropical forest plantations if management is switched towards more sustainable forestry. We first conducted a literature review to explore possible management factors that contribute to the potentials in ecosystem C in tropical and subtropical plantations. We found that broadleaves plantations have significantly higher ecosystem C than conifer plantations. In addition, ecosystem C increases with plantation age, and reaches a peak with intermediate stand densities of 1500–2500 trees ha^{-1}. We then used the FORECAST model to simulate the regional implications of switching from traditional to sustainable management regimes, using Chinese fir (*Cunninghamia lanceolata*) plantations in subtropical China as a study case. We randomly simulated 200 traditional short-rotation pure stands and 200 sustainably-managed mixed Chinese fir – *Phoebe bournei* plantations, for 120 years. Our results showed that mixed, sustainably-managed plantations have on average 67.5% more ecosystem C than traditional pure conifer plantations. If all pure plantations were gradually transformed into mixed plantations during the next 10 years, carbon stocks could rise in 2050 by 260.22 TgC in east-central China. Assuming similar differences for temperate and boreal plantations, if sustainable forestry practices were applied to all new forest plantation types in China, stored carbon could increase by 1,482.80 TgC in 2050. Such an increase would be equivalent to a yearly sequestration rate of 40.08 TgC yr^{-1}, offsetting 1.9% of China's annual emissions in 2010. More importantly, this C increase can be sustained in the long term through the maintenance of higher amounts of soil organic carbon and the production of timber products with longer life spans.

Citation: Wei X, Blanco JA (2014) Significant Increase in Ecosystem C Can Be Achieved with Sustainable Forest Management in Subtropical Plantation Forests. PLoS ONE 9(2): e89688. doi:10.1371/journal.pone.0089688

INTRODUCTION

Reforestation efforts are promoted globally to meet the growing demand of forest products, especially in developing countries. Subtropical forests are therefore promoting the maintenance of forest cover in the region and the production of goods that improve the livelihoods of local communities. Subtropical forests cover 7–10% of the global land area, and store 40–50% of carbon present in terrestrial vegetation [1]. However, subtropical forests also account for most of current deforestation, with accumulated losses in the 1996–2010 period of ~100 million ha[2]. Because forests are a major carbon pool, their management is crucial to develop successful policies for climate change mitigation. Carbon storage versus production of timber and non-timber forest products are seen sometimes as incompatible objectives to be implemented at the same time in a forest, but this view is being challenged [3]. Clarifying if subtropical forest plantations can support multi-objective forestry and estimating how much carbon could be stored in working forest plantations would help to maximize the outcome of forest management plans while helping to reach a more sustainable development of the subtropical regions.

New planted forests are typically established on sites that have rapid growth and access to processing facilities and growing markets. Hence, very large areas of new forest plantations have been established in subtropical countries, particularly in Latin America and parts of Asia. In spite of such rapid growth, only 3% of the world forest land is covered with productive forest plantations. However, this area expanded by 2 million ha annually in the 1990s and by 2.8 million ha in the 2000s [4].

While much attention and discussion have been devoted on controlling and reverting deforestation, how to enhance forest carbon stocks by improving forest ecosystem condition has received less emphasis until recently [5]. Reforestation has been suggested not only as an effective way to restore degraded ecosystems, but also as a way to mitigate elevated atmospheric CO_2, hence contributing towards the reduction of climate change [3], [6]. Whether and how these efforts can maximize and sustain forest C storage largely remains unexamined[7], [8], [9]. Recent decisions by the U.N. Framework Convention on Climate Change have encouraged consideration of other ecosystem services while implementing REDD+ (Reducing greenhouse gas Emissions from Deforestation and forest Degradation) projects for forest carbon sequestration and storage. Therefore, selecting the most appropriate approaches for implementing multipurpose forest management in planted forests is just as important as growing more forests [10]. However, few incentives and policy guidelines are in place to encourage growing better forests globally, perhaps due to the lack of a clear understanding of long-term benefits from growing better forests as well as operational difficulties.

Traditional forest management practices in forest plantations, particularly in developing countries, usually involve application of short rotations, usage of monoculture and exotic species, and high-level removal of biomass, with the main purpose of maximizing short-term economic gains [11]. The ecological problems associated with those traditional forestry practices have been well documented: declines in long-term productivity [12], [13], biodiversity loss [14], increased susceptibility to insects and diseases [15], physicochemical changes in forest soils [16], and losses of other ecological services [17]. In the context of climate change mitigation, traditional forestry practices may lead to short-term carbon sequestration in wood products, but those gains are unlikely sustainable due to losses in ecosystem C linked to the above-mentioned ecological problems. Sustainable forestry practices, however, could have an important role in mitigating climate change impacts, if they are well designed and implemented[3]. For example, transformation of conifer monocultures into mixed conifer-broadleaved plantations has been considered as an efficient strategy to sustain forest productivity and restore degraded forests [18].

China has both subtropical and temperate forests, and aggressive reforestation policies have turned it into one of the five most forest-rich countries in the world, accounting for the largest gain in forested areas globally [12], [19]. However, most of ecosystems reforested during the past decades in China have been by monocultures dominated by coniferous species. It has been widely reported that timber yield declines and degradation in soil fertility in coniferous plantations over several continuous rotations (see [20], and references therein). On the other hand, in subtropical China, evergreen broadleaved forests are the main native ecosystem type in the region, with a clear potential capacity for carbon sequestration [11]. Among the native species, the broadleaf *Phoebe bournei* (Hemsley) Yang is one of the most valuable tree species in this subtropical region because of its high-quality wood properties, with significant economic and ecological benefits [21]. Sharing the same distribution area, the conifer Chinese fir (*Cunninghamia lanceolata* (Lamb.) Hook. is the most important commercial forest species in subtropical China. It has been widely planted in the southeast provinces of China [20], [22], becoming one of the most abundant planted species in the world [23].

The traditional management of Chinese fir plantations is representative of the type of management usually carried out in many other tropical and subtropical forest regions around the world. After harvesting and slash-burning the native mixed evergreen broad-leaved forests, pure plantations of fast-growing conifers were established. The plantation sites used to be abandoned after one or two rotations and allowed to regenerate naturally to mixed species stands by stump sprouting and natural seeding, which then acted as a fallow period to restore the site [24]. However, since the 1950s, the plantation area of Chinese fir has been enlarged, and this species has been repeatedly planted on the same sites without intercropping or periods of fallow. Local foresters have generally used a 25-year rotation, with variation from 20 to 30 years depending on soil nutrient abundance [25]. A trend towards rotation length shortening is being observed in the latest years, driven by increasing demand for timber products fueled by economic development and population increase in China. Continuous cultivation of Chinese fir at the same site has resulted in a well-known ecological problem: yield decline over consecutive rotations [20], [26]. With these conditionings, an important question has recently been raised in Chinese forestry with implications for the management of forest plantations in other subtropical regions of the world: can forest plantations be managed to produce forest products but in a way that also maintains or even increases ecosystem C? If so, can the differences in ecosystem C between traditional and sustainable management be quantified?

To answer these questions, in this work we tested the hypothesis that subtropical tree plantations accumulate less carbon under traditional exploitative management than under sustainable forest management, which is more similar to natural stand development. We also aim to quantify the potential increase in ecosystem C at a regional level. To test this hypothesis we first reviewed available information on C pools in subtropical forest plantations and related them to factors defining their management (species type, rotation length, stand density). Also, the literature review allowed us to identify the management prescriptions with the highest carbon storage, and to define realistic sustainable management scenarios. Finally, we estimated potential C sequestration gains through time at a regional level by comparing ecosystem C values under traditional and sustainable management regimes in subtropical plantations in China, simulating plantation growth with the forest ecosystem model FORECAST[27].

MATERIALS AND METHODS

We first carried out a literature review to test if differences in ecosystem C among subtropical plantation types exist, and also to identify the management factors that have maximized ecosystem C in those plantations in the past. Then, we used ecological modelling to estimate the average differences between traditional forest management and sustainable forest management, with consideration of the factors identified in the literature review. As ecological models need to be calibrated for specific ecosystem types [28], we used Chinese fir plantations (*Cunninghamia lanceolata* (Lamb.) Hook.) located in the subtropical region of China as a case study. Continuous repetitions of traditional practices over several rotations (monoculture, short rotations, whole-tree harvesting, and slash burning) have resulted in severe yield decline [26]. Among various possible reasons, N limitation by nutrient exports and resource competition between tree seedlings and understory are the most important factors [20]. Therefore, there is a need to improve the management of this species to establish a more sustainable and multi-purpose forestry in China.

Literature Review of Carbon in Subtropical Plantations

We conducted a structured search of published scientific documents to define the observed range in carbon storage in the ecosystem, as well as in the belowground and aboveground fractions, and their relationships with plantation type (broadleaf or coniferous), plantation age, and stand density. We searched in the on-line databases Google Scholar (by Google Inc.http://scholar.google.com), ISI Web of Knowledge (by Thomson Reuters,http://www.isiwebofknowledge.com), China Academic Journal Database (by China National Knowledge Infrastructure, http://cnki.en.eastview.com/kns50), CAB Abstracts (by the Centre for Agricultural Bioscience International, http://www.cabdirect.org), and the databases hosted by the Canadian Forest Service Bookstore (http://bookstore.cfs.nrcan.gc.ca), the University of British Columbia's library (http://www.ubc.ca/library), and the Public University of Navarre's library (http://www1.unavarra.es/biblioteca).

We divided the population of collected documents into two groups: 1) documents reporting C pools in broadleaf or other non-coniferous forests, and 2) documents describing coniferous or mixed conifer-broadleaf plantations. The final number of documents and sites used for further analysis after a standardized selection process [29] were, for broadleaves 38 documents and 138 sites; whereas for conifers 51 documents and 160 sites (Figure 1). For each site, we gathered data on plantation age, density, aboveground C density (MgC ha^{-1}), belowground C density, and ecosystem C density. Aboveground C was calculated as tree C plus understory C (if reported). Belowground C carbon was calculated as C in tree roots (and in plant roots if reported), litter C and mineral soil C. Values from graphs were extracted using DataThief III[30]. Original values for mineral soil C were often not directly comparable, as they were measured for different soil depths. To standardize soil carbon values, we modified the method by [31] and used bulk density and soil C concentration to calculate and plot soil C content versus soil depth, obtaining a significant correlation between both variables (Figure 2). We then estimated the standardized mineral soil carbon content at 60 cm depth as the value that would correspond if soil carbon distribution in each soil followed a relationship with soil depth parallel to the regression line.

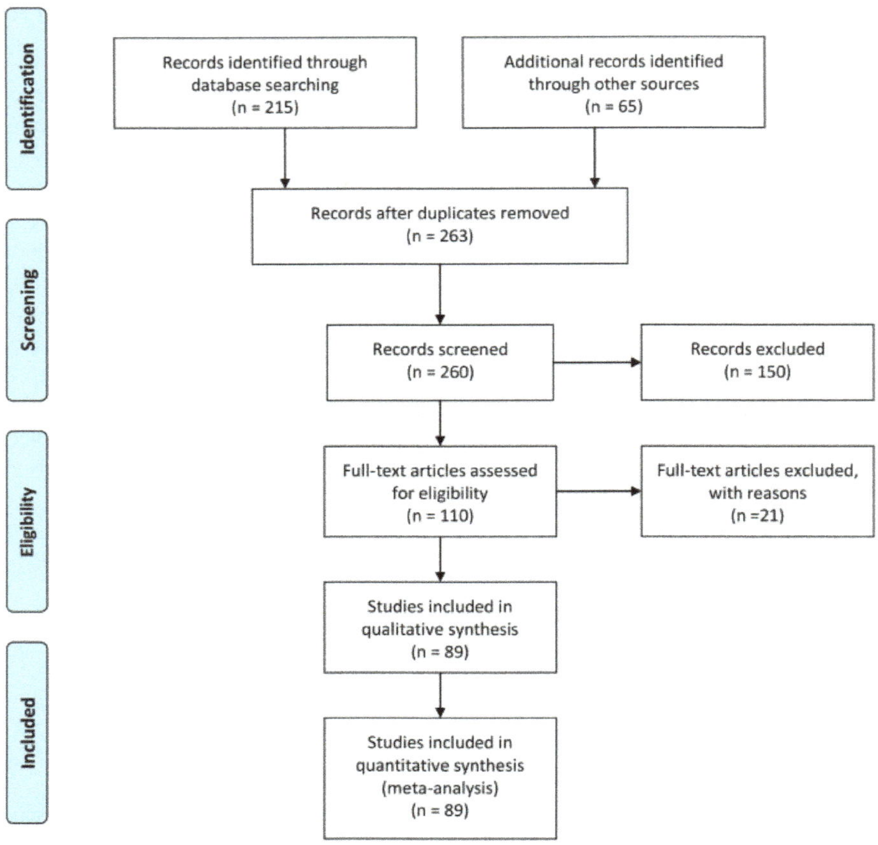

Figure 1. PRISMA flow diagram for the creation of the database of carbon pools in tropical and subtropical forest plantations.

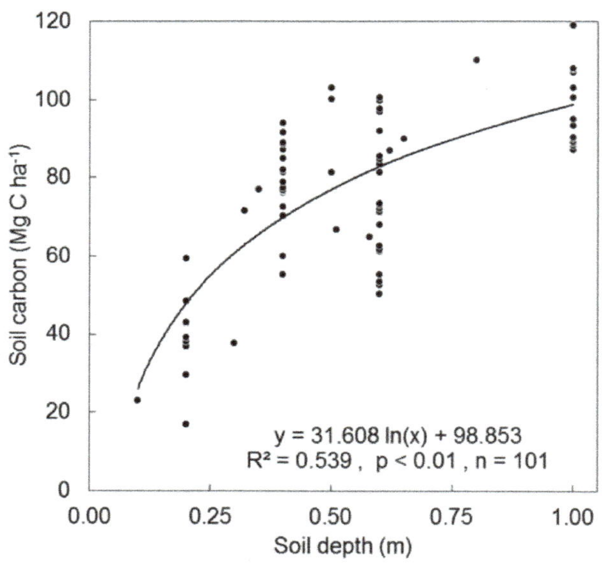

Figure 2. Values of soil carbon *vs.* soil depth in the reviewed literature.

We tested the equality of means for aboveground, belowground and ecosystem C between the two types of plantations using Welch's modified *t*-test, (a robust test that does not assume equality of variances; [32]), setting significance level at 0.05. We also carried out regressions (weighted by the error associated to each measure) of carbon density vs. plantation age and stand density, lumping together data from conifer and broadleaf plantations as initial analysis did not show improvement of the regression results by splitting the data into species groups or location (data not shown). All the statistical analyses where carried out with Sigmaplot 10.0 (Systat Software Inc.), JMP v.5.0.1 (SAS Institute, NC, USA), and S-PLUS 6.1 (Insightful Corp.).

Simulating Ecosystem C in Planted Forests with FORECAST

In the second part of our research, we used the results from the literature review to define probability distributions for the two main factors defining forest plantation management (stand density and rotation length) that maximized ecosystem C, and we assumed that they are indicators of more sustainable forest management. Probability distribution of site quality was defined using the distribution map of site qualities for Chinese fir in south-eastern China [33]. Probability distributions of planting density, rotation length and utilization level for traditional management were also created for both sets of simulations after consultations with panels of local experts composed by academics, forest officers, and forest managers (Figure 3). The panel of local experts indicated the need for creating different management scenarios based on site quality for sustainable management. Hence, probability distributions of tree density and rotation length were different in each site quality for sustainable management, but they were the same in all sites under traditional management.

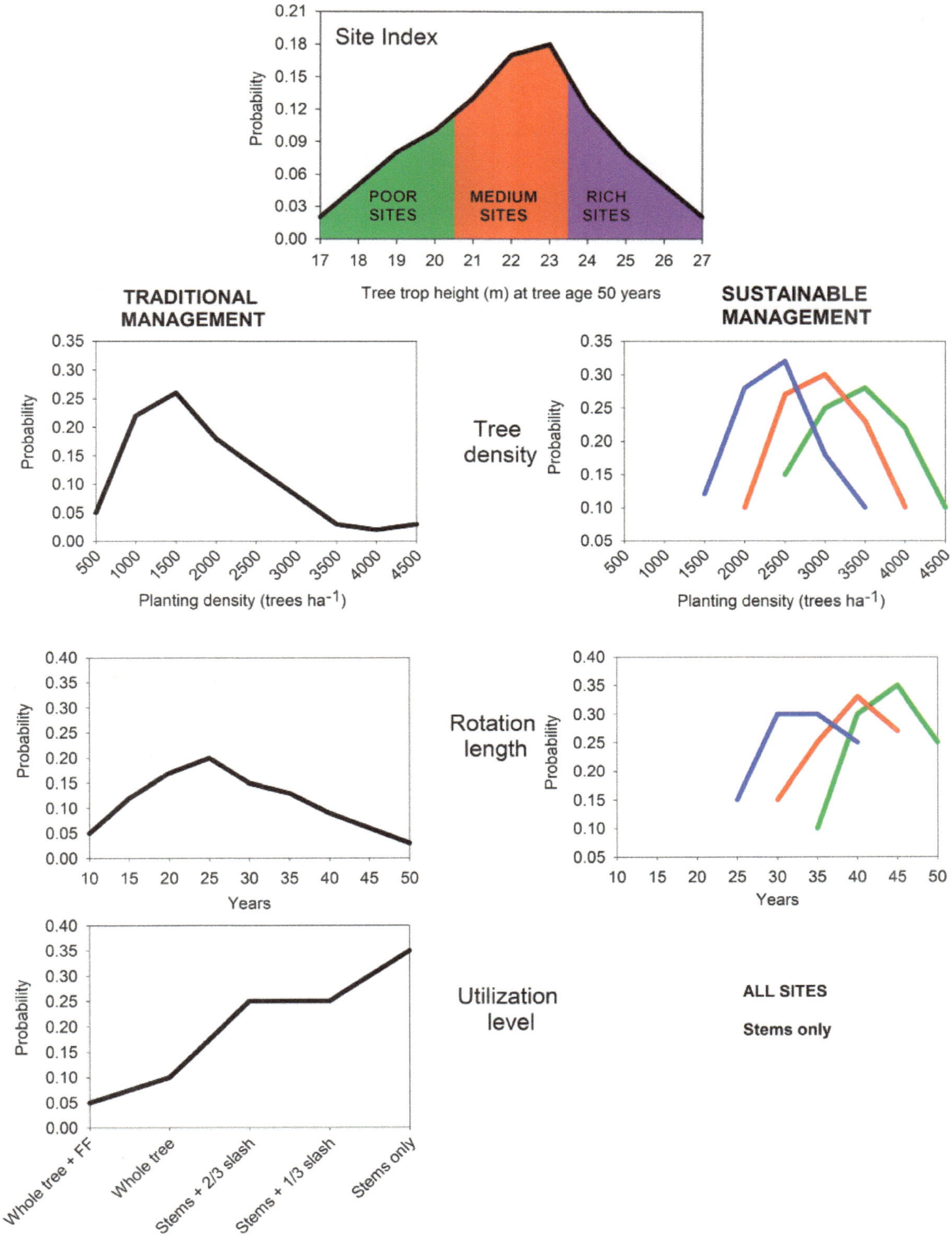

Figure 3. Probability distributions of the four variables defining the management scenarios. Different color lines correspond to the different site qualities as defined in the top graph. FF: Forest floor removal.

Using the probability functions, we constructed 200 Monte Carlo projections for each set of simulations (traditional management and sustainable forest management). Each run simulated a management scenario (a random combination of site quality, rotation length, and stand density). In traditional management, all trees planted were Chinese fir. In sustainable management, probability distributions show total tree density, which was composed by Chinese fir and *Phoebe bournei* in a 1:1 ratio, the most suitable species mix for carbon sequestration[21].

We then simulated each scenario for 120 years with the ecosystem model FORECAST, which is a stand-level model specially designed for assessing the effects of various forest management practices on forest carbon storage and sequestration. FORECAST has been successfully tested and applied in many world forests before, including tropical and subtropical forests [28].

FORECAST is a management-oriented, deterministic, non-spatial stand-level forest growth and ecosystem dynamics simulator that operates at annual time steps. The model has been described in detail before [27], [28], and therefore only a summary of the main driving function used to calculate tree growth is provided. The model uses a mass balance approach to estimate ecosystem nutrient circulation, and how nutrient availability limits tree growth together with available light in the canopy. FORECAST has three application stages: 1) assembling calibration data and generating historical rates of key ecosystem processes; 2) model initialization by establishing the ecosystem condition for the beginning of a simulation run; and 3) simulation of tree and plant growth.

Input data came from sites covering the observed range of Chinese-fir plantation growth, qualitatively described as poor, medium and good (Table 1). The sites were scaled quantitatively as 17, 21 and 27 (from poor to good, based on canopy top height in m at age 50 years) to represent a relative index of tree growth required for extrapolation within the model. Tree data on biomass, mortality, stand density, tree height, canopy height, nutrient concentrations of live tissues and other data were based on reported values for similar Chinese-fir plantations [25], [34], [35], [36], [37], [38], [39], [40], [41], [42], [43].

Table 1. Values used to calibrate FORECAST parameters related to Chinese-fir

Parameter	Unit	Rich site	Poor site
Chinese fir parameters			
Nitrogen concentration in needles young/old/dead	%	1.53/1.36/1.13	1.21/1.11/0.93
Nitrogen concentration in stem sapwood/heartwood	%	0.14/0.03	0.12/0.03
Nitrogen concentration in bark live/dead	%	0.44/0.27	0.37/0.24
Nitrogen concentration in branches live/dead	%	0.67/0.52	0.55/0.47
Nitrogen concentration in root sapwood/heartwood	%	0.37/0.06	0.35/0.06
Nitrogen concentration in fine roots live/dead	%	1.17/0.97	0.96/0.79
Shading by maximum foliage biomass	% of full light	8	30
Soil volume occupied at maximum fine root biomass	%	100	95
Efficiency of N root capture	%	98	100
Retention time for young/old foliage/dead branches	years	1/2/40	1/2/40
Fine roots turnover	$year^{-1}$	0.95	1.35
Decomposition rates			
Sapwood (by litter age)	% $year^{-1}$	1–5 years (2.0); 6–10 years (10.0); 11–15 years (30.0); 16–20 years (20.0); > 20 years (4.0)	
Heartwood	% $year^{-1}$	1–10 years (0.4); 11–15 years (10.0); 16–25 years (15.0); 25–40 years (10.0); >40 years (2.0)	
Bark	% $year^{-1}$	1–5 years (2.0); 6–20 years (12.0); 20–40 years (20.0); >40 years (4.0)	
Branches and large roots	% $year^{-1}$	1–5 years (10.0); 6–10 years (45.0); 11–15 years (35.0); >15 years (4.0)	
Needles (poor site)	% $year^{-1}$	1–2 years (20.0); 3–5 years (30.0); 6–10 years (40.0); >10 years (2.0)	
Needles (good site)	% $year^{-1}$	1–2 years (27.0); 3–5 years (30.0); 6–10 years (40.0) >10 years (3.0)	
Fine roots	% $year^{-1}$	1–2 years (30.0); 3–4 years (50.0); >4 years (9.0)	

Decomposition rates indicate the mass loss in one year as a fraction of the initial mass at that year. See text for list of bibliographical sources for model calibration.
doi:10.1371/journal.pone.0089688.t001

Model calibration data for *Phoebe bournei* forests (Table 2) came from literary sources and field data [21]. They included data of biomass accumulation and stand density [44], [45], [46],[47], [48], [49], [50], decomposition rates [51], [52], [53], [54], soil nutrients under different site conditions [55], [56], [57], and photosynthetic efficiency [58], [59], [60], [61]. Understory was simulated as grass and shrub complexes [20]. Published data were used to characterize shrub biomass, height, tissue nutrient concentrations and other relevant data [62], [63], [64], [65](Table 3).

Table 2. Values used to calibrate FORECAST parameters related to *Phoebe bourne*

Parameter	Unit	Rich site	Poor site
Chinese fir parameters			
Nitrogen concentration in leaves young/old/dead	%	1.35/0.92/0.69	1.15/0.80/0.67
Nitrogen concentration in stem sapwood/heartwood	%	0.18/0.04	0.16/0.04
Nitrogen concentration in bark live/dead	%	0.55/0.47	0.49/0.33
Nitrogen concentration in branches live/dead	%	0.33/0.29	0.28/0.26
Nitrogen concentration in root sapwood/heartwood	%	0.37/0.29	0.34/0.27
Nitrogen concentration in fine roots live/dead	%	0.81/0.57	0.65/0.56
Shading by maximum foliage biomass	% of full light	8	25
Soil volume occupied at maximum fine root biomass	%	100	100
Efficiency of N root capture	%	100	100
Retention time for young/old foliage/dead branches	years	1/0.5/30	1/0.5/30
Fine roots turnover	$year^{-1}$	1.00	1.15
Decomposition rates			
Sapwood (by litter age)	% $year^{-1}$	1–3 years (0.1); 4–15 years (2.0); 15–20 years (12.0); >20 years (9.0)	
Heartwood	% $year^{-1}$	1–3 years (0.1); 4–15 years (2.0); 15–20 years (12.0); 20–40 years (9.0); > 40 years (2.0)	
Bark	% $year^{-1}$	1–10 years (20.0); 11–20 years (40.0); 21–30 years (35.0); >30 years (2.0)	
Branches and large roots	% $year^{-1}$	1–5 years (15.0); 6–15 years (35.0); 16–20 years (45.0); >20 years (15.0)	
Leaves (poor site)	% $year^{-1}$	1–2 years (45.0); 3–5 years (35.0); >6 years (20.0)	
Leaves (good site)	% $year^{-1}$	1–2 years (50.0); 3–5 years (40.0); >6 years (20.0)	
Fine roots	% $year^{-1}$	1–5 years (45.0); >6 years (35.0)	

Decomposition rates indicate the mass loss in one year as a fraction of the initial mass at that year. See text for list of bibliographical sources for model calibration.
doi:10.1371/journal.pone.0089688.t002

Table 3. Values used to calibrate FORECAST parameters related to the shrub complex and soil processes

Shrub complex parameters	Unit	Rich site	Poor site
Nitrogen concentration in leaves live/dead	%	1.68/1.38	1.12/0.76
Nitrogen concentration in stems live/dead	%	0.48/0.14	0.32/0.10
Nitrogen concentration in rhizomes live/dead	%	1.14/1.05	0.76/0.70
Nitrogen concentration in roots live/dead	%	1.10/0.84	0.73/0.56
Shading by maximum foliage biomass	% of full light	0.20	0.45
Soil volume occupied at maximum fine root biomass	%	60	50
Efficiency of N root capture	%	99	99
Transfer from live to dead stem/rhizomes/roots	% $year^{-1}$	20/20/30	20/20/40
Retention time for foliage	years	1	1
Decomposition rates		Litter age in years (decomposition rate in %)	
Foliage (poor site)	% $year^{-1}$	1 year (80.0); 2–3 years (60.0); 4–5 years (50.0); >5 years (2.0)	
Foliage (rich site)	% $year^{-1}$	1 year (95.0); 2–3 years (60.0); 4–5 years (50.0); >5 years (2.0)	
Stems & roots	% $year^{-1}$	1 year (20.0); 2–3 years (30.0); 4–5 years (40.0); >5 years (2.0)	
Soil parameters		**Rich site**	**Poor site**
Nitrogen concentration in slow/fast humus	%	3.00/1.40	3.0/1.40
Decomposition rate slow/fast humus	% $year^{-1}$	0.15/2.00	0.15/2.00
CEC soil (CEC humus)/AEC	kg N ha^{-1}	80.0 (0.2)/20.0	40.0 (0.2)/5.0
Atmospheric deposition/non-symbiotic fixation	kg N ha^{-1} $year^{-1}$	4.9/1.0	4.9/1.0
Non-symbiotic N fixation rate	Kg N ha^{-1}	2.5	3.0

Decomposition rates indicate the mass loss in one year as a fraction of the initial mass at that year.
doi:10.1371/journal.pone.0089688.t003

Data describing soil processes and nutrient inputs in precipitation and slope seepage, mineral soil cation and anion exchange capacities, humus mass, nutrient concentrations in litterfall, litter decomposition rates and others. were derived from literature [38], [39], [40], [66], [67], [68],[69], [70], [71] (Table 3). To establish initial site conditions the model was run in set-up mode, forcing the model to match the observed site conditions [72], [73]. Initial conditions were created for each site quality, by running several cycles of tree growth ending with stand-replacing windthrow, as typhoons are the most common natural disturbance in this region [74]. A library of initial conditions was created with initial soil organic matter values (humus+litter) ranging from 90.9 to 296.3 Mg ha^{-1} for the poorest and richest sites, respectively.

Published field data on several chronosequences at different site qualities were used to evaluate FORECAST performance for Chinese fir and *Phoebe bournei* plantations. After obtaining a satisfactory model performance, we run the model for the 400 randomly-generated scenarios described above. A sub-set of documents gathered during the literature review dealing with Chinese fir plantations was also used to evaluate the model performance for ecosystem C.

RESULTS

From our review, 23.9% of the studies on bradleaves used local native species, but *Populus*spp. (19.6%), *Tectionia sp.* (13.1%) and *Eucalyptus* spp. (8.10%) were the most common commercial species used. For conifers, 81.4% of the studies reported data from *Cunninghamia* sp. plantations, and 14.3% came from *Pinus* spp. plantations. Our results showed that subtropical broadleaves plantations can, on average, sustain higher carbon densities than subtropical conifer plantations (Figure 4). The differences, however, are mostly caused by differences in belowground C, whereas differences among plantations types in aboveground C are non-significant. Not surprisingly, the trend in subtropical forest plantations is to increase ecosystem carbon storage as plantation age increases. This is basically the result of increase in tree size, and accumulation of litter and coarse woody debris biomass. This trend, however, seems to slow down over time, with little expected gains beyond plantation age ~90, as described by the significant exponential model (Figure 5). Ecosystem C initially increases as plantation density increases, reaching a peak at about 1500–2500 stems ha^{-1}. For higher densities the trend is to reduce ecosystem C (Figure 5).

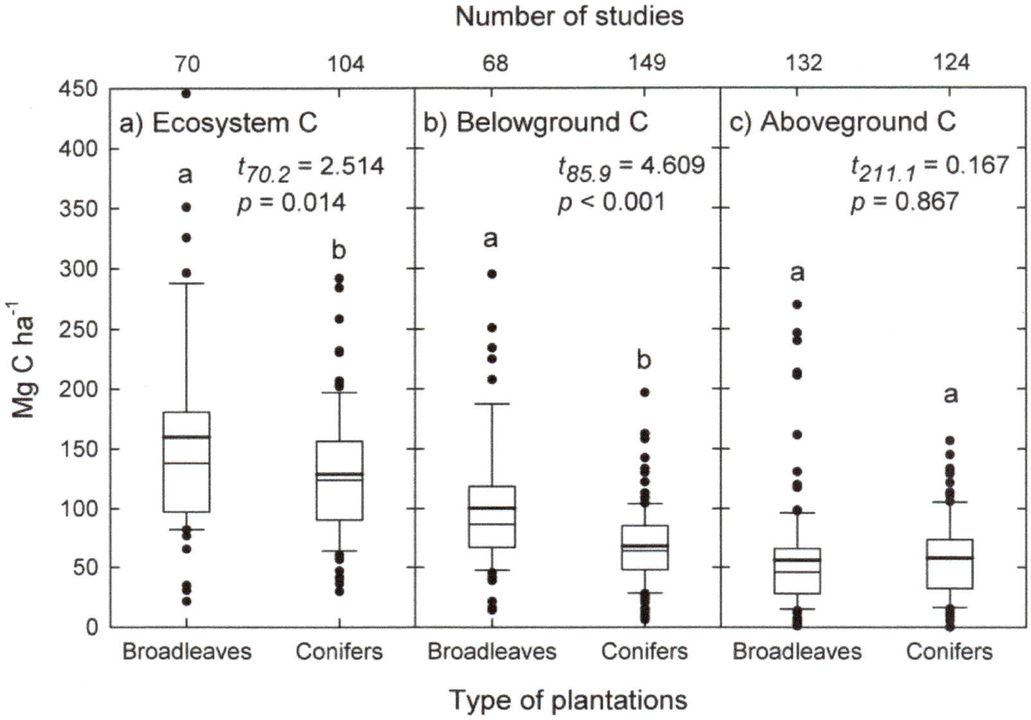

Figure 4. Distribution of C storage in published bibliography for different subtropical plantation types. The thick horizontal line indicates the average of all observations. Letters indicate significant differences with Welch's t at P<0.05.

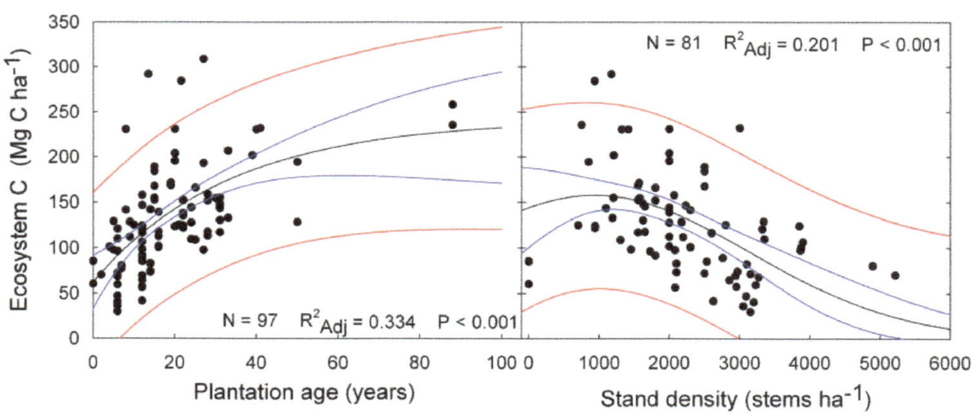

Figure 5. Carbon stocks in subtropical plantations for different A) rotation ages, and B) stand densities. Black lines indicate the best-fit model, blue lines 95% confidence interval and red lines 95% prediction intervals.

During the simulation runs, the most frequent scenario for sustainable forest management was 3000 trees ha^{-1} (1500 Chinese fir, 1500 *P. bournei*), rotation length of 40 years and stem-only harvesting. For traditional management, the most frequent scenario was 1500 trees ha^{-1}(Chinese fir only), rotation length of 25 years, stem plus slash removal, and site preparation by understory burning. Comparing these two scenarios when run for 240 years, we can see how plantations managed for sustainable, long-term goals, maintain ecosystem productivity higher and during longer periods that traditional intense management.

Yearly net primary productivity under traditional management quickly drops in each consecutive rotation, and as a consequence tree biomass (and the associated merchantable volume) also decreases rapidly over time (Figure 6). In addition, there is an increasingly stronger competition with understory. Although with traditional management higher peaks of available soil N are achieved than in the sustainable management, the total amount of ecosystem N decreases over time. On the other hand, under sustainable management higher levels of soil organic matter are sustained (Figure 6).

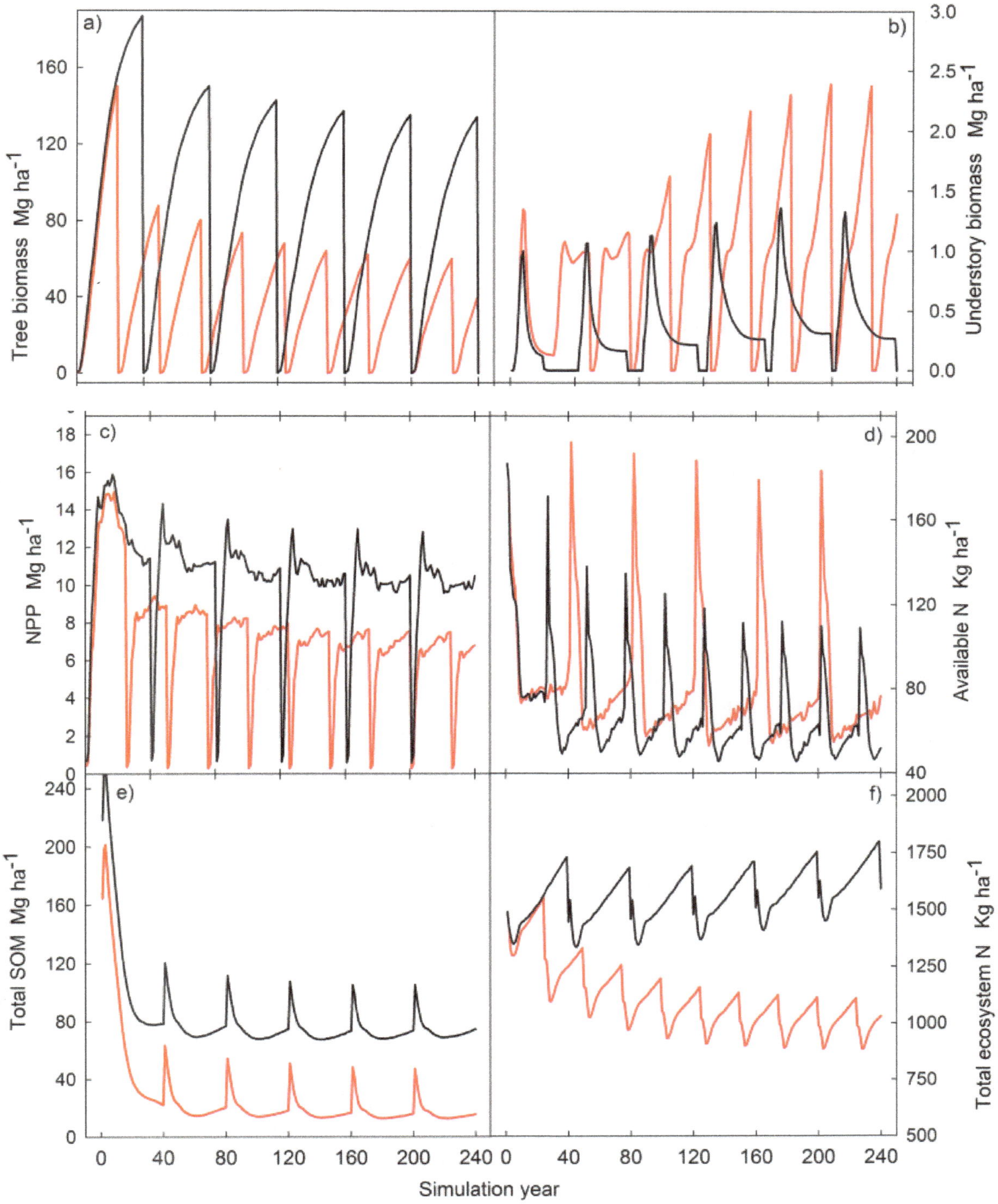

Figure 6. Change in ecological variables in Chinese fir plantations in 240 years of simulation under the average plan two types of management. Current (red line) 25-year rotation, whole-tree harvesting, 1500 trees ha^{-1} Chinese fir planted; and close-to-nature management (black line) 40-year rotations, stem-only harvesting, 1500 trees ha^{-1} Chinese fir +1500 trees ha^{-1} *Phoebe bournei.*

When all the inputs and outputs are accumulated through the 240 years of simulation, it is clear how current management is depleting ecosystem N reserves. On the other hand, sustainable management can maintain and even increase N pools (Figure 7). Traditional management removes N from the ecosystems during the site preparation (directly through slash-burning and indirectly by erosion caused by heavy machinery and losses of plant cover), with the export of harvested products, and by N leaching. Under sustainable forest management, slash-burning is not used and erosion is greatly reduced as practices affecting soil stability and plant cover are minimized. N removal in harvested products is also lower as only stems (with low N concentrations) leave the plantations. On the other hand, as N reserves under sustainable management are higher, leaching losses are also higher, just because the soil N pool is bigger than in traditional management. All flows considered, N outputs with current practices are not only higher, but N inputs are also smaller, as there is less N fixation due to the reduced biological activity in sites of lower productivity. Also, because any given stand gets less N inputs by seepage as it is surrounded by stands that also have lower N capitals.

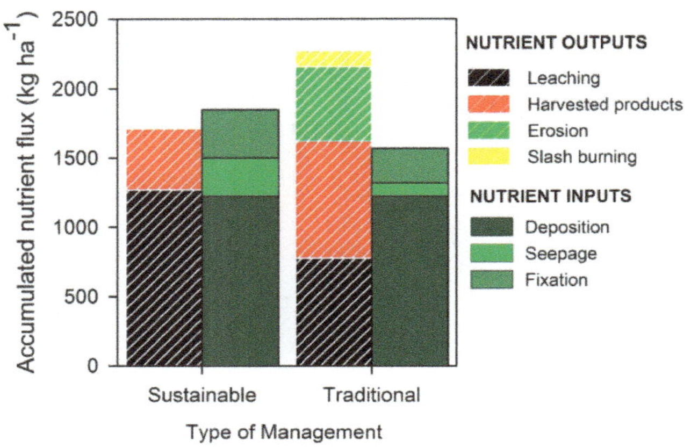

Figure 7. Accumulated N inputs (solid bars) and outputs (dashed bars) in Chinese subtropical plantations after 240 years of management under two most typical management regimes for traditional and sustainable plantations.

When simulating how the C pools in Chinese fir stands would evolve over the 21st century, our results showed that if sustainable forestry practices were introduced at a rate of 10% of the total planted area, the region would keep on average a higher and sustained level of carbon storage over the long term, compared to traditional management (Figure 8). At the beginning of our scenarios a large overlapping range for ecosystem C between traditional and sustainable plantations can be seen (grey area). Such fact indicates that some time is needed for the sustainable management to significatively reduce C loses compared to traditional management. This is not surprising as productivity losses are accumulative and occur mostly at the change of rotation. However, by the middle and until the end of the simulated 120 years the overlap has been greatly reduced. By that time, only the sites with the highest C contents and managed traditionally would be comparable to the poor sites under sustainable management. Such increase in carbon density could be 12% by 2020. This difference would grow over time, reaching 17%, 52%, 79%, and 67% greater carbon density under sustainable forest management after 25, 50, 75 and 100 years, respectively. The average weighted carbon density increase over the 100-year simulation period was 67.5%.

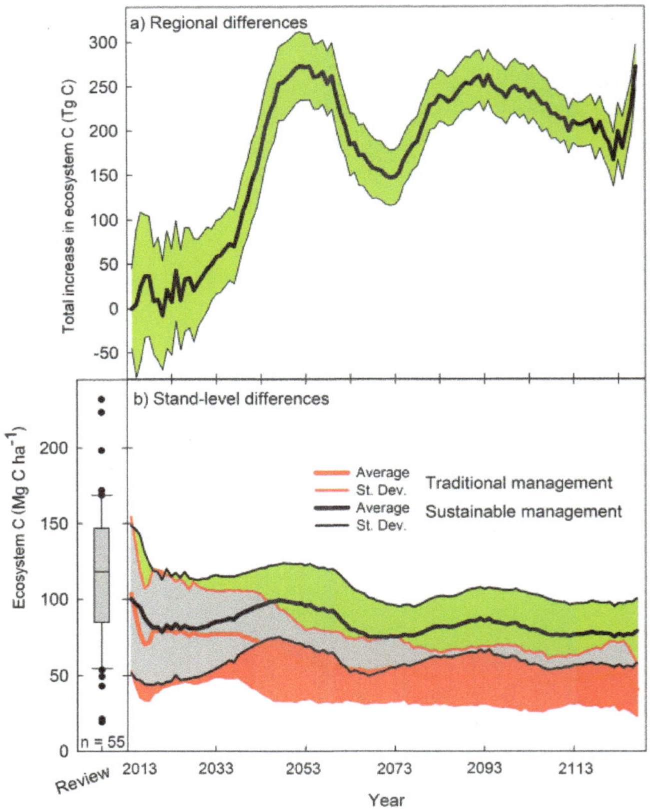

Figure 8. Total carbon gains when switching from current traditional intensive forestry to sustainable forest management at an annual rate of 10% of total area dedicated to Chinese-fir plantations (panel A). Ecosystem carbon density in Chinese fir plantations under current intensive management (red lines), and sustainable management (green lines) (panel B). Thick lines represent average and color areas represent one standard deviation. The grey area indicates values of ecosystem C that can be achieved by either sustainable or traditional management plans. The box-plot in the bottom-left summarizes the values obtained from a review of 55 published peer-reviewed papers on carbon stocks in Chinese fir plantations.

As a way to validate our estimations at the stand level, when analyzing the average and 95% confidence intervals for the whole set of projected ecosystem C values they were always inside the range of reported values for Chinese-fir plantations (Figure 8), providing confidence on our estimations.

Stand-level differences between management regimes would have large regional implications. Extrapolated to the whole area dedicated to Chinese fir plantations (7,046,520 ha [11]), together with the plan of substituting traditional by sustainable management implemented at a ratio of 10% of the land every year, the increase in ecosystem C stored in the mixed Chinese fir plantations could reach 261.2 TgC by 2050 (Figure 8). After that year, the differences in ecosystem carbon storage between management options show a temporal decline. This is as a consequence of the harvesting of sustainable forest plantations primarily clustered around the years 40 to 60. The second rotation growth would again produce an increase in stored C, which would remain above 200 TgC after 120 years of simulation.

DISCUSSION

Subtropical plantations have been described as the most effective for C sequestration [75],[76], [77]. In subtropical regions, local broadleaved species are usually planted, but the available data on C stocks in subtropical plantations came mostly from the five genuses more widely planted around the globe: *Eucalyptus, Populus, Tectonia, Pinus* and *Cunninghamia*. These are the most valuable among the fast-growing species currently used, and our results show that selecting among the species type (broadleaves or conifers) can produce significant differences in C pools depending on plantation type and management regime. Ecosystem C increases with tree age in subtropical plantations, as these systems develop ecological structures that make them closer to natural forests over time [78]. Older plantation age has also been associated to higher soil carbon levels [79]. Therefore, increasing rotation length should be encouraged when implementing sustainable forestry plans for increasing carbon stocks, as well as generating more biophysical structure for higher niche diversity and its associate biodiversity.

[80] have pointed out that density plays an important role in the productivity of mixed stands, with higher productivities in mixed stands than in monocultures at medium to high densities (>1500 trees ha^{-1}). Our results also indicate that density should be maintained at medium values, avoiding excessive inter-tree competition for resources such as light, water and nutrients that could reduce stand productivity, but also avoiding low densities that may leave some resources unused by trees. Our estimates also fit with several field experiments in mixed plantations of Chinese fir with other trees species, which have consistently shown higher biomass and C density in mixed stands with a tree density ratio of 1:1 for Chinese fir/companion species, for total densities in the 2000–3000 trees ha^{-1} range [81], [82], [83], [84].

Our results indicate that managing plantations designed for multiple values could sustain higher carbon storage potentials if they include broadleaves, are managed for rotations lengths of 25–50 years, and have planting densities of 1500–2500 stems ha^{-1}. Our estimations corroborate previous field research that have reported maximum biomass values in mixed Chinese fir plantations for similar rotation and density ranges, reporting little gain or even biomass decrease with plantation densities above 3000 trees ha^{-1} [82], [85]. In this regard, forest plantations seem to have important functional similarities with native forests, where intermediate tree densities have been reported as the ones reaching the carrying capacity of the ecosystem, without surpassing it [86].

Differences in C density among management types can be better understood by comparing the typical (most frequent in our set of simulations) management regimes for traditional and sustainable forest management practices. Depletion in ecosystem N reserves reduces ecosystem productivity, which in turn produces a long-term decrease in tree biomass [87]. This productivity reduction is much more pronounced under traditional practices, as shorter rotation lengths accumulate higher and faster nutrient exports. In addition, short rotations and lower stand densities produce an under-canopy environment with more light available and therefore more suitable for understory growth (Figure 6). In these plantations, understory grass and shrubs usually sprout from rhizomes and other underground structures, rather than from seeds[20]. This phenomenon provides a competitive advantage against planted tree seedlings, which is exacerbated by the advantage of understory's lower nutrient needs in increasingly nutrient-poor sites as a consequence of intensive organic matter extraction after harvesting [88], slash burning [89], and N leaching [90] (Figure 7).

Reductions in soil organic matter and its associated sources of nutrients, water retention and physical properties would also be more severe under traditional management [16], leading to the already reported yield decline in Chinese fir plantations [26]. On the other hand, our results show that using longer rotations, moderately higher stand densities, and mixing conifers and broadleaves would prevent site productivity decline. With more sustainable management, the ecosystem would maintain and even can increase N reserves by keeping a positive N balance, an indicator of ecological sustainability of forest management in the long-term [26], [91].

Theoretical forest ecology indicates that in tree species mixtures, if each species has different growing patterns (differential shade tolerance, physical separation of canopies, phenological differences, successional separation, and differences in soil resource utilization), inter-specific competition may be reduced. Growing evidence indicates a generalized pattern in mixed forests, in which niche separation at both canopy and root levels leads to a more complete use of site resources [80], [92], [93], [94]. This phenomenon has already been reported in mixed tropical plantations [95], [96], [97]. Therefore, resources in the mixed stand are used more efficiently, hence total stand productivity increases. This phenomenon is due to two facts: complementary resource use between species and facilitative improvement in nutrition in mixtures with a nitrogen-fixing species [93], [95], [98].

Our results fit with this theoretical background, showing that in Chinese fir plantations, implementing mixtures with *Phoebe bournei* will not only increase productivity, but it will also prevent SOM and N losses and their associated yield decline over time. The additional increase in sequestered ecosystem C is another positive effect of species mixtures in these sites. Similar results have also been reported when simulating boreal [99], temperate [73] and tropical [100]forests.

Forest management can greatly affect net C exchange with the atmosphere, not only because it changes ecosystem C pools, but also because the C removed from the ecosystem during harvesting is stored in wood products, such as paper, cardboard, furniture, structural timber, etc. [101]. We have shown that subtropical plantations managed under sustainable forest management can store more ecosystem carbon. However, their net effect on C sequestration depends on forest types, their management regimes, the type of wood products produced, and the efficiency of biomass conversion, as well as assumptions about how the wood and wood residues will substitute for other products with greater GHG emissions and for fossil energy[102].

In Chinese fir plantations under traditional management, most of the production goes to the market as small diameter trunks. Over time, harvested sizes become increasingly smaller as plantation sites lose productivity due to the nutrient losses and competition with understory. However, using a more sustainable forest management, harvesting rates are reduced as it takes more time to complete the harvesting turn, but on the other hand harvested tree sizes are bigger and stay bigger through several rotations. These wood products are expected to reach better prices and last longer because they will be used more as structural timber and high quality furniture [103]. The increased ecosystem C storage, combined with the longer lifespan of higher quality wood products, could compensate the reduced harvesting rates compared to traditional management. Although a life-cycle analysis is needed to test this hypothesis, previous work has shown that secondary forests allowed growing rather than being harvested and their wood products stored could result in lower net CO_2 emissions [104].

Effects of forest management actions on carbon should be examined for large areas and over long time periods [105]. In the case of China, government's commitment to increase forest cover by 40 million ha with new plantations may not have an impact on carbon storage as important as intended unless better forest management is also implemented [106]. In this context, applying a more ecologically-based forestry could make a significant difference. Assuming those new 40 million ha were evenly distributed among all Chinese forest types, there could be 17.55 million ha of new subtropical plantations by 2020 [8]. If the differences between sustainable and traditional forestry were similar to the ones we have estimated for Chinese fir plantations, a more ecologically-based forestry implemented in Chinese subtropical plantations could add 650.8 TgC by 2050.

Previous research on long-term C storage under sustainable forest management regimes in Chinese temperate forests have estimated ecosystem C densities of 600–750 MgC ha^{-1} in poplar (Dai *et al.* under review), 450–800 MgC ha^{-1} in spruce [107], and 200–325 MgC ha^{-1}larch plantations [108]. In all cases, estimates C densities for temperate plantations were higher than our estimates for subtropical plantations. Therefore, it seems reasonable to assume that differences between traditional and sustainable forest management regimes in temperate plantations are at least similar to the ones calculated for subtropical plantations. Under this assumption, C storage in Chinese plantations could potentially be increased by 1,482.8 TgC by 2050. These estimations could even be conservative, as temperate plantations have on average higher levels of ecosystem carbon stocks [109].

In terms of emissions, applying sustainable forestry to all new Chinese plantations could potentially lead to an average net increase in stored C of 0.04 PgC y^{-1} in 2050, equivalent to offsetting 1.91% of China's annual carbon emissions in 2010 [110]. More importantly, this carbon increase would be sustainable with opportunities to maintain other ecological services and functions linked to increased biodiversity and less degraded forest soil [3].

In conclusion, our results show that applying a more ecologically-based forestry would have positive impacts not only on long-term plantations productivity, but it can also be a better strategy to implement climate change mitigation programs. Carbon stored in plantations could be further increased if sustainable forestry practices were combined with other practices not examined here, such as genetic improvement of trees, landscape design, etc., although the ecological sustainability of such practices should also be assessed. Finally, socio-economic incentives for farmers to extend rotations and to market non-conifer timber and other products would also be important.

AUTHOR CONTRIBUTIONS

Conceived and designed the experiments: XW JAB. Performed the experiments: JAB. Analyzed the data: XW JAB. Contributed reagents/materials/analysis tools: XW JAB. Wrote the paper: XW JAB.

REFERENCES

1. Lewis SL, López-González G, Sonké B, Affum-Baffoe K, Baker TR, et al. (2009) Increasing carbon storage in intact African tropical forests. Nature 457: 1003–1007. doi: 10.1038/nature07771
2. Food and Agriculture Organization (2012) State of World's Forests 2012. Rome: Food and Agriculture Organization.
3. Paquette A, Messier C (2010) The role of plantations in managing world's forests in the Anthropocene. Fron Ecol Environ 8: 27–34. doi: 10.1890/080116
4. Kirilenko AP, Sedjo RA (2007) Climate change impacts on forestry. PNAS 104, 19697–19702.
5. Rhemtulla JM, Mladenoff DJ, Clayton MK (2009) Historical forest baselines reveal potential for continue carbon sequestration. PNAS 105: 6082–6087. doi: 10.1073/pnas.0810076106
6. Kimmins JP, Blanco JA, Seely B, Welham C, Scoullar K (2008) Complexity in Modeling Forest Ecosystems; How Much is Enough? For Ecol Manage 256: 1646–1658. doi: 10.1016/j.foreco.2008.03.011
7. Niu XZ, Duiker SW (2006) Carbon sequestration potential by afforestation of marginal agricultural land in the Midwestern U.S. For Ecol Manage. 223: 415–427. doi: 10.1016/j.foreco.2005.12.044
8. State Forestry Administration People's Republic of China (2009). People's Republic of China forestry outlook study. FAO Working Paper No. APFSOS II/WP/2009/11. Bangkok.
9. Chen X, Zhang X, Zhang Y, Wan C (2009) Carbon sequestration potential of the stands under the Grain for Green program in Yunnan Province, China. For Ecol Manage 258: 199–206. doi: 10.1016/j.foreco.2008.07.010
10. Nabuurs GJ et al.. (2007) Forestry. In: Metz, B., Davidson, O.R., Bosch, P.R., Dave, R., Meyer, L.A. (eds) Climate Change 2007: Mitigation. Contribution of Working Group III to the Fourth Assessment Report of the Intergovernmental Panel on Climate Change. Cambridge: Cambridge University Press.
11. Wei X, Blanco JA, Jiang H, Kimmins JP (2012) Effects of nitrogen deposition on carbon sequestration in Chinese fir forests. Sci Tot Environ 416: 351–361. doi: 10.1016/j.scitotenv.2011.11.087
12. Food and Agriculture Organization (2001) Biological Sustainability of productivity in successive rotations. Report based on the work of J. Evans. Forest Plantation Thematic Papers, Working Paper 2. Forest Resources

Development Service, Forest Resources Division. Rome: Food and Agriculture Organization.

13. Carson KM, Curran LM, Asner GP, Pittman AM, Trigg SM, et al. (2013) Carbon emissions from forest conversion by Kalimantan oil palm plantations. Nature Clim Change 3: 283–287. doi: 10.1038/nclimate1702
14. Gibson L, Lee TM, Koh LP, Brook BW, Gardner TA, et al. (2011) Primary forests are irreplaceable for sustaining tropical biodiversity. Nature 478: 378–381. doi: 10.1038/nature10425
15. Jactel H, Brockerhoff EG (2007) Tree diversity reduces herbivory by forest insects. Ecol Let 10: 835–848. doi: 10.1111/j.1461-0248.2007.01073.x
16. Liao C, Luo Y, Fang C, Chen J, Li B (2012) The effects of plantation practice on soil properties based on the comparison between natural and planted forests: a meta-analysis. Global Ecol Biogeogr 21: 318–327. doi: 10.1111/j.1466-8238.2011.00690.x
17. Constanza R, D'arge R, De Groot R, Farber S, Grasso M, et al. (1997) The value of the world's ecosystem services and natural capital. Nature 387: 253–260. doi: 10.1038/387253a0
18. Lo YH, Lin YC, Blanco JA, Yu CH, Guan BT (2012) Moving from ecological conservation to restoration: an example from central Taiwan, Asia. In: Blanco JA, Lo YH (Eds.) Forest Ecosystems: more than just trees. Rijeka: InTech. 339–354.
19. Lewis SL (2006) Tropical forests and the changing Earth system. Phil Trans R Soc Lond B 261: 195–210.
20. Bi J, Blanco JA, Kimmins JP, Ding Y, Seely B, et al. (2007) Yield decline in Chinese Fir plantations: A simulation investigation with implications for model complexity. Can J For Res 37: 1615–1630. doi: 10.1139/x07-018
21. Wang W, Wei X, Liao W, Blanco JA, Liu Y, et al. (2013) Evaluation of the effects of forest management strategies on carbon sequestration in evergreen broad-leaved (*Phoebe bournei*) plantation forests using FORECAST ecosystem model. For Ecol Manage 300: 21–32. doi: 10.1016/j.foreco.2012.06.044
22. Lu ZH, Wu G, Ma X, Bai GX (2002) Current situation of Chinese forestry tactics and strategy of sustainable development. J For Res 13: 319–22. doi: 10.1007/bf02860099
23. Food and Agriculture Organization (2006) Global planted forests thematic study: results and analysis, by D. Del Lungo, J. Ball, J. Carle. Planted Forests and Trees Working Paper 38. Rome: Food and Agriculture Organization.
24. Yu XT (1997) (Ed.) Silviculture of Chinese Fir. Fuzhou: Science and Technology Press of Fujian. 250 pp. (in Chinese).
25. Wu ZL (1984) Chinese-fir China Forestry. Beijing: Publishing House. (in Chinese).
26. Tian DL, Xiang WH, Chen XY, Yan WD, Fang X, et al. (2011) A long-term evaluation of biomass production in first and second rotations of Chinese fir plantations at the same site. Forestry 84: 411–418. doi: 10.1093/forestry/cpr029
27. Kimmins JP, Mailly D, Seely B (1999) Modelling forest ecosystem net primary production: the hybrid simulation approach use in FORECAST. Ecol Model 122: 195–224. doi: 10.1016/s0304-3800(99)00138-6
28. Kimmins JP, Blanco JA, Seely B, Welham C, Scoullar K (2010) Forecasting Forest Futures: A Hybrid Modelling Approach to the Assessment of Sustainability of Forest Ecosystems and their Values. London: Earthscan. 218 p.
29. Moher D, Liberati A, Tetzlaff J, Altman DG (2009) The PRISMA Group (2009) Preferred Reporting Items for Systematic Reviews and Meta-Analyses: The PRISMA Statement. PLoS Med 6(6): e1000097 doi:10.1371/journal.pmed1000097.
30. Tummer B (2006) DataThief III v1.6. Available: http://datathief.org. Accessed: 22 October 2013.
31. Guo LB, Gifford M (2002) Soil carbon stocks and land use change: a meta analysis. Global Chan Biol 8: 345–360. doi: 10.1046/j.1354-1013.2002.00486.x
32. Quinn GP, Keough MJ (2002) Experimental design and data analysis for biologist. Cambridge: Cambridge University Press.
33. Yu XT (1999) Chinese fir: In Chinese Forests Forests Editorial Committee (Eds) (China Forestry Publishing House, Beijing) 1–29 (in Chinese).
34. Pan WC, Tian DL, Lei ZX, Kang WX (1983) Studies on the nutrient cycling in the Chinese-fir plantations: (II) Contents, accumulation rate and biological cycling of nutrient elements in the fast-growing Chinese-fir forest in the hill regions. J Central South For Inst 3: 1–12 (in Chinese)..
35. Shao L (1992) Chinese fir. In: Sheng W, Chen B, Xu H, Xu X (Eds) Research on Site Degradation of Timber Plantation. Beijing: Chinese Science & Technology Press. 87–92 (in Chinese).
36. Yang YS, He ZM, Qiu RH, Yu XT, Huang BL (1999) Growth pattern of 29-year-old Chinese-fir grown from seed in different rotations. Sci Silvae Sin 35: 32–36 (in Chinese)..
37. Xiao W, Nie D, Zhang J (1999) Study on biomass and energy use efficiency of the stands of *Cunninghamia lanceolata* in China. For Res 3: 253–258.
38. Liu F, Luo R, Jiang J (1991) Quantities and effects of various phosphorus compounds in soil under Chinese-fir plantation. J Nanjing For Univ 4: 7–12 (in Chinese)..
39. Tian DL, Zhao K (1989) Studies on the litter in a Chinese-fir plantation ecosystem: (I) Amounts, composition and dynamics of litter. J Central South For Inst 9(Sup): 38–43 (in Chinese)..
40. Tian DL (1994) Study on bio-geochemistry cycling in Chinese-fir ecosystem. In: Forestry Ministry of China (Ed.) Studies on Forest Ecosystems in China. Harbin: Publishing House of Northeast Forestry Univ. 136–145 (in Chinese).
41. Zhong AL, Hsiung WY (1993) Evaluation and diagnosis of tree nutritional status in Chinese-fir (*Cunninghamia*

lanceolata (Lamb) Hook) plantations, Jiangxi, China. For Ecol Manage 62: 245–270. doi: 10.1016/0378-1127(93)90053-p

42. Zhou X (1994) Long-Term Research On China's Forest Ecosystems. Harbin: Northeast Forestry Univ. Press. 320 pp. (in Chinese).
43. Zhou H (1999) Study on productivity of mixed forest planted on site of continuously planted Chinese-fir. J Fujian College For 2: 165–169 (in Chinese)..
44. Du J, Lu Y (2009) Study on growth regularity of *Phoebe zhennan* plantation. J Zhejiang For Sci Technol 29: 9–12 (in Chinese)..
45. Ma M, Jiang H, Liu Y (2008) Biomass, carbon content, carbon storage and their vertical distribution of *Phoebe bournei* artificial stand. Scien Silv Sin 44: 34–39 (in Chinese)..
46. Peng L (2003) Biomass and productivity in 35 year old *Phoebe bournei* plantation. J Fujian College For 23: 128–131 (in Chinese)..
47. Peng L (2008) Preliminary study on the biomass of *Phoebe bournei* plantation with different density and site condition. J Fujian For Sci Technol 35: 15–23 (in Chinese)..
48. Wu W (2009) Study on stand structure and growth law of near mature *Phoebe bournei*plantation. Develop. For Sci Technol 23: 75–78 (in Chinese)..
49. Lu M (2006) Study On Biomass And Nutrient Of Phoebe Bournei Plantation In Heterogeneity Spatiality Of Density And Site. PhD Thesis. Fujian: Fujian Agriculture and Forestry Univ (in Chinese).
50. Liao H, Di D, Zhang C, et al. (1989) Planting density management to *Cinnamoum camphora* and *Phoebe bournei.* J. Fujian College For. 9: 80–84 (in Chinese)..
51. Cheng Y (2003) Study On Litter Fall Decomposing Dynamic And Nutrients Releasing Rule Of Phoebe Bournei. PhD Thesis. Fujian: Fujian Agriculture and Forestry Univ. (in Chinese).
52. Guo Y (2007) Decomposition and the nutrient dynamics of *Phoebe bournei* leaf litter. J Fujian College For 27: 199–202 (in Chinese)..
53. Lin KM, Zhang ZQ, Cao GQ, He ZM, et al. (2006) Decomposition characteristics and its nutrient dynamics of leaf litter mixtures of both Chinese fir and *Phoebe bournei*. Acta Ecol Sin 26: 2732–2738 (in Chinese)..
54. Lin KM, Zhang ZQ, Ye FM, Lin Y, Li QS (2010) Dynamic analysis of decomposition characteristics and content change of nutrient elements of leaf litter of *Cunninghamia lanceolata, Phoebe bournei* and *Schima superba* under *C. lanceolata* artificial forest. J Plant Resour Environ 19: 34–39 (in Chinese)..
55. Peng L (2008) Study on nutrient elements of Phoebe bournei plantation in the different site conditions. J Fujian For Sci Tech 35: 10–15 (in Chinese)..
56. Chen S (2010) Effect of different site conditions on the growth of *Phoebe bournei.* J. Fujian College For. 30: 157–160 (in Chinese)..
57. Liu Q (2009) Study of soil fertility of *Phoebe bournei* planted forest. J Jiangxi For Sci Technol 1: 31–33 (in Chinese)..
58. Du J, Fan ZX, Ye DY, Lu CT (2010) The predicting models of crown volume and LAI for *Phoebe zhennan* plantation. J Zhejiang For Sci Technol 30: 37–41 (in Chinese)..
59. Hu J, Liu G (2010) Preliminary investigation on photosynthetic characteristics of saplings of *Phoebe zhennan* and *Machilus leptophylla.*. J Anhui Agri Univ 37: 541–546 (in Chinese)..
60. Ren CH, Chen H, Zhou LH (1990) Study on L.A.I. (Leaf Area Index) of *Phoebe bournei*stand in Youxi county. J Fujian College For 10: 67–71 (in Chinese)..
61. Wu Z, Chen S (2004) The effect of light on growth of *Phoebe bournei* plantation. J Fujian College For 24: 371–373 (in Chinese)..
62. Fan X, Ma X, Fu R, Liu A (2001) Comparative study on underground vegetation develop of different generation plantations of Chinese-fir. For Res 14: 11–19 (in Chinese)..
63. Lin KM, Yu XT, Hong W, Huang BL (2001) Effect of minor vegetation on soil fertility in Chinese-fir plantation. Scient Silv Sin 37: 94–98 (in Chinese)..
64. Xiang WH, Tian DL, Yan WD, Kang WX, Fang X (2003) Biomass and nutrient content of vegetation on fallow site after clear-cutting of Chinese-fir plantation. Acta Ecol Sin 23: 695–702 (in Chinese)..
65. Yan W, Tian D, Jiao X (2003) A study on biomass dynamics and distribution of undervegetation in the secondary generation of Chinese-fir plantation in Hui Tong. For Res 16: 82–86 (in Chinese)..
66. Zhou X, Luo R, Ye J (1991) Effect of continuous cropping with Chinese-fir upon soil nutrients and its feedback. J Nanjing For Univ 3: 47–52 (in Chinese)..
67. Tian DL, Zhu XP, Cai BY, Gao YM (1989) Studies on the litter in a Chinese-fir plantation ecosystem: (II) Nutrient contents and decomposition rate of litter. J Central South For Inst 9(Sup): 45–55 (in Chinese)..
68. Liao LP (2000) Decomposition of leaf litter of Chinese-fir in mixture with major associated broad-leaved plantation species. Acta Phytoecol Sin 24: 27–33 (in Chinese)..
69. Ding YX, Tian Y, Qi L (1999) A testing simulation with FORECAST on long-term productivity of Chinese-fir plantations. For Studies China 1: 34–38 (in Chinese)..
70. Ding YX, Chen JL (1995) Effect of continuous plantation of Chinese-fir on soil fertility. Pedosphere 5: 57–66.
71. Yang YS, Chen GS, Huang BL (2000) Variation in the soil water and nutrients between different rotation stands of Chinese-fir. J Nanjing For Univ 24: 25–28 (in Chinese)..
72. Seely B, Welham C, Kimmins JP (2002) Carbon sequestration in a boreal forest ecosystem: results from the ecosystem simulation model, FORECAST. For Ecol Manage 169: 123–135. doi: 10.1016/s0378-1127(02)00303-1

73. Blanco JA, Seely B, Welham C, Kimmins JP, Seebacher TM (2007) Testing the performance of FORECAST, a forest ecosystem model, against 29 years of field data in a *Pseudotsuga menziesii* plantation. Can J For Res 37: 1808–1820. doi: 10.1139/x07-041

74. Guan BT, Wrigth WE, Chung CH, Chang ST (2012a) ENSO and PDO strongly influence Taiwan spruce height growth. For Ecol Manage 267: 50–57. doi: 10.1016/j.foreco.2011.11.028

75. Masera OR, Garza-Caligaris JF, Kanninen M, Karjalainen T, Liski J, et al. (2003) Modeling carbon sequestration in afforestation, agroforestry and forest management projects: the CO2FIX V.2 approach. Ecol Model 164: 77–199. doi: 10.1016/s0304-3800(02)00419-2

76. Cannell MGR (2003) Carbon sequestration and biomass energy offset: Theoretical, potential and achievable capacities globally, in Europe and the UK. Biomass Bioener 24: 97–116. doi: 10.1016/s0961-9534(02)00103-4

77. Van Minnen JG, Strngers BJ, Eickhout B, Swart RJ, Leemans R (2008) Quantifying the effectiveness of climate change mitigation through forest plantations and carbon sequestration with an integrated land-use model. Carbon Balance Manage 3: 3. doi: 10.1186/1750-0680-3-3

78. Blanco JA (2012) Forests may need centuries to recover their original productivity after continuous intensive management: an example from Douglas-fir. Sci Total Environ 437: 91–103. doi: 10.1016/j.scitotenv.2012.07.082

79. Marín-Spiotta E, Sharma S (2013) Carbon storage in successional and plantation forest soils: a tropical analysis. Global Ecol Biogeogr 22: 105–117. doi: 10.1111/j.1466-8238.2012.00788.x

80. Amoroso MM, Turnblom MM (2006) Comparing productivity of pure and mixed Douglas-fir and western hemlock plantations in the Pacific Northwest. Can J For Res 36: 1484–1496. doi: 10.1139/x06-042

81. Wu Z (2005) Study on growth effect of the mixed forest of *Cunninghamia lanceolata*and *Phoebe bournei*.. J Fujian College For 25(2): 142–146.

82. Duan W, Zhu B, Lu JL, Lu R, Yang CJ, et al. (2008) Carbon Storage Effects in Artificial Mixed Forest of *Cryptomeria fortune - Cunninghamia lanceolata* in MontaneAreas of Western Sichuan Province of China. Sichuan Environment 27(5): 57–63 (in Chinese)..

83. Zhang G (2009) Growth and biomass in the Chinese fir mixtures. Anhui Agri. Sci. Bull. 15(21): 131–132.

84. Lin T (2011) Mixed Forests of *Michelia macclurei* and *Cunninghamia lanceolata* in Coastal Mountains Land of Southern Fujian. Jiangxi Forestry Sci Tech 3: 16–19.

85. Chen XY (2009) Stand biomass and its distribution of mixed forests of *Alnus crematogyne* and *Cunninghamia lanceolata*. Jiangxi Forest Science and Technology 1, 14–17.

86. Keith H, MaKey B, Berry S, Lindenmayer D, Gibbons P (2010) Estimating carbon carrying capacity in natural forest ecosystems across heterogeneous landscapes: addressing sources of error. Glob Change Biol 16: 2971–2989. doi: 10.1111/j.1365-2486.2009.02146.x

87. Berthrong ST, Jobbagy EG, Jackson RB (2009) A global-meta-analysis of soil exchangeable cation, pH, carbon and nitrogen with afforestation. Ecol Appl 19: 2228–2241. doi: 10.1890/08-1730.1

88. Minghe L, Ritchie GA (1999) Eight hundred years of clonal forestry in China: I. Traditional afforestation with Chinese fir (*Cunninghamia lanceolata* (Lamb.) Hook.). New For 18: 131–142.

89. Zhang XQ, Kirschbaum MKU, Hou Z, Guo Z (2004) Carbon stock changes in successive rotations of Chinese-fir (*Cunninghamia lanceolata* (Lamb.) Hook) plantations. For Ecol Manage 202: 131–147. doi: 10.1016/j.foreco.2004.07.032

90. Blanco JA, Wei X, Jiang H, Jie CY, Xin ZH (2012) Impacts of enhanced nitrogen deposition and soil acidification on biomass production and nitrogen leaching in Chinese fir plantations. Can J For Res 42: 437–450. doi: 10.1139/x2012-004

91. Blanco JA, Zavala MA, Imbert JB, Castillo FJ (2005) Sustainability of forest management practices: Evaluation through a simulation model of nutrient cycling. For Ecol Manage 213: 209–228. doi: 10.1016/j.foreco.2005.03.042

92. Kelty MJ (1992) Comparative productivity of monocultures and mixed-species stands. In: Kelty, M.J., Larson, B.C., Oliver, C.D. (Eds.), The Ecology and Silviculture of Mixed-Species Forests. Kluwer Academic Publishers, Dordrecht, Boston, 125–141.

93. Pretzsch H, Bielak K, Block J, Bruchwald A, Dieler J, et al. (2013) Productivity of mixed versus pure stands of oak (*Quercus petraea* (Matt.) Liebl. and *Quercus robur* L.) and European beech (*Fagus sylvatica* L.) along an ecological gradient. Eur J For Res 132: 263–280. doi: 10.1007/s10342-012-0673-y

94. Primicia I, Camarero JJ, Imbert JB, Castillo FJ (2013) Effects of thinning and canopy type on growth dynamics of *Pinus sylvestris*: inter-annual variations and intra-annual interactions with microclimate. Eur J Forest Res 132: 121–135. doi: 10.1007/s10342-012-0662-1

95. Bauhus J, Khanna PK, Menden N (2000) Aboveground and belowground interactions in mixed plantations of *Eucalyptus globulus* and *Acacia mearnsii*. Can J For Res 30: 1886–1894. doi: 10.1139/x00-141

96. Erskine PD, Lamb D, Bristow M (2006) Tree species diversity and ecosystem function: Can tropical multi-species plantations generate greater productivity? For Ecol Manage 233: 206–210. doi: 10.1016/j.foreco.2006.05.013

97. Forrester DI, Bauhus J, Cowie AL, Vanclay JK (2006) Mixed-species plantation of*Eucalyptus* with nitrogen fixing trees: a review. For Ecol Manage 233: 211–230. doi: 10.1016/j.foreco.2006.05.012

98. Kelty MJ (2006) The role of species mixtures in plantation forestry. For Ecol Manage 233: 195–204. doi: 10.1016/j.foreco.2006.05.011

99. Seely B, Welham C, Blanco JA (2010) Towards the application of soil organic matter as an indicator of ecosystem productivity: Deriving thresholds, developing monitoring systems, and evaluating practices. Ecol Ind 10: 999–1008. doi: 10.1016/j.ecolind.2010.02.008

100. Blanco JA, González E (2010) Exploring the sustainability of current management prescriptions for Pinus caribaea plantations in Cuba: a modelling approach. J. Trop. For. Sci. 22: 139–154.

101. Fahey TJ, Woodbury PB, Battles JJ, Goodale CL, Hamburg SP, et al. (2010) Forest carbon storage: ecology, management, and policy. Front Ecol Environ 8: 245–252. doi: 10.1890/080169

102. Eriksson E, Gillespie AR, Gustavsson L, Langvll O, Olsson M, et al. (2007) Integrated carbon analysis of forest management practices and wood substitution. Can J For Res 37: 671–681. doi: 10.1139/x06-257

103. Lippke B, Oneil E, Harrison R, Skog K, Gustavsson L, et al. (2011) Life cycle impacts of forest management and wood utilization on carbon mitigation: knowns and unknowns. Carbon Manage 2: 303–333. doi: 10.4155/cmt.11.24

104. Perez-Garcia J, Lippke B, Comnick J, Manriquez C (2005) An assessment of carbon pools, storage, and wood products market substitution using life-cycle analysis results. Wood Fiber Sci 37: 140–148.

105. McKinley DC, Ryan MG, Birdsey RA, Giardina CP, Harmon ME, et al. (2011) A synthesis of current knowledge on forests and carbon storage in the United States. Ecol Appl 21: 1902–1924. doi: 10.1890/10-0697.1

106. Liu P, Yin R (2012) Sequestering carbon in China's forest ecosystems: potential and challenges. Forests 3: 417–430. doi: 10.3390/f3020417

107. Jie C, Jiang H, Zhou G, Wei X, Blanco JA, et al.. (2011) Simulating the carbon storage of spruce forests based on the FORECAST model and remotely sensed data. Proceedings - 2011 19th International Conference on Geoinformatics, Geoinformatics 2011. Article number 5981581. ISBN: 978–161284848–8. DOI: 10.1109/GeoInformatics.2011.5981581.

108. Qi G (2011) Carbon pool and sequestration potential of larch plantations in Northeast China. PhD thesis. Shenyang: Chinese Academy of Sciences.

109. Keith H, Mackey BG, Lindenmayer DB (2009) Re-evaluation of forest biomass carbon stocks and lessons from the world's most carbon-dense forests. PNAS 106: 11635–11640. doi: 10.1073/pnas.0901970106

110. Guan D, Liu Z, Geng Y, Lindner S, Hubacek K (2012b) The gigatonne gap in China's carbon dioxide inventories. Nature Clim Change 2: 672–675. doi: 10.1038/nclimate1560

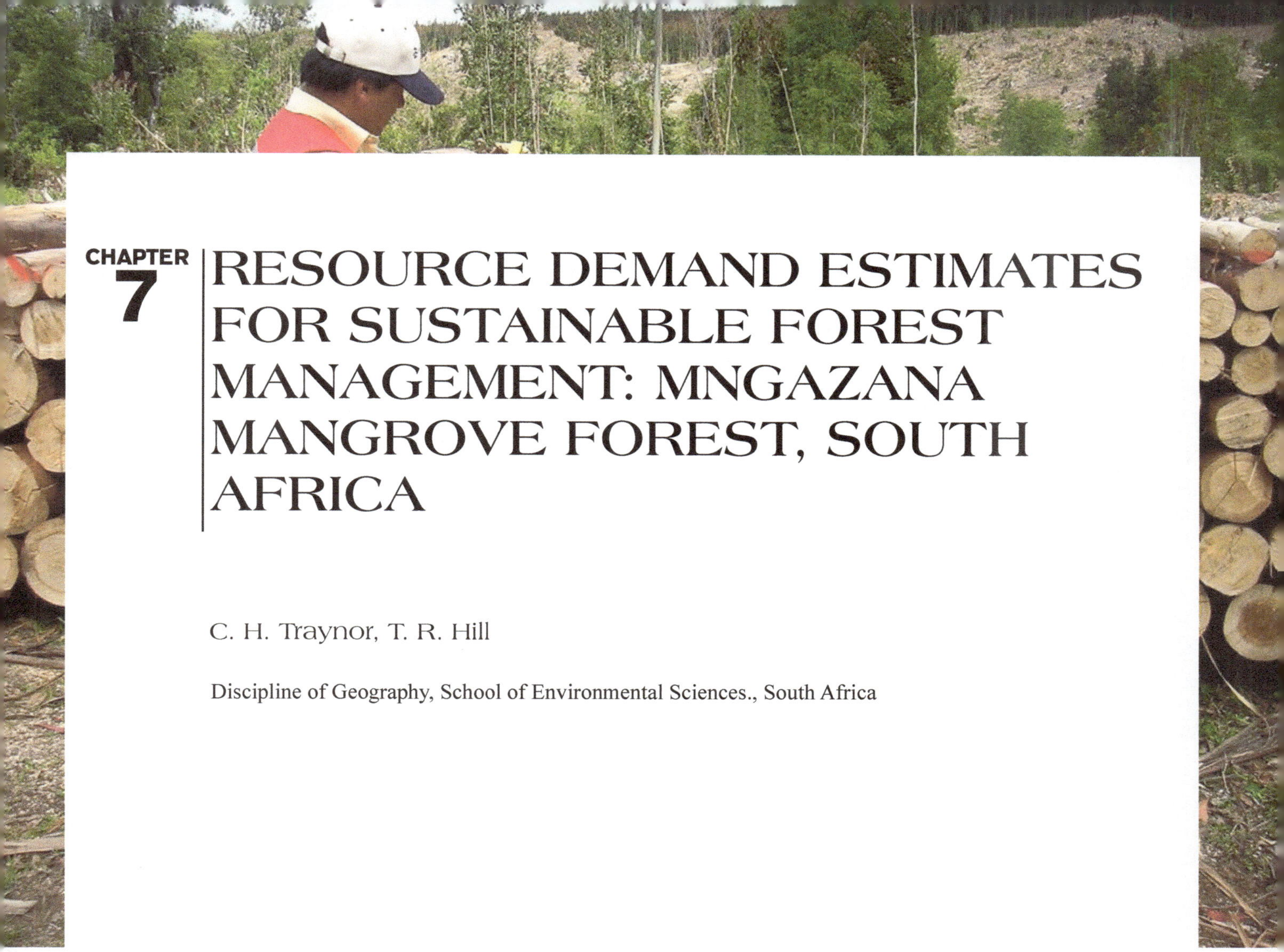

CHAPTER 7

RESOURCE DEMAND ESTIMATES FOR SUSTAINABLE FOREST MANAGEMENT: MNGAZANA MANGROVE FOREST, SOUTH AFRICA

C. H. Traynor, T. R. Hill

Discipline of Geography, School of Environmental Sciences., South Africa

ABSTRACT

Since democratization in 1994, South African forest policies have promoted sustainable forest management. However, implementation has been problematic due to limited information concerning forest product utilization. This paper investigates and quantifies timber use from the Mngazana Mangrove Forest. Eastern Cape Province. South Africa. Three local communities utilize stems of the mangrove species *Rhizophora mucronata* Lam. and *Bruguiera gymnorrhiza* (L.) Lam. for building construction. There were two distinct building shapes, circular and rectangular. On average. 155 stems were used for circular buildings and 378 stems for rectangular buildings. Most buildings were constructed using mangroves as well as indigenous timber from coastal scarp forests. The proportion of mangrove stems in buildings varied from 0 to 95%. The annual demand for mangroves was estimated to be 18 400 stems. Due to the high annual demand, projected human population growth rates have a minor influence upon future demand values. For effective sustainable forest management, the standing stock at Mngazana should be restricted to the two mangrove species utilized for building construction, and a forest inventory performed so that demand for building can be compared to supply.

INTRODUCTION

Over the last decade, global advancements in forest management have progressed towards Sustainable Forest Management (SFM), a term first incorporated into Agenda 21 and the Forest Principles which were both outputs of the United Nations Conference on Environment and Development (UNCED) 1992. Agenda 21 called for enhanced sustainable management of all forests, whereas the Forest Principles stated that 'forest resources and forest lands should be sustainably managed to meet the social, economic, ecological, cultural and spiritual needs of present and future generations'. It also specified that these needs include forest products such as wood and wood products (United Nations General Assembly 1992: 291, 292). Since its inception, the SFM concept has evolved through international forest policy dialogue, as well as country and eco-regional initiatives. These activities have resulted in global alterations to forest policies and forest management strategies (FAO 1994).

Citation: C. H. Traynor, T. R. Hill, "Resource demand estimates for sustainable forest management: Mngazana Mangrove Forest, South Africa," Bothalia; Vol 38, No 1 (2008), 103-110. doi: 10.4102/abc.v38i1.274.

Although South Africa did not participate in UNCED, global developments have influenced its policies and since democratization in 1994, forest policy has striven to broaden access and sustainable use of indigenous forests (Lawes et al. 2004a). The new South African Constitution promotes conservation, ecologically sustainable development and the use of natural resources within justifiable social and economic development (RSA 1996a). Furthermore, the White Paper on Sustainable Forest Development in South Africa (RSA 1996b), and the National Forest Act (RSA 1998), both acknowledge the vital role of forests in the livelihoods of rural communities. These policies recognize the obligation w hich the country has, to use its natural resources to further the development of the poor (Willis 2004). Thus, South African policy promotes the sustainable use of forests. Forest biomes and their management have been well researched (Muir 1990; Oribi et al. 2002; Lawes et al. 2004a), but implementation of management plans sometimes poses a problem. There is also a paucity of information on the usefulness of the forests, especially to local people and on actual removal of w ood and other products (Lawes et al. 2004).

SFM approaches have been applied to mangrove forests (International Tropical Timber Organization 2002), but no examples exist for South Africa's mangroves. even though their forest products are utilized by local communities (Bruton 1980; Ward et al. 1986; Steinke 1999; Rajkaran et al. 2004). Mangroves in South Africa are located along the Indian Ocean coastline in the prov inces of KwaZulu-Natal and Eastern Cape. Over 17 years, a reduction of 7% in mangrove area has been reported for the Eastern Cape, and easily accessible forests that lie outside of protected areas are under great pressure from resource users (Adams et al. 2005). Mangrove ecosystems have a positive influence on coastal protection, nutrient cycling and export, sediment trapping, and serve as breeding and nursery grounds for fish (Lugo & Snaedakar 1974; Hogarth 1999; Mumby et al. 2004). South African mangrove habitats are threatened and as a result, many authors have called for greater conservation and management efforts (Branch & Gnndley 1979; Day 1981; Ward et al. 1986; Berjak et al. 1997; Steinke et al. 1995).

The Mngazana Mangrove Forest (MMF) in the Eastern Cape was selected for study, as mangrove trees have been harvested throughout the forest (Rajkaran et al. 2004) and local communities use them with other indigenous wood species as building construction materials. In the villages surrounding the mangroves there are limited employment and income opportunities. People have restricted means to purchase building construction materials and thus over 90% of dwellings are traditional huts or structures made from natural materials (Statistics South Africa 2001). Every South African citizen has the right to access adequate housing (RSA 1996a) and the communities at Mngazana are dependent upon the local natural resources, including the mangrove forest, to meet their housing needs. At Mngazana the buildings are typical of rural styles constructed from indigenous wood; a wooden framework is made to which stone and mud walls and roofing materials are added (Liengmel983; Cunningham 1985; Shackleton et al. 2002).

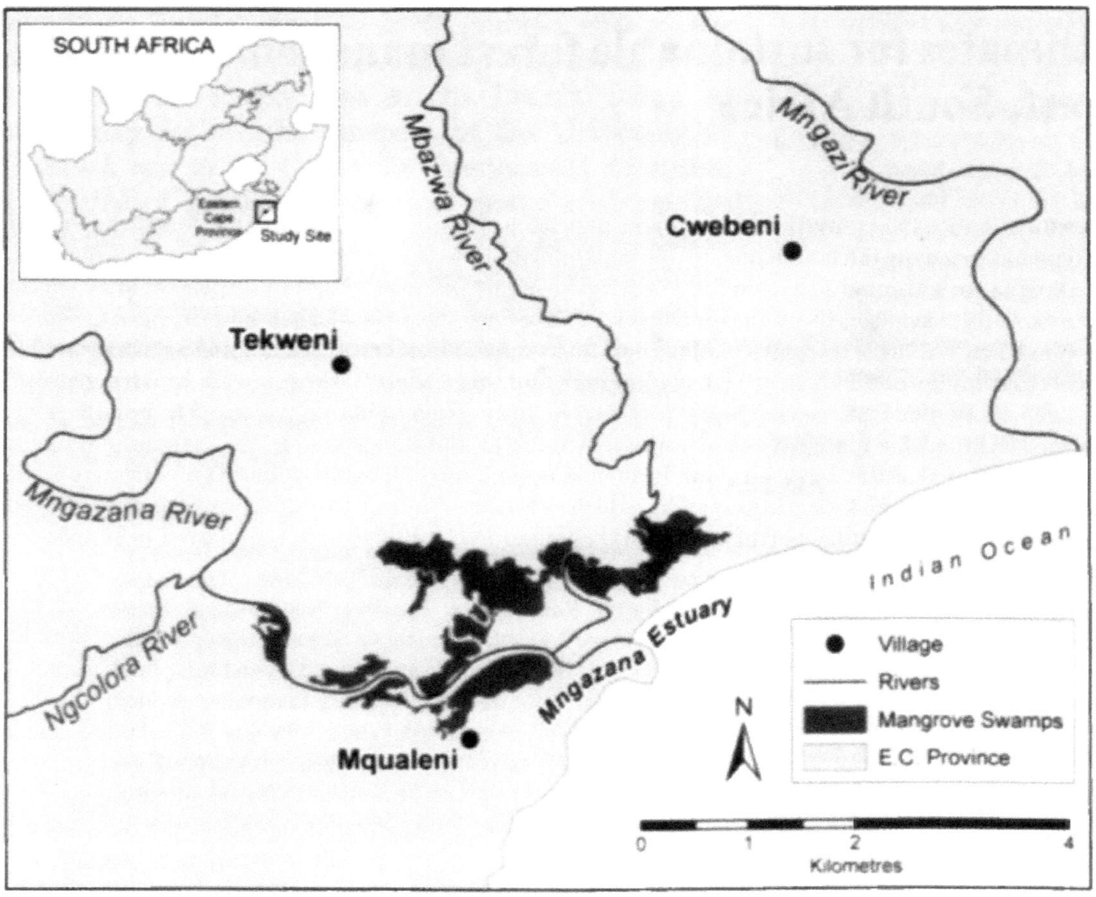

Figure 1. Location of the Mngazana Mangrove Forest and study villages. Eastern Cape, South Africa.

This paper investigates the SFM concept focusing on the estimation of empirical values for forest product utilization. Through the case of the MMF, the intricacies of precisely calculating demand for housing construction timber is presented and the usefulness of these estimates to SFM considered. Firstly, the total numbers of mangrove stems within standing structures is quantified. Secondly, the current annual demand for mangrove stems for building construction is evaluated. Thirdly, future demand for mangrove stems is predicted under different scenarios. Finally, the contribution the demand estimates can make to a mangrove SFM strategy and the wider relevance of the case is discussed.

The context of the Mngazana Mangrove Forest

The topography of the Mngazana area is a tidal flat, up to 1 500 m wide in places, attributed to the geological formation, the Ecca and Beaufort Groups of the Karoo Sequence (Harrison et al. 1999). These groups are composed of conglomerates, sandstones and mudstones which produce weakly structured, shallow soils when weathered (Nicolson 1993). The average annual rainfall is 1 200 mm, and 70% occurs in summer. The vegetation includes a coastal strip of dune forest, mangroves, saltmarsh and inland grassland and forest patches.

The MMF is a riverine mangrove forest that is situated within the Mngazana Estuary (31°42'S 29°25'E), Eastern Cape. It is the third largest mangrove forest in South Africa (Ward & Steinke 1982) covering an area of ± 145 ha (Adams et al. 2005) and is composed of the mangrove species Avicennia marinia (Forssk.) Vierh., Bruguiera gymnorrhiza (L.) Lam. and Rhizophora mucronata Lam. The estuary is biologically important with the highest recorded invertebrate diversity for east coast estuaries (Branch & Grindley 1979) and rated first in the region using a botanical importance rating system (Colloty 2000) and 22nd in South Africa for its biodiversity importance (Turpie et al. 2002). The MMF has been heavily disturbed, and Rajkaran et al. (2004), using GIS analysis, estimated that over 40% of the forest showed a ratio of one adult tree to two harvested stumps, indicating that the harvesting intensity was extremely high.

The area is inhabited by Xhosa-speaking Mpondos. Prior to democracy in 1994, the area was classified as a homeland by the apartheid government and consequently is poorly developed (Ashley & Ntshona 2003). There are three villages surrounding the Mngazana Estuary; Mqualeni, Cwebeni and Tekweni (Figure I). In these villages, ± 30% of the population have received no schooling, less than 10% of the labour force are employed and over half of the homesteads are female-headed (Statistics South Africa 2001). The land is in trust to the Tribal Authorities but owned by the state, and community members may utilize the land subject to legislation and local rules. The communities at Mngazana and a nonprofit organization hope to devise a SFM plan for the mangrove forest that will permit mangrove harvesting.

METHODS

Investigations focused upon Rhizophora mucronata and Bruguiera gy mnorrhiza as these species are utilized exclusively for building construction. They are harvested by cutting the main stem at its base; this entire stem minus branches is used in building construction. It has been suggested that R. mucronata can regenerate after cutting 'if cut high enough on the stem that live branches (with leaves) are spared' (Walters 2005a: 344), but coppicing of cut stems has never been observed at Mngazana for R. mucronata and B. gymnorrhiza (Adams et al. 2005) and thus, under the current cutting regime, one stem equates with one tree destructively harvested. Therefore, demand (and supply) calculations can use numbers of individual trees rather than weight and volume measurements of wood.

Mangrove utilization for building construction purposes has been determined through empirical observations supplemented by information from semi-structured interviews with homestead occupants and group discussions. As house construction activities only occur every few years, a respondent's recollections concerning quantities of materials used may not be very precise, so empirical observations are generally more reliable (Shackleton & Shackleton 2004). The villages of Mqualeni, Cwebeni and Tekweni were selected for study. There are 574 homesteads in these villages (Statistics South Africa 2 0 0 1) and of these, 108 (19%) were randomly selected and surveyed, 32 in Cwebeni, 48 in Tekweni and 28 in Mqualeni. The semi-structured interviews with homestead occupants were in Xhosa (the vernacular language spoken in the area) with English translation. The interviews were designed to elicit information regarding mangrove characteristics, utilization and building longevity. Within each homestead the number and shape of individual buildings were recorded. Buildings constructed w ith mangroves were identified and the number of mangrove poles utilized in each structural unit (wall poles, roof poles and roof laths) was counted. The diameter of mangrove stems used in each structural unit was determined by sampling a random ten exposed stems and measuring the diameter using metal calipers to the nearest millimeter. The length of stems used for different units was obtained through group discussions. The presence of alternative building materials was also recorded. Three group discussions with community members were held from October 2003 to October 2004 and relevant information from these has been included for comparative purposes. Groups were composed of members of the Mngazana Mangrove Management Forum and interested villagers from all three villages, between 11 to 2 0 people attended each meeting.

Human population growth rates were estimated from the 2001 South African census (Statistics South Africa 2(H) 1) and the population estimates of the United Nations (United Nations 2005). The following terminology will be used throughout this paper: homestead: an area allocated for a family's dwellings, usually consisting of several different individual buildings; mangrove used as a prefix to structural unit, building or homestead denotes use of mangroves in construction. Utilization was calculated at different scales of structural unit, building, homestead and village using the procedure summarized in Figure 2 .

RESULTS

Mangrove Utilization

Group discussions and direct observations indicated that the main use of mangroves is for building construction: *Rhizophora mucronata* and *Bruguiera gymnorrhiza* are the only species utilized for this purpose. An extremely small minority of homesteads used these species for non-building construction purposes such as fencing. *Avicennia marinia* was collected by some homesteads for fuel wood; dead wood was preferred to green wood. There was a low frequency of collection because non-mangrove tree species were also utilized for fuel wood. This study focuses upon construction use because it is the predominant use of green mangrove wood.

Mangrove Stems in Standing Structures

Buildings constructed from mangroves are circular or rectangular. Both of these shapes are constructed from structural units of wall poles, wall laths and roof poles. Wall poles are positioned vertically and then laths laid at ninety degrees to these to produce a framework.

The average number of mangrove stems used in building units constructed entirely from mangroves was calculated for all three villages (Table 1). A circular building if constructed entirely from mangroves would utilize 155 stems. ± 1 10 of these would be wall laths and the remainder wall or roof poles. Rectangular buildings composed of only mangrove stems utilize more than double the number of stems (± 378 stems), because the wall laths require 335±151 stems. Requirements for wall and roof poles are approximately similar. The average diameter of poles was 6.1 ±2.0 cm (n = 787) and the average diameter of laths was 3.3±1.0 cm (n = 550). Group discussions revealed that poles and laths over 2 m long were utilized.

However, mangroves are utilized in conjunction with other timbers, the percentage of mangroves within these structural units shows variation (Figures 3, 4). Mangroves compose a high percentage of building materials in the roof poles of circular buildings (up to 1 0 0%) and the wall laths of rectangular buildings (75% or more). In other structures, mangroves account for 50% or less of materials. The percentage of mangroves utilized showed variation between the villages studied. Mqualeni Village showed the most consistent use of mangroves and they were utilized in all structural units composing ± 30% to 80% of all stems. In Cwebeni and Tekweni, the percentage of mangroves used shows wider variations. All the roof poles in circular buildings in these villages are mangroves, but mangroves make up 10% or less of wall poles and laths. Rectangular buildings in these two villages display different patterns; mangroves make up 75% or more of wall laths but only up to 25% of wall and roof poles.

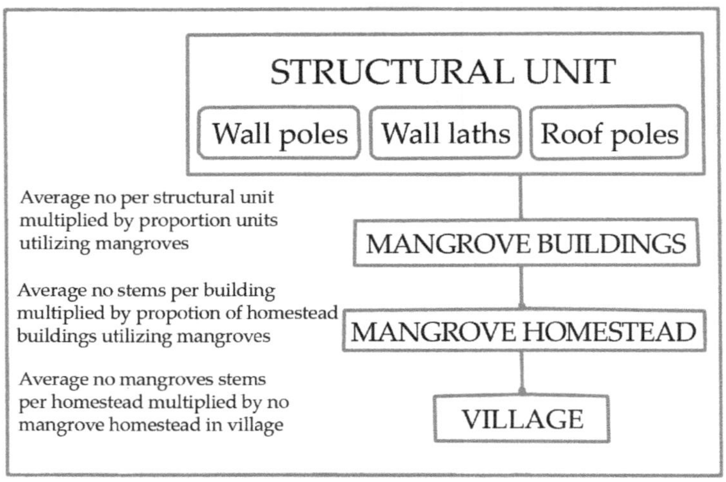

Figure 2. Summary of method used to calculate demand for mangrove stems within each village.

Table 1. Average no. mangrove stems (± std dev.) per building unit constructed entirely from mangroves in three villages studied

	Tekweni	Cwebeni	Mqualeni	Average
Circular building				
Wall poles	22 (n = 1)		24±10 (n = 3)	23±8 (n = 4)
Wall laths		96 (n = 2)	125±7 (n = 2)	111±17 (n = 4)
Roof poles	19±6 (n = 37)	24±5 (n = 21)	22±2 (n = 2)	21±6 (n = 60)
Extrapolated				**155**
Rectangular building				
Wall poles		30±6 (n = 2)		25±9 (n = 10)
Wall laths	336±69 (n = 3)	372±172 (n = 19)	281±118 (n = 13)	335±151 (n = 35)
Roof poles	10±1 (n = 2)	15±6 (n = 2)	44 (n = 1)	18±15 (n = 5)
Extrapolated				**378**

A one-way between-groups analysis of variance was conducted to explore the impact of village location upon the number of mangrove stems used in different-shaped buildings. There was a statistically significant difference at the p = < 0.05 level for the number of mangroves stems per circular building between villages ($F(2, 59) = 10.7$, $p = 0.00$). Post-hoc comparisons using the Turkey HSD test indicated the mean score for Maqualeni (M = 72.80, SD = 64.76) was significantly different from Cwebeni (M = 33.00, SD = 28.70) and Tekweni (M = 20.17, SD = 7.50). There was no statistically significant difference at the p = < 0.05 level in the number of mangrove stems used in rectangular buildings between villages ($F(2,36) = 2.32$, $p = 0.11$).

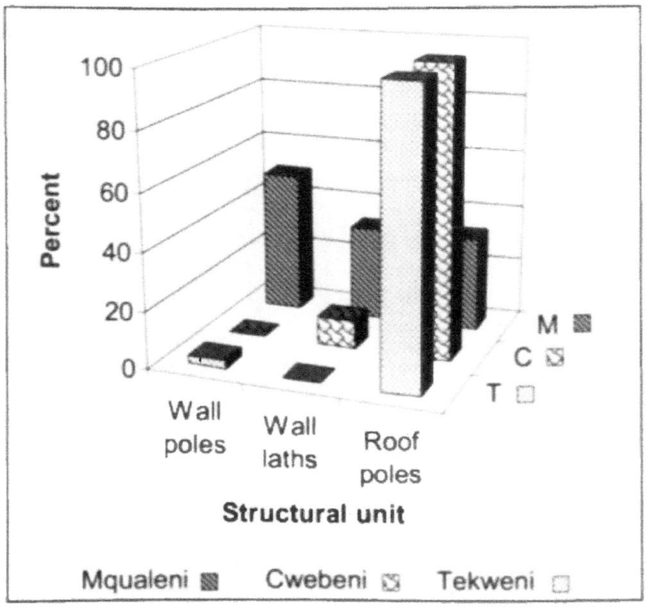

Figure 3. Percentage of mangrove stems used within structural units in circular mangrove buildings.

Typical homesteads in the villages studied consist of circular and rectangular buildings, constructed from a variety of materials including indigenous timber and blocks.

Thus a conversion factor was utilized to estimate the number of mangrove stems utilized per homestead (Table 2). This conversion factor was the average number of mangrove stems in mangrove buildings multiplied by the average number of buildings containing mangroves in homesteads (that utilized mangroves). Accounting for these variations, mangrove homesteads used an average of 89 stems in Tekweni, 346 stems in Mqualeni and 397 stems in Cwebeni. The percentage of homesteads utilizing mangroves was 54% in Tekweni, 79% in Mqualeni and 8 8 % in Cwebeni.

At the village scale, the total number of homesteads using mangroves was 67 in Mqualeni, 96 in Tekweni and 272 in Cwebeni and the estimated total number of mangrove stems utilized was 8 581 stems in Tekweni, 23 108 stems in Mqualeni, 108 034 stems in Cwebeni. Thus in total, 139 723 mangrove stems were utilized in standing structures within buildings at the time of the study.

Current Annual Demand for Mangrove Stems (2004)

The percentage of new buildings compared to the total number of buildings was 3% in Tekweni, 16% in Cwebeni and 18% in Maqualeni. As the different-shaped buildings utilize different amounts of mangrove stems, the shape of buildings constructed has an important influence upon demand for mangrove stems. The ratio of circular to rectangular mangrove buildings in the villages studied was 1:1, thus the current demand for stems can be calculated using this building shape ratio. Circular buildings utilize fewer mangrove stems than rectangular buildings, thus demand from circular buildings represents the minimum number of stems necessary to fulfil housing requirements and demand from rectangular buildings represents maximum demand. Using these demand scenarios, the minimum demand would be for 3 700 stems, the 1:1 demand would be for 18 400 stems and the maximum demand for 33 500 stems (Figure 5).

Table 2. Conversion of mangrove utilization figures from buildings to homestead level

	Tekweni		Cwebeni		Mqualeni	
	C	R	C	R	C	R
Average no. stems per building (a)	21	255	33	358	61	254
Average no. buildings containing mangroves per homestead (b)	0.37	0.32	0.43	1.07	0.55	1.23
Average no. mangrove stems per homestead by building shape (a × b)	8	81	14	383	34	312
Total average no. mangrove stems per homestead	89		397		346	

C, circular building; R, rectangular building

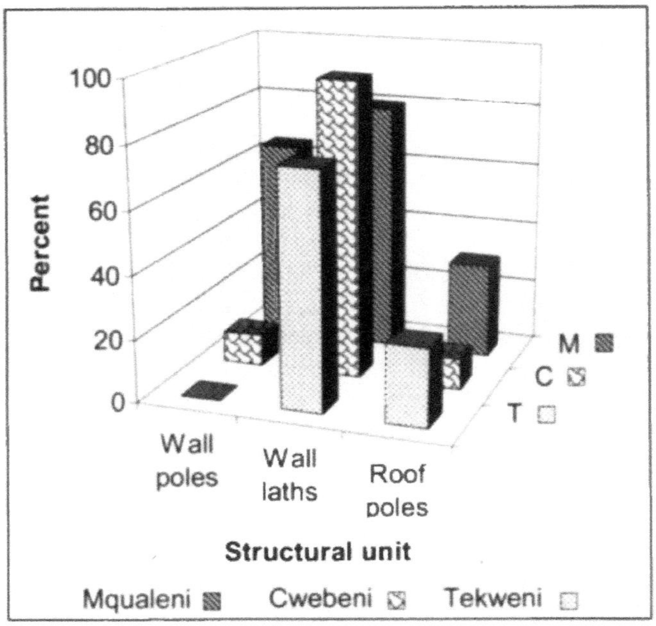

Figure 4. Percentage of mangrove stems used within structural units in rectangular mangrove buildings.

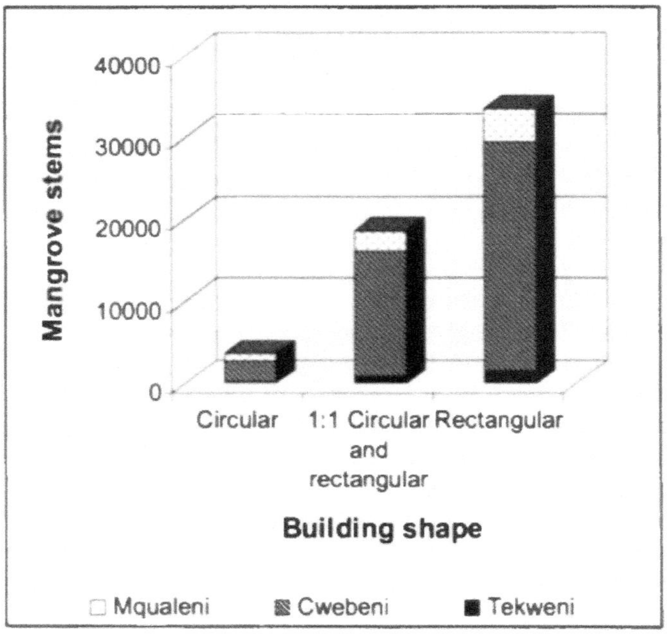

Figure 5. Demand for mangroves in 2004 under different building shape scenarios.

Group discussions suggested that the longevity of mangrove buildings was 10 years and homestead surveys suggested an average of 2 0 years, thus, the average is 15 years.

Predicting Future Demand Scenarios

Future demand for mangrove stems can be estimated using human population growth predictions derived from past growth rates or demographic models. Past rates (1996-2001) of annual human population growth in all three villages was 3% (Statistics South Africa 2001). However, this census data may underestimate the impact of HIV/AIDS upon the population and as the most direct demographic consequence of HIV/A IDS is an increase in mortality (Whiteside 2001) demographics will change. The United Nations demographic models take account of the effect of HIV/AIDS and in South Africa annual human population growth rates were estimated to be 0.78% for the period 2000 to 2005 (United Nations 2005).

The influence of a decade of human population growth on mangrove stem demand can be estimated using the different human population growth forecasts and various building-shape scenarios (Figure 6). These estimates suggest that under the 0.78% human population growth rate, demand in all building scenarios over the next ten years will increase gradually but it will not exceed more than 0.5% of the current demand figure. Under the 3% human population growth forecast, demand for mangrove stems increases by 6% of current estimated values, and in the maximum demand scenario of rectangular buildings, an additional 2 628 stems per year would be required in 2015 compared to current values.

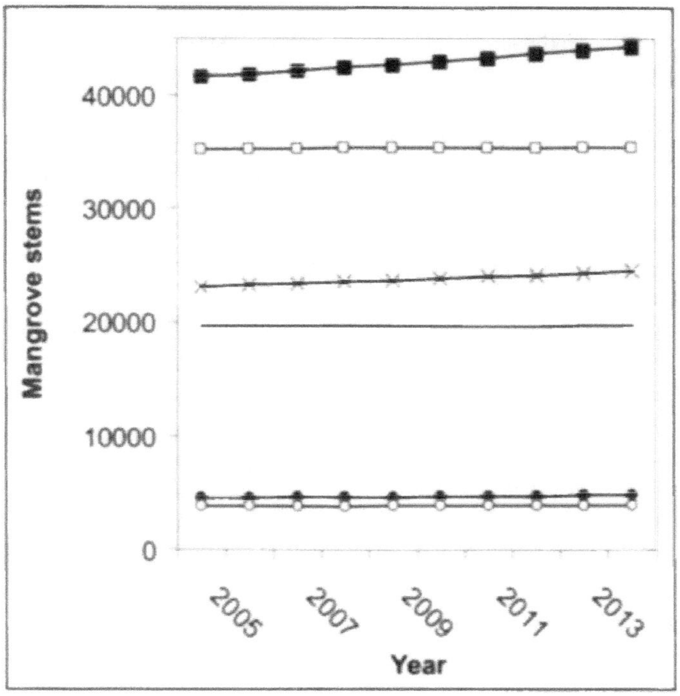

Figure 6. The predicted demand for mangrove stems under different population growth forecasts and building-shape scenarios —■—3% growth (rectangular*; —□— 0 78% growth (rectangular); —✕—3% growth (circular and rectangular); — 0 78% growth (circular and rectangular); —▲— 3%growth (circular); —○— " 0.78%growth (circular).

DISCUSSION

The investigations have demonstrated that different-shaped buildings require a different total number of resource stems. In buildings constructed entirely from mangroves, rectangular buildings require ± 1 0 0% or over 200 more stems than circular buildings. Liengme (1983) found that rectangular buildings used ± 50% more poles than circular buildings, the main difference attributable to rectangular buildings having two rather than one row of roof poles. In the current study, the major variation between building shapes was in the number of wall laths required.

Buildings in the current study are rarely constructed entirely from mangroves and species combinations have also been documented in other parts of South Africa (Liengme 1983). The proportion of mangroves used varies with structural unit and building shape. Mangroves are the preferred materials for circular building roof poles and rectangular building wall laths. Shackleton et al. (2004) reported that proportions of species could alter within walls and roofs. In the current study, mangroves composed a higher percentage of rectangular than circular building stems.

The average distance from surveyed homesteads to the mangrove boundary was 393 m in Mqualeni, 1 733 m in Tekweni and 2 040 m in Cwebeni. Mangrove use varied spatially between the villages; mangroves are used for all structural units in both circular and rectangular buildings in Mqualeni Village, however in Tekweni they are rarely used as roof poles and never used as circular building wall laths. Circular buildings have curved walls and group discussions revealed that indigenous species from the coastal scarp forests were preferred for this purpose because they were more flexible than mangroves. In Mqualeni Village, mangroves were often used for circular wall laths as the mangrove resource was nearby, ± 400 m from homesteads. Similar examples of resource selection for less preferred species due to higher availability and ease of collection have been documented by Nomtshongwana (1999) and Walters (2005b). In the current study, the relationship betw-een mangrove use and distance to the mangroves was confounded by the fact that mangrove use was also affected by utilization of alternative building materials such as indigenous timbers and sand blocks.

Utilization evaluation at the homestead scale shows a variety of mangrove and net-mangrove materials used for both circular and rectangular buildings. Homesteads in Tekweni use approximately a quarter the amount of mangroves compared to homesteads in the other villages studied. Tekweni is the furthest distance from the mangrove forest and also has proximity to the coastal scarp forest that can provide building construction materials (Obiri 1997). Analysis of use at the village scale indicates that observed differences at the homestead scale are generally enlarged at the village scale. Tekweni Village which has homesteads with the lowest utilization values, the lowest percentage and actual number of mangrove homesteads, is estimated to have used less than 9 000 mangrove stems. Cwebeni Village at the other extreme, has high homestead utilization values and a high percentage and actual number of mangrove homesteads. These factors combine to produce utilization figures of approximately 1 0 0 0 0 0 stems.

Demand for a resource may constantly change and the dynamic needs to be understood (Ellery et al. 2004). Both the annual and predicted demand for mangrove stems is influenced greatly by building shape. The annual demand for mangrove stems is based upon the percentage of sampled homesteads that were observed constructing new mangrove buildings, the value in Mqualeni was over 50%. This high value may indicate that this category of homestead was over-sampled. However, as Mqualeni is a small village, its influence on the overall demand for mangroves from Mngazana Forest is relatively limited. The high annual demand indicates that many homesteads are constructing new buildings on their land. These tend to be required due to a change in individual circumstance often with young adults requiring the privacy of a separate building.

The influence of human population growth rates of 0.78% or 3% upon demand is limited as the current demand rates are high when compared to the demand changes due to population growth which are relatively small. The differences in values between current demand in 2004 and predicted demand in 2015 due to human population growth suggest that there are factors other than human population growth which are influencing the construction of new buildings. These factors may include maturing children who desire their own private room or marriage of family members who remain on the homestead and family members returning to live at the homestead such as migrant workers. Although demand is predicted to show minor changes due to human population growth rates, it could change drastically due to the shape of the new buildings being constructed.

In the past, fuel wood models were devised that predicted future demand scenarios and an impending 'fuel wood crisis', however this never materialized; the models failed partly because they did not appreciate the complexity of rural energy and focused upon supply without accounting for demand changes (Shackleton et al. 2004). Humans demonstrate adaptability to resource decline and increasing scarcity and Shackleton (1993) reported that where demand for the preferred fuel wood—dead wood—outstripped supply, harvesters responded by selecting the non-preferred live wood resources and even developed strategies to circumvent legislation. Factors such as these may well influence the demand scenarios presented for Mngazana, particularly as although mangroves are generally preferred for building construction, the inhabitants are not strictly reliant upon them and can obtain alternative timber resources from the indigenous terrestrial forests or use mud or sand blocks for construction.

A previous study concerning wood utilization for building construction focused upon quantities used per building (Liengme 1983). This study demonstrates that the number of stems required per building, the proportion of a species used in different structural units, the building shape and the proportion of buildings constructed from different materials must all be evaluated so that a realistic estimate of wood use can be determined.

Sustainable Forest Management, at Mngazana must recognize that the local communities utilize the mangroves for building construction and that the people have a right to access mangrove wood products. Given the socio-economic status of the local communities at Mngazana, the demand for low-cost building materials will persist. Although there are alternatives to mangrove wood, it is highly valued and will probably continue to be used preferentially for building in the immediate future. Mangrove forests in other developing countries are under similar pressures to those of Mngazana (Semesi 1992; Dahdouh-Guebas et al. 2000) and the most immediate value of mangroves is placed on their wood products (Alongi 2002). Additionally, natural forests as opposed to plantation forests are often viewed as open access resources and as a result may be intensively harvested (Walters 2005b). Given these conditions, the extent to which these demands can be met needs to be analyzed based upon a forest inventory.

SFM implies that there should be no decrease in wood products and that the capacity to regenerate be maintained. However, studies have demonstrated that 'small-scale, local woodcutting can be a significant form of ecological disturbance in mangroves' (Walters 2005a: 345), and it has been shown that forest structure was dramatically altered.

A desktop extrapolation, to determine the sustainability of current extraction rates, was made using published data from Adams et al. (2005). They reported that the density of trees greater than one meter high in the Mngazana Estuary was 230 Bruguiera gymnorrhiza trees and 489 Rhizophora mucronata trees per hectare. As the Mngazana Estuary covers an area of 145 ha this suggests that there are ± 104 255 trees > 1 m height of B. gymnorrhiza and R. mucronata. With an estimated annual demand of 18 400 stems, the annual off-take would be approximately 17.6%. The resource would last for five and a half years at the current demand, without any recruitment. This is only sustainable in the very short term. Such a desktop extrapolation can be used as a guideline within an adaptive management context. Further investigations concerning recruitment and growth rates are required to accurately assess sustainability. As Mngazana has already been subjected to past harvesting, the forest must be assessed to determine whether the present community structure is functioning in a desirable state or if current stands need to be improved. Harvesting needs to be planned so that the ecosystem functioning and biodiversity are maintained. Studies suggest that mangroves can be managed sustainably for their wood, the Matang Mangrove Forest in Malaysia has been managed since 1906 for commercial purposes and continues to be productive (Hogarth 1999).

SFM should also aim to broaden the range of benefits derived from the mangroves, particularly of nonconsumptive uses. Activities such as mangrove honey production, could demonstrate that non-consumptive mangrove uses have the potential to generate incomes. External factors that may affect demand also need to be considered: at Mngazana these include the link between mangrove use and use of other indigenous timbers and alternatives to mangroves such as sand blocks. In the village of Cwebeni, a group was formed to produce and locally sell sand blocks for building construction. These blocks are durable, highly regarded as building materials and, if competitively priced, could reduce demand for mangrove stems. Other factors may indirectly affect demand for mangroves, for example clearing of the coastal scarp forest for agriculture will reduce the supply of construction timber from these forests, builders may turn to the mangrove forests to meet their needs and thereby increase demand for mangrove stems. In the past, forest clearing for agriculture has reduced the supply of indigenous hardwood poles from coastal scarp forest in northern KwaZulu-Natal (Cunningham & Gwala 1986).

ACKNOWLEDGEMENTS

This project was funded by USAID, the National Department of Environmental Affairs and Tourism (Marine and Coastal Management) and the Southern African Consortium of

Universities for Development and Environment—Sustainable Land Use Project: SLUSE-SA. Thanks to the Institute of Natural Resources, particularly F. Lewis and A. Msimang who are facilitating the Mangrove Management Plan. The Discipline of Geography, University of KwaZulu-Natal (Pietermaritzburg Campus) provided logistical support and B. Gijsbertsen of the Carto-graphic Unit produced the location map. Thanks to Eamoch Mtambeki for translation and guidance in the field. Special thanks to members of the Mqualeni. Tekweni and Cwebeni villages who participated in the study.

REFERENCES

1. ADAMS. J.B., COLLOTY. B.M. & BATE. G.C. 2005. The distribution and state of mangroves along the coast of Transkei. Eastern Cape Province. South Africa. Wetlands. Ecology and Management 12: 531-541.
2. ALONGI. D M. 2002. Present state and future of the world's mangrove forests. Environmental Conservation 29: 331-349.
3. ASHLEY. C. & NTSHONA. Z. 2003. Transforming roles but not reality? Private sector and community involvement in tourism and forestry de\ elopment on the Wild Coast. South Africa. Sustainable Livelihoods in Southern Africa Research Paper 6. Institute of Development Studies. Brighton.
4. BERJAK. P., CAMPBELL. G.K.. HUCKETT. B I. & PAMMENTER. N.W. 1997. In the mangroves o f southern Africa: 64. Wildlife and Environment Society of South Africa. KwaZulu-Natal. South Africa.
5. BRANCH. G.M. & GRINDLEY. J.R 1979. Ecology of southern African estuaries. Part XI. Mngazana: a mangrove estuary in Transkei South African Journal o f Zoology 14: 149-170.
6. BRUTON. M.N. 1980. An outline of the ecology of the Mgobezelem lake system at Sodwana. with emphasis on the Mangrove Community. In M.N. Bruton & K.H. Cooper. Studies on the ecology o f Maputaland: 408-426. Rhodes University and Wildlife Society of Southern Africa. Grahamstown. South Africa.
7. COLLOTY. B.M. 2000. The Botanical Importance Rating of the estuaries in former Ciskei Transkei Ph.D. thesis. University of Port Elizabeth
8. CUNNINGHAM. A B 1985. The resource value of indigenous plants to rural people in a low agricultural potential area Ph.D. thesis. University of Cape Town.
9. CUNNINGHAM. A.B & GWALA. B.R. 1986. Building methods and plant species used in Tembe-Thonga hut construction. Annals o f the Natal Museum 27: 491-511.
10. DAHDOUH-GUEBAS. F.. MATHENGE, C.. KAIRO, J.G. & KOEDAM. N. 2000. Utilization of mangrov e wood products around Mida Creek (Kenya) amongst subsistence and commercial users Economic Botany 54: 513-527. DAY. J.H. 1981. Estuanne ecology with particular reference to southern Africa. 289 Balkema. Cape Town.
11. ELLERY. W . TOOLEY. J . BILL. C,, DAHLBERG. A.. HEINSOHN. D . MCKENZIE. A & MURPHEY. M 2004 Resource use and conflict in and around Thukela Biosphere Reserve: woodcarvers seeing the wood, farmers the trees In M.J Lawes. H.A.C. Eeley. C M . Shackleton & B G.S Geach. Indigenous forests and wood lands in South Africa: 367-397. University of KwaZulu-Natal Press, Pietermaritzburg. FAO. 1994. Mangrove forest management guidelines. Food and Agricultural Organisation of the United Nations, Forest Resources Development Branch, Forest Resources Division, FAO Forestry Paper 117: 1-319. Rome. Italy.
12. HARRISON, T.D., COOPER, J.A.G., RAMM, A.E.L. & SINGH, R.A. 1999. Application o f estuarine health index to South Africa 's estuaries. Transkei. Technical Report, CSIR, Congella, South Africa.
13. HOGARTH, PJ. 1999. The biology o f mangroves'. 174. Oxford University Press, Oxford.
14. INTERNATIONAL TROPICAL TIMBER ORGANISATION. 2002. Mangrove Work Plan 2002-2006. International Tropical Timber Organisation, Yokohama, Japan.
15. LAWES, M.J., EELEY, H.A.C., SHACKLETON, C M. & GEACH, B.G.S. 2004a. Indigenous forests and woodlands in South Africa. University of KwaZulu-Natal Press, Pietermaritzburg.
16. LAWES, M.J., EELEY, H.A.C., SHACKLETON, C M. & GEACH. B.G.S. 2004b. South African forests and woodlands: recurring themes in integrating policy, people and practice. In M.J. Lawes, H.A.C. Eeley, C.M. Shackleton & B.G.S. Geach. Indigenous forests and woodlands in South Africa: 815—831. University of KwaZulu-Natal Press, Pietermaritzburg.
17. LIENGME, C.A. 1983. A study of wood use for fuel and building in an area of Gazankulu. Bothalia 13: 501-518. LUGO, A.E. & SNEDAKER. S.C. 1974. The ecology of mangroves. Annual Review o f Ecology and Systematics 5: 39-64. MUIR. D. 1990. Forest utilisation in Kw aZulu: a case study of Hlatikulu Forest Reserve, Maputaland. Institute of Natural Resources. Pietermartitzburg.
18. MUMBY, P.J., EDWARDS, A.J., ARIAS-GONZALEZ , E., LINDERMAN, K.C., BLACKWELL, P.G., GAIL, A., GORCZYNSKA. M.I., HARBORNE, A.R., PESCOD, C.L.. RENKEN. H.. WABNITZ, C.C.C. & LLEWELLYN, G. 2004. Mangroves enhance the biomass of coral reef fish communities in the Caribbean. Nature-. 427: 533-536.
19. NICHOLSON, G. 1993. Transkei coastal development plan. Transkei Development Corporation, South Africa.
20. NOMTSHONGWANA, N. 1999. Indigenous plant use in Gxalingenw a and Kwa Yili Forests in the southern Drakensberg. KwaZuluNatal. M.Sc. thesis. School of Environment and Development. University of Natal. Pietermaritzburg. OBIRI. J.A.F. 1997. Socio-economic and environment impacts on the utilisation ofumSimbithi tree (Milletia grandis) in Eastern Cape: a case study o f Mt Thesiger Forest, Pondoland. M.Sc. thesis. School of Environment and Development. University of Natal, Pietermaritzburg.
21. OBIRI. J.A.F., LAWES, M.J.L. & MUKOLWE. M. 2002. The dynamics of sustainable use of high-value

tree species of the coastal Pondoland forests of the Eastern Cape Province. South Africa. Forest Ecology and Management 166: 131-148.

22. RAJKARAN. A., ADAMS, J.B. & DU PREEZ. D R. 2004. A method for monitoring mangrove harvesting at Mngazana Estuary. South Africa. African Journal o f Aquatic Science: 29: 57-65. RSA. 1996a. Constitution of the Republic of SA (Act No. 108 o f 1996). Pretoria. RSA. 1996b. The White Paper on sustainable forest development in South Africa. Department of Water Affairs and Forestry, Pretoria. RSA. 1998. National Forests Act No. 84 o f 1998. Government Gazette, vol. 400: no. 19408. Pretoria. SEMESI, A.K. 1992. Developing management plans for the mangrove forest reserves of mainland Tanzania. Hydrobiologia 247: 1-10.

23. SHACKLETON, C.M. 1993. Fuelwood harvesting and sustainable utilisation in a communal grazing land and protected area of the eastern Transvaal Lowveld. Biological Conservation 63 : 247-254.

24. SHACKLETON, C.M. & SHACKLETON, S.E. 2004. Use of woodland resources for direct household provisioning. In M.J. Lawes, H.A.C. Eeley, C.M. Shackleton & B.G.S. Geach, Indigenous forests and woodlands in South Africa'. 195-225. University of KwaZulu-Natal Press, Pietermaritzburg.

25. SHACKLETON, C.M., GRUNDY, I.M. & WILLIAMS, A. 2004. Use of South Africa's woodlands for energy and construction. In M.J. Lawes, H.A.C. Eeley, C.M. Shackleton & B.G.S. Geach, Indigenous forests and woodlands in South Africa: 337-363. University of KwaZulu-Natal Press, Pietermaritzburg.

26. SHACKLETON, S.E., SHACKLETON, C M., NETSHILUVHI. T.R., GEACH, B.S., BALANCE. A. & FAIRBANKS, D.H.K. 2002. Use patterns and value of savanna resources in three rural villages in South Africa. Economic Botany 56: 130-146.

27. STATISTICS SOUTH AFRICA. 2001. Census 2001. Pretoria. STEINKE. T. 1999. Mangroves in South African estuaries. In B. Allanson & D. Baird, Estuaries o f South Africa'. 119-140. Cambridge University Press, Cambridge. STEINKE, T,, WARD, C.J. & RAJH, A. 1995. Forest structure and biomass of mangroves in the Mgeni Estuary. South Africa. Hydrobiologia 295: 159-166.

28. TURPIE, J.K., ADAMS, J.B., JOUBERT. A.. HARRISON. T.D., COLLOTY, B.M., MAREE, R.C., WHITFIELD, A.K., WOOLDRIDGE, T.H., LAMBERTH. S.J., TALJAARD, S. & VAN NIEKERK, L. 2002. Assessment of the conservation priority status of South African estuaries for use in management and water allocation. Water SA 28: 191-205.

29. UNITED NATIONS. 2005. The world population prospects. The 2004 rex ision: highlights. Department of Economic and Social Affairs, Population Division. United Nations, New York.

30. UNITED NATIONS GENERAL ASSEMBLY 1992. Non-legally binding Authoritative Statement on Principles for a global consensus on the management, conservation and sustainable development of all types of forest. Report o f the United Nations Conference on Environment and Dex elopment. Rio de Janeiro, 3-14 June 1992, Annex III. A/CONF. 151/26 (Vol. III).

31. WALTERS, B.B. 2005a. Ecological effects of small-scale cutting of Philippine mangrove forests. Forest Ecology and Management 206: 331-348.

32. WALTERS, B.B. 2005b. Patterns of local wood use and cutting of Philippine mangrove forests. Economic Botany 59: 66-76.

33. WARD, C.J. & STEINKE, T.D. 1982. A note on the distribution and approximate areas of mangroves in South Africa. South African Journal of Botany 1: 51-53.

34. WARD, C.J., STEINKE, T.D. & WARD, M.C. 1986. Mangroves of the Kosi System, South Africa: their re-establishment since mass mortality in 1965/66. South African Journal o f Botanv 52: 501 — 512.

35. WHITESIDE, A. 2001. Demography and economics of HIV/AIDS. British Medical Bulletin 58: 73-88.

36. WILLIS, C.B. 2004. Policy frameworks pertaining to the conservation and sustainable use of forests and woodlands in South Africa. In M.J. Lawes, H.A.C. Eeley, C.M. Shackleton & B.G.S. Geach, Indigenous forests and woodlands in South Africa'. 77-107. University of KwaZulu-Natal Press, Pietermaritzburg.

CHAPTER 8

SUSTAINABLE FOREST MANAGEMENT IN POLAND– THEORY AND PRACTICE

Hanna Kruk[1], Bożena Kornatowska[2]

[1]Gdynia Maritime University, Faculty of Entrepreneurship and Quality Science, Department of Economics and Management, Morska 81–87, 81-225 Gdynia, Poland

[2]Institute of Environmental Protection-NRI, Department of Nature and Landscape Conservation, Kolektorska 4, 01-692 Warsaw, Poland

ABSTRACT

The conception of sustainable development has been implemented into practice in numerous economic sectors, including forestry. Forest ecosystems are extremely important in the global ecological system, therefore maintenance and appropriate management of forest resources according to sustainable development principles have en-gaged a great deal of attention. The concept of sustainable forest management (SFM) encompasses three dimensions: ecological, economic and social. A powerful tool to promote SFM are criteria and indicators. The aim of the article was evaluation of SFM in Poland, using one of the methods proposed by the Food and Agriculture Organization (FAO). According to data available, Polish forestry has a number of advantages: Poland has avoided the problem of deforestation, forest area has been permanently increasing, there has been observed improvement of forest health and vitality as well as a significant share of forests has carried out protective functions with no impact on timber production. Poland's model of SFM is an adaptive process of balancing the ever-changing set of economic, environmental and social expectations. Such a complicated undertaking requires constant assessing and adjusting forest practices, in response to new circumstances, scientific advances and societal input.

Key words

Sustainable forest management, forestry, forest functions, Poland

INTRODUCTION

The concept of sustainable development (Keiner 2004) is taken into consideration in numerous economy sec-tors, with forestry being no exception. Forests are extremely important in the ecology of the Earth. They produce multiple environmental benefits including bio-diversity conservation, carbon sequestration, and protection against desertification. Appropriately managed forests can enhance the provision of wood and non-wood forest products for millions of people. At the same time, forest ecosystems play a key role in mitigation of climate change effects.

Citation: Hanna Kruk and Bożena Kornatowska, "Sustainable forest management in Poland – theory and practice," DOI: 10.2478/ffp-2014-0005, July 2014.

Sustainable forest management (SFM) is the management of forests according to the principles of sustainable development. The most far and wide respected definition of sustainable development was presented in "The World Commission on the Environment and Development (the Brundtland Commission) Report on Our Common Future in 1987": "Development that meets the needs of the present without compromising the ability of future generations to meet their own needs" (Keiner 2004). Five years after publication of the Brundtland Report, the UN General Assembly convened the United Nations Conference on Environment and Development (UNCED) with the objectives built upon the achievements of the Report so as to respond to environmental problems and to agree major treaties on biodiversity, climate change and forest management (Report of the UN Conference on environment and development 1992).

There is no globally agreed-on definition of sustainable forest management (SFM). Gusstavson et al. 2005 (Panagopoulos 2009) defined SFM as "long-term management of complex habitats in order to perform their multiple functions into the future". Appropriate forest management is crucial for maintenance of forest resources for present and future generations. Forest management encompasses a broad scope of issues, namely: exploiting of wood and non-wood forest products at the same time avoiding deforestation, keeping forest ecosystems in good conditions, preserving bio-diversity, protecting rare endangered species and other values of nature including landscape, improving welfare of local societies as well as enhancing cultural values and tourism. The concept of SFM includes ecological, economic and social aspects. The ecological aspect encompasses not only inner- and inter-ecosystem dependences but also equal treatment of all values which forests provide. The economic dimension concerns improving welfare and the use of forest products. The latter denotes that needs of various groups of interest like local societies, hunters, farmers, tourists, landowners should be taken into account during planning forest management strategies (Szaro et al. 2000). Managing forests with and for the people means listening to and learning from them. The social aspect of SFM encompasses both technical and political dimensions and has been the major challenge in forest management for the last decades. Other concepts related to SFM include among others nature-oriented silviculture (Meyerhof et al. 2009) and the idea of functional areas in forestry (Drozdowski 2008).

The goals for SFM elaborated at a global level by the United Nations (United Nations Global Objectives to Forests towards 2015) comprise the following (Forestry in the EU and the world – a statistical portrait 2011):

- reversing deforestation, protecting forests and preventing forest degradation,
- improving the use of economic, social and environmental benefits from forests,
- enhancing the areas of protected forests and other areas of SFM (aiming at better proportion between productive forests and those sustainably managed),
- strengthening official assistance for SFM (including financial support).

The main aim of this paper was to present essential information on the state of forests in Poland evaluated in the context of their sustainable management. The analyses were carried out against the background of forest situation in the European Union. The evaluation of the latter was possible only partially due to the lack of relevant data at a European level or its inaccessibility.

Sustainable Forest Management in Europe

Awareness on the vital role that forests play in socio-economic development and multiplicity of forest values as well as knowledge about complex relationships within forest ecosystems has contributed to reshaping timber forest management into sustainable management of forests worldwide. Broadly accepted definition of sustainable forest management (SFM) at a Pan-European level was adopted at the Second Ministerial Conference on the Protection of Forests in Europe (MCPFE – known as Forest Europe, re-named in 2009) convened in Helsinki, Finland in 1993. According to MCPFE sustainable forest management "means the stewardship and use of forests and forest lands in such a way, and at a rate, that maintains their biodiversity, productivity, regeneration capacity, vitality and their potential to fulfill, now and in the future, relevant ecological, economic and social functions, at local, national, and global levels, and that does not cause damage to other ecosystems" (Forestry in the EU and the world – a statistical portrait 2011; Baycheva at al. 2013). A definition of SFM was developed by MCPFE (FOR-EST EUROPE) has since been adopted by the European Union as well as Food and Agriculture Organization

(FAO). In Poland, SFM definition provided in the Act on Forests (art. 6, § 1a, http://isap.sejm.gov.pl) is fully compliant with the one established by MCPFE.

According to the three dimensions embraced in sustainable development concept, the principal elements of SFM in Europe include:

- ecological dimension – maintaining and increasing biological diversity of forests, sustaining natural processes, preserving the most precious elements of nature;
- economic dimension – treating forests as a source of numerous product types (including timber) and work-place provider (of high importance in neglected regions), manufacturing based on forest products, monitoring (especially tree healthiness), conducting scientific research towards forest management enabling maintenance of their capacity to continuously restore;
- social dimension – improving the quality of life including poverty reduction, strengthening social participation, empowerment and activity of local societies, enhancing ecological awareness, fostering co-operation among forest owners.

The Forestry Strategy for the European Union adopted by the European Council in 1998 encompasses among others: supporting sustainable forest management in connection with multifunctional role of forests, respecting the principle of subsidiarity, accenting the role of the SFM for maintaining biodiversity and nature conservation, promoting timber and non-timber products used as a part of environmentally friendly production, enhancing cooperation and participation of stakeholders (private and public), emphasizing contribution of the forestry sector to national economy and the quality of life

(European Council 1998). Milestones for European forestry were established in 2006 by the EU Forest Action Plan. Principal foundations of the document include, among others, enhancing competitiveness of the forestry sector, amending governance of European forests, ameliorating coherence and coordination of the EU's cross-country forest policy, protecting forest eco-systems, improving the health condition of forests, bettering the quality of life and fostering communication and coordination (COM(2006) 302 final).

In 2013, the new EU Forest strategy was elaborated. The main principles of SFM were identified as follows (COM(2013) 659 final):

- sustainable forest management and the multifunctional role of forests, delivering multiple goods and services in a balanced way and ensuring forest protection,
- resource efficiency, optimizing the contribution of forests and the forest sector to rural development, growth and job creation,
- global forest responsibility, promoting sustainable production and consumption of forest products.

Assessment Methods and Data Sources

The level of sustainability is measured with numerous indicators in various arrangements. Their lists are usu-ally extensive, that is why they receive strong criticism from many authors. One must bear in mind that not all elements, values and services (especially environmental and social aspects) may be expressed by quantitative measures. Therefore, the estimates are sometimes ambiguous and the results – not always credible and objective (Bartelmus 2003). Nevertheless, a variety of indicators for SFM has been applied. In general, two groups of tools are distinguished: criteria and indicators that allow to implement consistently the principles of sustainable development into forestry and receive information on SFM achievements. The term criterion reflects the conditions of the process used for SFM assessment, whereas indicators allow measurement of given criterions (Wijewardana 2008).

The complex tool for assessment was prepared by MCPFE. The process has developed a pan-European set of criteria and indicators (C&I), which consists of 6 criteria and 35 quantitative indicators (describing the forest status and changes) as well as 17 qualitative indicators (describing national forest policies, institutions and instruments towards SFM). C&I set has served as the basis for State of Europe's Forests assessments in 2003, 2007 and 2011, and provided the foundation for regional and national policy formulation, analysis and monitoring. Criteria embrace the following aspects: enhancement of forests and their role in the global carbon cycle (C1), maintenance of forest ecosystem health and vitality (C2), maintenance and encouragement of productive functions of forests (C3), maintenance, conservation and appropriate enhancement of biological diversity in forest ecosystems (C4), maintenance and appropriate enhancement of protective functions in forest management (notably soil and water) (C5), and enhancement of other forests socio-economic functions (C6) (Baycheva at al. 2013).

Another comprehensive means for forest evaluation was proposed by FAO (Global Forest Resources Assess-ment 2005, 2010). Seven principal groups of indicators were separated, namely (Global Forest… 2010):

- the extent of forest resource [area of forests and other wooded land (ha), growing stock of forests (m^3) and carbon stock per hectare in forest biomass (tons/ha)];
- biological diversity [area of primary forests, area of forests with major protective functions (designated for conservation of biodiversity) and total forest area alongside with plantations (all measured in ha)];
- forest health and vitality [area of forests affected by fire and forest cover damaged by insects, fun-gus diseases and other factors (quantified in ha per year)],
- productive functions of forest resources [area of forests with productive primary function (ha), area of productive forest plantations (ha), commercial growing stock (m^3), total wood removals (m^3/year) and total non-wood forest products (NWFP) removals (tons/year)];
- protective functions of forest resources [area of protected forests (ha) and area of protective forest plantations (ha)];
- socio-economic functions [value of total timber products and non-timber products (US dollars/year), employment (number of workers in a given year), area of forests privately owned (ha) and area of forests with social services as a dominant function (ha)];
- legal, policy and institutional framework [policy, forest law forest national programme, institutional structure of forestry and human resources (staff, share of highly graduated in forestry, etc.)].

The tool developed by FAO allows for complex assessments of SFM implementation. However, some economic and social aspects of SFM are taken into consideration only in a limited scope, e.g. the quality of life, preservation of cultural heritage, tradition related with forests and hunting, social participation and empowerment, relationship between forests ownership and forests conditions are not included. Besides, in case of some categories, data are not commonly available.

Therefore, for the purpose of this paper the main criteria were modified.

- In order to evaluate the state of forests in Poland and a level of their sustainability, the following categories were selected and used in further analyses as criteria of SFM achievements:
- forest area, tree species composition and changes in them – total forest area, forests area per capita, rate of afforestation (or deforestation), share of coniferous and broadleaved trees;
- state of health in forests and its changes – threats posed by fungi, insects, animals, and damage due to air pollution (defoliation);
- forest production, main products and services wood and non-wood products, eco-certification;
- preserving biodiversity – forest age structure, habitat constitution and species composition (tree species and forest type habitat suitability), potential and existent threats to biodiversity (monotypic forests, neophytisation, pinetisation) and forest protection (protective forests, protected areas in forests and NATURA 2000 sites);

- forests ownership and its influence on forest estate and management;
- social functions of forests – utilization by different groups of interest (recreation, tourism, hunting), education (including Promotional Forest Complexes PFC, arboreta), preserving cultural heritage.

The analyses based on the above approach were performed with the use of statistical data and publications available from Eurostat, FAO, the Central Statistical Office of Poland (GUS), the State Forests National Forest Holding and other institutions. There were found some discrepancies among the data obtained, most probably due to different methodologies used by data providers. The year 2010 was assumed as the reference in comparisons carried out. Depending on data availability the changes were examined be-tween 2010 and previous years (2000 or 2005 or even earlier). In a few cases there were data available only for 2005.

The comparisons between the state of Polish and European forests were also conducted. As some statistical data concerning the European Union (EU-27) were not available, international comparisons were not performed for all the aforementioned categories (e.g. in point 4 – preserving biodiversity).

State of Poland's Forests

Poland is one of the European countries with the biggest forest area (app. 9, 4 million ha, Global Forest… 2010). The majority of Poland's forests belong to the State Treasury, on behalf of which state owned forests are managed by the National Forest Holding "State Forests" (State Forests) headed by the General Director. The General Directorate is in charge of 17 Regional Directorates with altogether 431 forest districts divided into Forest-ranges (The State Forests in figures 2011). Poland's forest owner-ship structure is summarized below (tab. 1).

Table 1. Forest ownership structure in Poland in 2010

Share	Public forests				Private forests
	Managed by the State Forests	National Parks	Treasury Agricultural Property Stock	Owned by municipalities and communities	
thousands of ha	7072	184	36	84	1686
%	78.04	2.03	0.40	0.93	18.60

Since UNCED held in Rio de Janeiro in 1992, new legislation concerning sustainable forestry in Poland has been adopted, including the Forest Act (1991), the National Forestry Policy (1997), Regulations on Pro-motional Forest Complexes (1994, 1999, 2002), Regulations by the General Director of State Forests: No. 11 (1995) and No. 11a (1999) on ecological basis of forest management; the Instruction for preparation of the Nature Protection Programme as an integrated part of the Forest Management Plan for each forest district (1996), the Instruction of Forest Management (1994) and so on. According to the provisions in the 1995 Act on the Protection of Agricultural and Forest Lands, the protection of forest lands is based "upon (...) prevention of degradation or devastation processes in forest land and of damage in tree stands (...)".

Forest Area, Tree Species Composition and their Changes

In Poland, forests and wood areas cover around 29.9% of land area, i.e. less than the average value for 27 countries of the European Union (EU-27) (Eurostat News release 85 2011). Basic data concerning the total area of forested land in Poland and that in the EU are presented in table 2.

Table 2. Forests areas and their changes in Poland and EU in 2010 (average for 27 countries)

Country	Forests and wood areas	Area of forests and other wooded land per capita (ha per capita)	Forest area: annual average change (2005–2010)
EU-27	41%	0,35	0,32%
Poland	29,9%	0,24	0,30%

The data presented demonstrate that Poland has avoided the problem of deforestation, however, the share of forests and wood areas is below the EU's aver-age. Poland's forest cover is planned to be consequently increased. The National Programme of Increasing Forest Cover (NPIFC, revised in 2002) is a key element of the Polish strategy on improving the country's space structure and developing rural areas relative to implementation of sustainable, balanced and multi-functional forest management. It embraces the principle to reach 30% forests cover in 2020 and 33% in 2050. However, the pace of increasing forests' extent in Po-land has been slackening. In 2006–2010 only about 1/3 of planned afforestation was realized – 36% within the state owned forests and 31% within private and com-munity owned forests, whereas in the years 2001–2005, 81% of planned afforestation was achieved (96% and 70%, respectively).

Assuring appropriate tree species composition and percentage share of tree stands in conformity with a type of a given forest site poses further important questions in performance of desirable and appropriate forest management. The changes in tree species com-position (percentage shares) and age classes in Poland's forests in the years 1945–2010 are presented in table 3.

Generally, in the forests managed by the State Forests, the share of coniferous trees (especially pine) in species composition is higher than that in privately owned (GUS 2010). The pine (*Pinus sylvestris)* is the most widespread species preferred by the timber industry. The share of broadleaved species has been rising constantly since the year 1945. However, the appraisal of species structure of forests in Poland with regard to forest site types (fig. 1) indicates that the share of broad-leaves trees is lower than potentially possible as well as that a part of tree stands grows in the conditions not consistent with their favorite habitats. Actions leading to correction of stands' structure have been undertaken during afforestation/reforestation activities, nevertheless the composition of elder tree stands cannot be altered promptly and without difficulty.

Table 3. Species and age composition of forests in Poland

Year	1945	1997	2006–2010*
Species structure of tree stands			
Coniferous trees	87.0%	77.3%	70.7%
Deciduous trees	13.0%	22.7%	29.3%
Age class of tree stand			
I (age 1–20 years and restocking class	23.2%	17.2%	14.0%
II (21–40 years)	22.8%	24.6%	16.0%
III (41–60 years)	18.5%	21.8%	27.1%
IV (61–80 years)	13.1%	17.6%	18.3%
V and higher (over 81 years)	14.3%	17.0%	20.9%
Blanks, feeling sites and irregularly stocked open stands	8.1%	1.8%	3.7%

* Due to changes in research methods, these data are not fully of the same kind with those from previous years.

In Poland, there also exist forest plantations of fast-growing tree species, mainly Populus sp. and Larix decidua (app. 87% of all plantations). However, the share of such plantations in the total forest area is not considerable (Kozioł and Matras 2011).

Another issue concerning forest stands is their age and its changes (tab. 3). The average age of all tree stands in Poland is app. 56 years (State Forests managed – 57 years and privately owned – 46). The age structure has been continually improved. Even though designated tree stands are harvested and new areas are continually afforested, all things considered – the share of older tree stands in the total forest area has been increasing for the last decades.

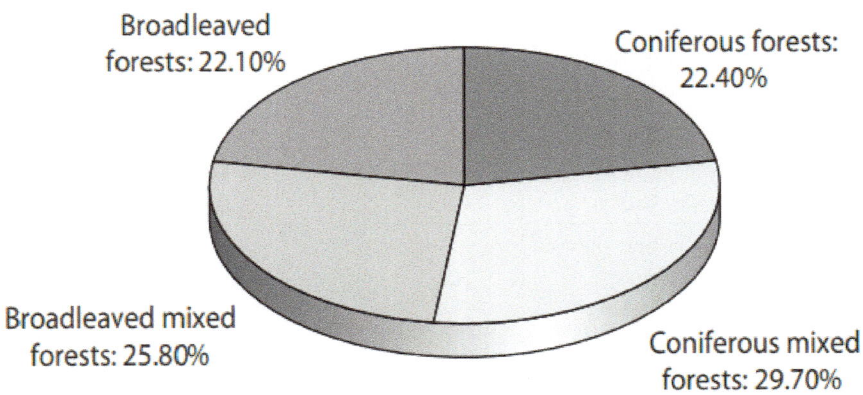

Figure 1. Share of coniferous and broadleaved stands in Polish forests.

State of Health in the Forests and its Changes

Health and vitality of forests may be assessed in line with the following criteria: defoliation (tab. 4) as well as leaf discoloration, diseases caused by fungi and insects, damages caused by animals and forest fires (tab. 5).

Table 4. Forest tress damaged by defoliation in Polish forests when compared with average within the EU

Defoliation	Poland	EU-27
Year 2000	in classes 2–4: 32%	in classes 2–4: 21.7%
Year 2010	in classes 2–4: 20.7% adequately: in class 0: 21.0%, in class 1: 58.3%, in class 2: 19.6% and in classes 3–4: 1.1%	in classes 2–4: 22.6% adequately: in class 0: 32.3%, in class 1: 45%, in class 2: 20.1% and in classes 3–4: 2.5%

Healthiness of trees is usually defined by the intensity of defoliation and a level of leaf or needle decolouration, and these illustrate damages due to air pollution (6.6% of Poland's forests are considered damaged by industry, GUS 2011b). Five classes of defoliation have been conventionally used in the EU including Poland: from class 0 (no defoliation) to class 4 (dead trees). The same classification has been also used to assess degradation of forests (GUS 2011b). Generally, the results of tree health evaluations in Polish forests carried out consistent with the abovementioned method on selected plots (the so called fixed observation areas) in the years 2000–2010, allow to draw the following conclusions with regard to forest tree defoliation (GUS 2011b; GUS 2008a; Forestry in the EU and the world – a statistical portrait 2011):

- there are no significant differences between health conditions of coniferous and broadleaved trees in Poland;
- comparison of spatial diversity of average defoliation shows, that forests in northern and west-northern parts of the country indicate the best conditions (average defoliation in 2010 for all these regions was below 20%);
- average defoliation does not depend on forest ownership – the percentage of trees suffering from the process is alike in public and private forests;
- the percentage of trees with severe defoliation in Poland is below the average in the EU;
- the results of damage evaluation in selected tree species (more than 20 -year-old specimens) show that both for coniferous and broadleaved trees the problem exists in very small fragments of forests: only 0.99% of coniferous trees and 1.63% broadleaved trees belong to damage classes 3–4.

Taking the above into account, healthiness of Polish forests measured by defoliation criterion can be considered as reasonably good and continually improving.

Table 5. Health of forests in Poland measured by means of other criteria

Question	Year 2000	Year 2010
Discolouration	98.67% of forests with no changes; 0.47% of forests in classes 3–4 (significant damage)	92.27% with no changes; 0.28% in class 3–4 (significant damage)
Forest area protected against insects	169,562 ha	12,825 ha
Forest area protected against parasitic fungal diseases	93,533 ha	23,747 ha
Forest area protected against damage caused by game animals	127,574 ha	79,842 ha
Total forest area endangered by fungi diseases	546,863 ha (7.9% of total forest area)	384,033 ha (5.5% of total forest area)
Number of forest fires	12,428	4,681
Area of burned forest	7,013 ha	2,126 ha
Reconstructed forest area	8,650 ha	10,641 ha

The analysis based on other criteria on forest health conditions[1] (tab. 5) allows the following conclusions concerning Poland's forests:

- the area of forests protected against harms caused by insects, fungal diseases and game animals has been continuously decreasing;
- emission of air pollutants and natural disasters cause the necessity of tree stand restoration; many weather anomalies have been recorded in the last decade and probably due to their impact there increased the area of reconstructed forests;
- the number of fires and the area of forests affected by them have decreased[2]; the main reasons for fires in forests are arsons and carelessness (GUS 2012).

Taking the above into account, one can state that for the last decade, the state of health in Polish forests has been improved.

Forestry, Production, Main Forest Products and Services

Timber is the main forest product. Poland's forests re-main one of the best stocked in Europe. In Polish conditions, average resources of wood in 2010 equaled 219 m^3/ha, whereas the corresponding figure for Europe (without Russian Federation) was 156 m ^{3}/ha. Timber falls into three main groups with regard to the thickness of removed wood: merchantable (large) timber, lumber (slash) and stump wood (GUS 2011b). The production of timber in Poland is characterized by a rising tendency. In 2000, 27,659 thousand m^3 of timber was harvested, in 2010: 34,567 th. m^3 (0.88 m ^{3} *per capita*), and this was 1.04% of global share (GUS 2012) and 9.28% of the EU-27 production (State of Europe's Forests 2011). The main group of timber was large timber (35,468 th. m^3), then slash (about 1,899 th. m^3) and stump wood (0.1 th. m^3) (GUS 2011b). In the same year, Poland took fifth place in the EU countries with reference to roundwood removals (Eurostat data). With regard to the species structure of tree stands in Poland, harvesting of coniferous trees is predominant (Forestry in the EU and the world – a statistical portrait 2011).

It is assumed that timber harvesting in Poland lies within the limits of securing ecosystems' ability to self-restoration and sustainability, and is performed without harming their capacity to sustain equilibrium along with maintaining a constant increase of forest area (Bernadzki 2006). According to the Report on the forests conditions in Poland 2010, forest resources (measured by log volume) have a rising tendency. However, in the case of the State Forests, about 18% of wood removals have come from tending cuts and those connected with the necessity of improvement felling after disasters so as to keep proper sanitary conditions of forests. It is also important to note that rising tendency of timber harvesting is related to improving the gross timber re-sources per hectare of the age group III (41–60 years) and older (Forests in Poland 2011).

Non-wood products of forest, such as forest fruits (mainly berries[3]), mushrooms[4] and game animals[5] have been gaining more and more importance. The Polish Central Statistical Office provides no more than figures on fruits and mushrooms procured for specialized enterprises. Due to free public access to Poland's forests, everyone has a right to collect forest fruits and mushrooms without limits (only protected species are restricted). Hence, the volume of forest fruits and mush-rooms harvested is impossible to estimate, likewise the volume of collected herbs[6].

Proper forests management including wood harvesting is subject to eco-certification. According to the data available at the State Forests' web page and those at web pages of Regional Directorates of the State Forests, in 2010, 15 out of all 17 regional directorates hold the Forest Stewardship Council (FSC) certificates. In case of the Regional Directorate of the State Forests in Bialystok, FSC certificates were awarded to 28 out of 31 forests districts[7].

Harmful activities in Polish forests comprise un-lawful use of forests, poaching, theft and property destruction and also timber theft (GUS 2011b). Although, in general, the number of harmful activities decreased, assessing economic costs due to the damages as well finding perpetrators create problems.

PRESERVING BIODIVERSITY

The concept of SFM encompasses the protection of forest ecosystems. Forests with productive, economic function constitute 47.48% of all forests in Poland (data from 2010), whereas those fulfilling protective functions comprise 21% forests with the aim of protecting water and 4.6% forests protecting soil. Forests existing within urban and around urban areas (non-productive) constitute 9.0% of the total forest cover in Poland, whereas 1.25% of forests are excluded from harvesting as nature reserves and 4.7% perform other protective functions (e.g. environmentally valuable areas, animal sanctuaries, permanent research areas, reproductive areas, health resort forests). Additionally, 1.87% of forests remain non-productive for military reasons. There exist considerable differences between public and private forests: in 2010, 46.6% of forests managed by the State Forests were marked out as protective, while the function in this scope was declared for 29.4% of forests owned by communities and 3.9% of those privately owned (GUS 2011b).

The structure of the system of protected areas which include forested lands is summarized in tab. 6. There are also designated other protected elements such as ecological sites, nature monuments, protective zones and selected seed stands. Numerous rare forest plant and animal species are protected under the law. Other protected forest areas are established in line with inter - national agreements, e.g. Man and Biosphere Baltic Sea Protected Area, UNESCO World Heritage Sites and NATURA 2000 areas.

NATURA 2000 areas were established within al-ready protected areas such as national and landscape parks or nature reserves or else were newly designated. Unfortunately, information about the whole share of forests in NATURA 2000 areas is not available. Exact, updated figures exist only for the forests managed by the State Forests. In 2010, in Poland there were 144 areas of special habitat protection and 823 areas of special bird protection. Within state owned forests, about 40% of the area is protected by the Natura 2000 network (GUS 2011a; The State Forests in figures 2011).

Table 6. Nature protection in Poland in 2010

Protected areas	Number	As a percentage of Poland's area	Total area [in ha]	% forests in protected areas
National system of protected areas				
National parks	23	1.0	314,474.5	61.9
Natural reserves	1463	0.5	164,202.1	60.4
Landscape parks	121	8.1	2,529,022.0	50.15
Protected landscape areas	386	22.3	6,990,047.7	31.5
Other*	7350	0.5	145,378.3	53.6***
Natura 2000**				
Areas of special habitat protection	823	11	3,791,720	21.4***
Areas of special bird protection	144	16	5,570,912	29.2***

In Polish forests, safeguarding of biological diversity is connected with decreasing monotypisation (one-species tree stands), pinetisation (introduction of coniferous trees, especially *Pinus sylvestris*) and neophytisation (introduction of foreign species). Negative effects of inappropriate forest management were due to focusing forest management on timber production in the past. In Poland, the process of reconstruction of tree stands has taken many years. Since 1991 monotyp-ic forests have not been planted. Moreover, deformed tree stands have been rebuilt progressively for the last dozens of years (Kozioł and Matras 2011). This has allowed to reduce pinetisation. As mentioned above, the pine has always been the most welcome species in the wood industry. Hence, coniferous monotypic forests have been before laid out. Pines used to be planted within the natural habitats of other tree species. Nowa-days, reconstruction of tree stands allows to increase the share of broadleaved trees in the total forest area, keeping in line with natural conditions of forest habitats (GUS 2011b). However, the share of tree stands not compatible with forest type sites is still notable (19% of the total forest area) (Kozioł and Matras 2011). In the case of Polish forests, the share of alien tree species is not higher than 0.5% of the total forest area and none of them is treated as an invasive species (Kozioł and Matras 2011).

FOREST OWNERSHIP AND ITS INFLUENCE ON FOREST ESTATE AND MANAGEMENT

The majority of Poland's forests is state owned (tab. 1). The share of public forests in Poland exceeds that aver - age for 27 countries of the European Union (approximately 40% of the total forest area in 2005) (Forestry in the EU and the world – a statistical portrait. 2011). The ownership structure as such secures state responsibility for coherent national forest policy, open access to forests for society, and consent for exploitation of non-wood forest products in addition to assurance of protective forest functions. For instance, in 2005, 35.5% of Poland's total forest area was designated as forests with protective functions, whereas those in the EU countries (27) constituted 12.7% (Eurostat 2012).

Owing to the high share of state owned forests, Po-land has avoided in great measure the problem experienced by other European countries that is forest frag-mentation. In Poland, forest fragmentation occurs when large continuous forests are divided into smaller blocks due to road-building (Environmental statistics and ac-counts in Europe 2010).

SOCIAL FUNCTIONS OF FORESTS

Social functions of forests are difficult to define and measure, but they are commonly connected with the so-called social services: recreation, tourism, education and various cultural aspects (e.g. tradition).

The State Forests puts a lot of efforts into rising public awareness based on recreational and educational infrastructure, including: parks and arboreta (17), centers of ecological education (over 20), educational rooms (50), points (318) and paths (over 150), the so-called green schools and Promotional Forest Complexes. Additionally, ecological NGOs, managements of municipal forests, national and landscape parks have their own educational and promotion activities, and they most often collaborate with the State Forests. Besides, kilometers of tourist routes have been traced and equipped in forests by the best known tourism organization in Po-land – PTTK (Polish Tourist Country-Lovers Society).

The Promotional Forest Complexes (PFCs) are special areas designated in Polish forests to demonstrate different forest habitats, changeability of forest ecosystems and their multifunctional role. PFCs also promote FSM, support scientific research (including testing new technologies in forestry) and public education. PFCs are also acknowledged as a forum of co-operation between forest administration, local authorities and communities. In 2010, in Poland, there existed 19 PFCs in various regions with the total area of over 1 million ha, that was about 14% of forests managed by the State Forests (Blichtarska et al. 2012).

CONCLUSIONS

Sustainable forest management concerns the use of forest resources that assures safeguarding and enhancement of forest functions. Sustainable forestry encompasses timber production biodiversity protection and fulfillment of a variety of human needs. Achieving balance between these aspects is very difficult, especially when human needs are constantly growing. Forests, as ecosystems, provide people not only for material goods such as timber or fruits and herbal plants or else animals but also for non-material benefits connected with rest recreation and culture. Forest are indispensable to maintain energy flows and matter circulation in the world. Consequently, proper management of forest re-sources is imperative.

The analysis the SFM achievements in Poland and general comparison with those in the EU[9] allowed the following conclusions:

- the high share of state owned forests in Poland seems to guarantee SFM implementation, even though not all activities undertaken are fully successful;
- the forest cover and stand age structure are increasing: Poland has avoided the problem of deforestation and the amount of forest resources has magnified, however it is yet less than the EU's average;
- tree stands are reconstructed towards adjustment to suitable habitats;
- forest health and vitality are maintained at a good level; one of the main problems which is tree defo-liation has been successfully overcome during last 10 years; based on criteria used it can be stated that the condition of Polish forests is in the main better than that in the EU's forests;
- Poland's forest biodiversity is well protected;
- productive and protective functions of Poland's forests are quite well-balanced; in fact the share of pro-tective forests is the biggest in the EU;
- Poland is one of the important wood producers in Europe: timber harvesting is above the average for Eu-rope (and has an increasing tendency); viable exploitation of non-timber products has been also achieved.

The results of the analyses carried out justify the conclusion that forestry in Poland has improved towards implementation of SFM. However, many factors, along with a number of conceptual and practical problems still remain a challenge to face by Polish forestry on its way to sustainability.

REFERENCES

1. Bartelmus P. 2003. Dematerialization and capital maintenance: two sides of the sustainability coin. *Ecological Economics*, 46, 61–81.
2. Baycheva T., Inhaizer H., Lier M., Prins K., Wolfslehner B. 2013. Implementing criteria and indicators for sustainable forest management in Europe, report. available at: http://www.ci-sfm.org/report. html. Accessed: 10.01.2014.
3. Bernadzki E. 2006. Forest and forestry in European Union countries. Centrum Informacyjne Lasów Państwowych, Warsaw.
4. Blichtarska M., Angelstam P., Elbakidze M., Axelsson R., Skorupski M., Węgiel A. 2012. The Polish Promotional Forest Complexes: Objectives, implementation and outocomes towards sustainable forest management? *Forest Policy and Economics*, 23, 28–39.
5. COM(2006) 302 final: Communication from the Commission to the Council and the European Parliament on an EU forest action plan {SEC(2006) 748}. 2006.
6. COM(2013) 659 final: Communication from the Commission to the European Parliament, the Council the European Economic and Social Committee and the Committee of the Regions. A new EU Forest Strategy: for forests and the forest-based sector {SWD(2013)342 final} {SWD(2013) 343 final}. 2013.
7. Drozdowski S. 2008. Functional areas in forestry – the theoretical background. *Studia i Materiały Centrum Edukacji Przyrodniczo-Le nej*, 3 (19), 11–20 [in Polish]. Environmental statistics and accounts in Europe. 2010. Eurostat, Statistical Books.
8. European Council. 1998. Council Resolution of 15 December 1998 on a forestry strategy for the European Union. *Official Journal of the European Com-munities*, 1999/C 56/01.
9. Eurostat Newsrelease 85. 2011. http://epp.eurostat. ec.europa.eu. Accessed: 23.11.2012.
10. Forestry in the EU and the world – a statistical portrait. 2011. Eurostat, Statistical Books.
11. Forests in Poland 2011. Centrum Informacyjne Lasów Państwowych, Warsaw.
12. Global Forest Resources Assessment 2005. Progress towards sustainable forest management. *FAO Forestry Paper*, 147, 1- 350.
13. Global Forest Resources Assessment 2010. Main report. *FAO Forestry Paper*, 163, 1–378.
14. GUS. 2008a. Environment 2008. Ochrona środowiska 2008. Zakład Wydawnictw Statystycznych, War-saw.
15. GUS. 2008b. Forestry 2008. Zakład Wydawnictw Statystycznych, Warsaw.
16. GUS. 2009. Forestry 2009. Zakład Wydawnictw Statystycznych, Warsaw.
17. GUS. 2010. Forestry 2010. Zakład Wydawnictw Statystycznych, Warsaw.
18. GUS. 2011a. Environment 2011. Ochrona środowiska 2011. Zakład Wydawnictw Statystycznych, War-saw.
19. GUS. 2011b. Forestry 2011. Zakład Wydawnictw Statystycznych, Warsaw.
20. GUS. 2012. Forestry 2012. Zakład Wydawnictw Statystycznych, Warsaw.
21. Gustavsson R., Hermy M., Konijnendijk C., Steidle-Schwahn A. 2005. Management of urban woodlands and parks – Searching for Creative and Sustainable Concepts. In: Urban Forests and Trees (eds.: C.C. Konijnendijk, K. Nilsson, Th.B. Randrup, J. Schip-perijn), Springer, 369–397.
22. Keiner M. 2004. Re-emphasizing sustainable development – the concept of evolutionability. *Environment, Development and Sustainability*, 6, 379–392.
23. Kozioł C., Matras J. 2011. Home Report on forests genetic resources. Available at: http://www.lbg. jgora.pl/Raport_FAO.pdf. Accessed: 27.11.2012 [in Polish].
24. Meyerhof J., Liebe U., Harje V. 2009. Benefits of biodiversity enhancement of nature -oriented silviculture: Evidence from two choice experiments in Germany. *Journal of Forest Economics*, 15, 37–58.
25. Panagopoulos T. 2009. Linking forestry, sustainability and aesthetics. *Ecological Economics*, 68, 2485–2489.
26. Regulation 11 of the State Forests' General Director. 1995.
27. Regulation 11A of the State Forests' General Director. 1999.
28. Report on the forests conditions in Poland 2010. 2011. Centrum Informacyjne Lasów Państwowych, Warsaw. Available at: www.lasy.gov.pl. Accessed: 12.11.2012.
29. Report on the forests conditions in Poland 2011. 2012. Centrum Informacyjne Lasów Państwowych, Warsaw. Available at: www.lasy.gov.pl. Accessed: 12.11.2012.
30. Report of the UN Conference on environment and development. 1992. Annex III, A/CONF.151/26.
31. State of Europe's Forests. 2011. UNECE, FAO.
32. Szaro C.R., Langor D., Yapi A.M. 2000. Sustainable forest management in the developing world: Science challenges and contributions. *Landscape and Urban Planning*, 47, 135–142.
33. The State Forests in figures. 2011. Centrum Informacyjne Lasów Państwowych, Warsaw.
34. Wijewardana D. 2008. Criteria and indicators for sustainable oforest management: The road travelled and the way ahead. *Ecological Indicators*, 8, 115–122.

CHAPTER 9

SILVICULTURAL ALTERNATIVES TO CONVENTIONAL EVEN-AGED FOREST MANAGEMENT - WHAT LIMITS GLOBAL ADOPTION?

Klaus J Puettmann[1], Scott McG Wilson[2] , Susan C Baker[3] , Pablo J Donoso[4] , Lars Drössler[5] , Girma Amente[6] , Brian D Harvey[7] , Thomas Knoke[8] , Yuanchang Lu[9] , Susanna Nocentini[10], Francis E Putz[11], Toshiya Yoshida[12] and Jürgen Bauhus[13]

[1] Department of Forest Ecosystems and Society, Oregon State University, Corvallis, OR 97330, USA

[2] Consultant Forester and Forest Ecologist

[3] Thorngrove Crescent, Aberdeen, Scotland AB15 7FH, UK. 3 School of Biological Sciences, University of Tasmania, Private Bag 55, Hobart, TAS 7001, Australia

[4] Department of Forests and Society, Faculty of Forest Sciences and Natural Resources, Universidad Austral de Chile, Isla Teja s/n, Valdivia, Chile

[5] Southern Swedish Forest Research Centre, Swedish Agricultural University, Alnarp SE-23053, Sweden

[6] Haramaya University, Diredawa, Ethiopia

[7] Institut de recherche sur les forêts, Université du Québec en Abitibi-Témiscamingue, 445, Boulevard de l'Université, Rouyn-- Noranda, Québec J9X 5E4, Canada

[8] Institute of Forest Management, TUM School of Life Sciences Weihenstephan, Technische Universität München, Hans-Carl-von-Carlowitz-Platz 2, Freising 85354, Germany

[9] Department of Forest Management and Statistics IFRIT, Chinese Academy of Forestry, Dongxiaofu 2, Xiangshan Road, Beijing, Haidian 100091, China

[10] Department of Agricultural, Food and Forestry Systems, University of Florence, v. S. Bonaventura 13, Florence I-50145, Italy

[11] Department of Biology, University of Florida, Gainesville, FL 32611-8526, USA

[12] Uryu Experimental Forest, Field Science Center for Northern Biosphere, Hokkaido University, Moshiri, Horokanai 074-0741, Japan

[13] Chair of Silviculture, Faculty of Environment and Natural Resources, University of Freiburg, Tennenbacherstr. 4, Freiburg 79085, Germany

ABSTRACT

Background

The development of forestry as a scientific and management discipline over the last two centuries has mainly emphasized intensive management operations focused on increased commodity production, mostly wood. This "conventional" forest management approach has typically favored production of even-aged, single-species stands. While alternative management regimes have generally received less attention, this has been changing over the last three decades, especially

Citation: Klaus J Puettmann, Scott McG Wilson, Susan C Baker, Pablo J Donoso, Lars Drössler, Girma Amente, Brian D Harvey, Thomas Knoke, Yuanchang Lu, Susanna Nocentini, Francis E Putz, Toshiya Yoshida and Jürgen Bauhus, Silvicultural alternatives to conventional even-aged forest management - what limits global adoption?, DOI 10.1186/s40663-015-0031-x.

in countries with developed economies. Reasons for this change include a combination of new information and concerns about the ecological consequences of intensive forestry practices and a willingness on the part of many forest owners and society to embrace a wider set of management objectives. Alternative silvicultural approaches are characterized by a set of fundamental principles, including avoidance of clearcutting, an emphasis on structural diversity and small-scale variability, deployment of mixed species with natural regeneration, and avoidance of intensive site-preparation methods.

Methods

Our compilation of the authors' experiences and perspectives from various parts of the world aims to initiate a larger discussion concerning the constraints to and the potential of adopting alternative silvicultural practices.

Results

The results suggest that a wider adoption of alternative silvicultural practices is currently hindered by a suite of ecological, economic, logistical, informational, cultural, and historical constraints. Individual contexts display their own unique combinations and relative significance of these constraints, and accordingly, targeted efforts, such as regulations and incentives, may help to overcome specific challenges.

Conclusions

In a broader context, we propose that less emphases on strict applications of principles and on stand structures might provide additional flexibility and facilitate the adoption of alternative silvicultural regimes in a broader set of circumstances. At the same time, the acceptance of alternative silvicultural systems as the "preferred or default mode of management" will necessitate and benefit from the continued development of the scientific basis and valuation of a variety of ecosystem goods and services. This publication is aimed to further the discussion in this context.

REVIEW

Much of the history of forest science and management in the last two centuries has focused on optimizing the efficiency of commodity production, mostly of wood for timber, pulp, and fuel. Based on the notion that homogenous products are cheaper to produce and manipulate, management practices have typically led to even-aged, mono-specific or species-poor stands (Puettmann and Ek 1999, Paquette and Messier 2009, FAO 2010, West 2014). Such practices can include planting or natural regeneration, e.g., through coppice or natural or sown seed, combined with weed control and timber stand improvement practices to control species composition, tree size and quality; hereafter called "conventional" forest management. Currently, stands under conventional forest management for commodity production cover about 30% of the global forest land base (FAO 2010). At the same time, silvicultural approaches that can be viewed as alternatives to conventional management have been developed in various parts of the world, but have received less attention at a global scale (e.g., Gayer 1886, Kimmins 1992, Gustafsson et al. 2012, Lindenmayer et al. 2012). However, interest in these alternative approaches has greatly increased in many regions over the last three decades (Figures 1 and 2), particularly in countries with developed economies (Brukas and Weber 2009). Motivations for this recent development have varied across regions, but generally involve increased public scrutiny (Brunson and Steel 1994, Bliss 2000), impetus from forest certification bodies (Baker 2011), and better understanding of the implications of conventional forest management practices on biodiversity and habitat quality (Harlow et al. 1997, Manolis et al. 2000, Bauhus and Schmerbeck 2010), soils, water and nutrient cycling (Swank et al. 2001; Little et al. 2009), landscape visual quality (Ribe 1989; Paquet and Bélanger 1997), and other dimensions of natural capital and ecosystem services provision. Especially in countries with developed economies, public landowners, and, increasingly, many small private landowners do not necessarily rely on forests as their primary income source. Instead, other management goals, including recreational opportunities and landscape amenities, have gained importance (Butler and Leatherberry 2004; Hugosson and Ingemarson 2004; Urquhart and Courtney 2011).

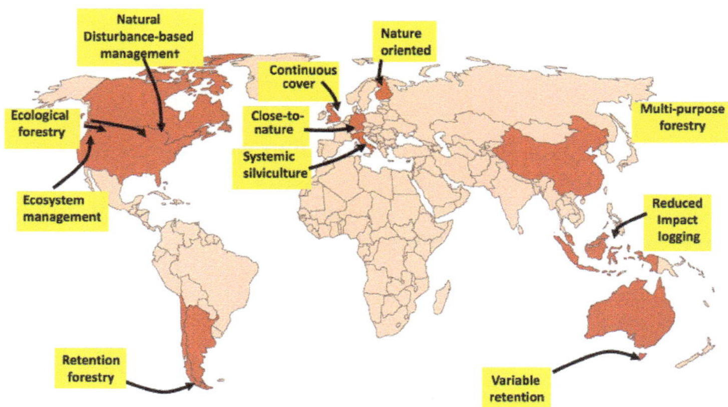

Figure 1: Examples of efforts to apply alternative silvicultural regimes. Countries with notable examples of alternative silviculture practices are in red. Associated text boxes display the common terminology used for alternative silvicultural practices in that region. *For more

information about close-to-nature forestry in central Europe see Jacobsen (2001), continuous cover forestry see Mason et al. (2003), ecological forestry in the USA, see Franklin et al. (2007), ecosystem management in the USA see Kohm and Franklin (1997), multipurpose forestry in China see Lu et al. (2013), natural disturbance-based management in Canada see Bergeron et al. (2002), nature oriented forestry in Finland see Kuuluvainen (2009), reduced impact logging in southeast Asia see Putz et al. (2008), retention forestry in Chile see Donoso and Promis (2013), systemic silviculture in Italy see Ciancio and Nocentini (2011), and variable retention in Tasmania see Baker and Read (2011).

Figure 2: Examples of alternative silvicultural practices in different locations. **(a)** Ethiopia, natural regeneration of*Juniperus escelsa* in gaps, **(b)** Chile, recent restoration thinning in second growth *Nothofagus dombeyi* forests in Llancahue, **(c)** Japan, single tree cutting in natural *Quercus crispula* forest with *Abies sachalinensis* regeneration, **(d)**Germany, single-tree selection in mixed-species (*Abies, Picea, Fagus*) mountain forests, **(e)** Australia, aggregated retention in *Eucalyptus* forests in Tasmania, **(f)** Canada, removal of overstory *Populus tremuloides* to release and favor establishment of *Abies* and *Picea* species **(g)** England, silvicultural development in species-enriched *Pinus* and *Larix* plantations, **(h)** USA, variable density thinning with gaps and leave islands in Douglas-fir stands in Oregon, and **(i)** and**(j)** southern China, *Eucalyptus* overstory with *Pinus* underplanted between *Eucalyptus* sprouts and dominant *Pinus*overstory with *Quercus griffithii* and *Castanopsis hystrix* understory, respectively. Photo credit for **(f)** Philippe Duval, all other photos by the authors.

The current variety of alternative silvicultural approaches (Figures 1 and 2) appears to have originated in two distinct contexts: in Europe and North America. Aesthetic and ecological critiques of conventional forest management during the period of 1850–1950 led to an emphasis on irregular size-class distribution and single-tree selection versus clearcuts, thus promoting the plenter forest (Plenterwald) model (Gurnaud 1882; Biolley 1901; Schütz 2002) and other "continuous cover" (Dauerwald) concepts (Möller 1923).

Other early European critiques highlighted concerns about the environmental effects of monocultures, and favored mixed-species stands (Gayer 1886). Similar concerns influenced management in other parts of the world, such as Japan (Marten 2005). Selection and shelterwood silvicultural systems suitable for the management of tropical moist forests were developed during the British and French colonial periods and were deployed in forests in India, Burma, Malaya, and West Africa until the 1960s. Technical development of these approaches has continued over recent decades, especially in Indonesia and Malaysia. The recent prominence in some European countries of Continuous Cover Forestry (e.g., UK and Ireland; Wilson 2013) and Close-to-Nature forestry (e.g., Germany, Switzerland, Slovenia, Italy) also reflects these historical influences (ProSilva 2014).

A second school of silvicultural approaches originated in the Pacific Northwest of North America under the banner of "New Forestry" (Franklin and Forman 1987) as a reaction to large-scale clearcutting in natural forest ecosystems and its impact on wildlife habitats, visual quality, and other ecosystem functions. Based on an increased understanding of how fire and other natural disturbances invariably leave patches of intact or partially affected forests in affected landscapes, this approach emphasizes the importance of residual structures and organisms (legacies), such as undisturbed vegetation patches, trees, and dead wood left behind after harvesting. These legacies are considered crucial for the speedy recovery of forest ecosystems after human-caused disturbances and thus for a continuation of ecosystem functioning and biodiversity and habitat values. Because this approach focuses on what is actually left behind after harvesting, it has spread to many regions of the world under the broad rubric of "retention forestry" (Lindenmayer et al. 2012).

A distinction between these different schools of alternative silviculture is the relative emphasis on management versus natural disturbances. For example, silviculturists in Central and Southern Europe tend to see a greater role for silvicultural interventions, reflecting the longer history of managed forests in their part of the world. By contrast, the lack of a long management history and an associated focus on natural disturbances in parts of North America and Australia are reflected in an emphasis on natural disturbance patterns as a template for alternative silviculture (Attiwill 1994; Drever et al. 2006; Long 2009). This is based on the notion that the historical range of variability in forest conditions can provide suitable guidance for management decisions (Higgs et al. 2014).

Despite their wide geographical distribution, in most contexts alternative silvicultural practices are not yet considered mainstream. They are often viewed and represented as less accepted alternatives to the conventional forest management paradigm in teaching and field practice (Puettmann et al. 2009). After long-standing arguments about alternative silvicultural systems, especially in the European literature of the last century, it appears that the debate is gaining momentum, as reflected by a number of recent summaries and overviews that discuss basic conceptual frameworks for these approaches, including complexity science (Puettmann 2011; Messier et al. 2013; Filotas et al.2014), natural disturbance-based systems (Bergeron et al. 2002; Franklin et al. 2007), and legacies and retention (Gustafsson et al. 2012; Lindenmayer et al. 2012).

In many regions, these silvicultural alternatives to conventional forestry are enacted under a variety of social, economic, and ecological conditions and a wide array of ownership objectives, as well as legal and regulatory frameworks. While less applicable to highly intensive, short-rotation plantations (3% of the total forest area are covered by "productive plantations"; FAO 2010) and areas with the primary objective of biodiversity conservation (12%), silvicultural alternatives are most likely to be of interest on land that is designated primarily for either "commodity production" (30%), "multiple use" (24%), "soil and water protection" (8%), or "social services" (4%) (FAO2010). These last four objectives may not be best achieved by adherence to any single management approach. Instead, a variety of local solutions have developed, including planted forests and forests that are naturally regenerated by seed or vegetative reproduction, reflecting the wide variety of ecological, economic, and social conditions and constraints of different jurisdictions. For an overview of the different efforts and discussions about labels, see Pommerening and Murphy (2004) and Evans (2006); also see Figure 1, including the notation with a list of references.

A basic distinction between conventional and alternative silvicultural approaches is the relative balance of selected values and objectives. Conventional approaches typically emphasize commodity production and view other objectives as constraints, e.g., intrinsic ecosystem values, accounting for natural processes, and maintaining species and structural diversity. In contrast, alternative silvicultural approaches place a unique set of emphases on each value. They regard all values, including non-commodity values, as a basic foundation necessary to achieve high levels and sustainable provision of ecosystem services, including product extraction (*sensu lato*) (Evans 2006).

This maturing view of forest management alternatives expresses itself in a set of five silvicultural principles (Jacobsen 2001; Schütz 2002; Mason et al. 2003; Bauhus et al. 2013):

- *Partial harvest*—use of partial harvesting and avoidance of large clearcuts
- *Natural regeneration*—where possible, preferential use of natural regeneration and native tree species
- *Structural diversity and small-scale variability*—varying management approaches across a range of spatial scales, with a special emphasis on diversity of stand structures at small scales, including single-tree and neighborhood conditions
- *Mixed species*—promotion of mixed-species stands
- *Avoidance of intensive operations*—minimization of intensive site preparation, fertilization, and weed

control practices, and reliance on natural process such as self-thinning of seedlings and small saplings.

These principles correspond closely with the "low" and "medium" silvicultural intensity categories described in Duncker et al. (2012), in which management approaches are aligned along a gradient of intervention intensity.

When applying silvicultural alternatives, the focus on variability in stand structures across spatial scales typically results in an emphasis on the development and harvest of individual trees or small patches of trees. This principle may appear on the surface to be similar to selective exploitation harvesting approaches, where activities are limited to harvesting single high-value trees. However, alternative silviculture approaches, as discussed in this paper, are philosophically and practically quite distinct from such practices. In contrast to early selective exploitation, harvesting decisions in alternative silvicultural approaches are driven by an appreciation of the economic and ecological value of retained trees in terms of ensuring the continuity of ecosystem processes and functions. Thus, the application of all five principles aims to maintain or increase the growth potential of the site, conforming to traditional sustained-yield paradigms.

The potential advantages of alternative silvicultural approaches have been hailed in various dimensions, including in terms of biodiversity (Steventon et al. 1998; Fedrowitz et al. 2014), maintenance of "undeniably better resilience" (Schütz et al. 2012), general resilience (O'Hara and Ramage 2013) or, specifically, resilience to climate change (Schütz 1999a). However, limitations to specific adaptation strategies may exist; for example, the introduction of selected or improved genetic material may not be allowed or difficult to achieve in forests managed under these principles (Brang et al. 2014). Alternative approaches have also been claimed to provide other benefits compared to even-aged monocultures, albeit under specific sets of conditions. These conditions include more uniform cash-flow of harvest revenues (Knoke et al. 2001), higher income resulting from the production of high-quality wood (Hanewinkel 2002), more diverse wildlife habitats, especially for late-seral species (Hyvaerinen et al. 2006; Rosenvald and Lohmus 2008), increased wind stability of individual trees (Dvorák et al. 2001, but see Cameron2002), and improved recreation value and public acceptance (Tönnes et al. 2004; Ribe 2005). This said, there is still much more work to be done in order to investigate and quantify the impacts under specific regional conditions (Gustafsson et al. 2010). Given these reported advantages, however, as well as the fact that many of the public and private landowners who have begun applying these principles are very satisfied with the results, "it is surprising that [alternative silvicultural approaches] are so seldom applied" (Schütz et al. 2012). In the following sections, we hope to provide a basis for a discussion about broader application of alternative silvicultural approaches and ask the following questions:

- What are the major factors that influence the decision to apply alternative silvicultural approaches?

What is currently limiting the wider application of alternative silvicultural approaches at a global scale?

- What can be done to remove such limitations?

METHODS

The authors have extensive experience dealing with research and teaching of alternative silviculture in various parts of the world. An initial collection of arguments by the authors was based on their own experience and familiarity with the relevant literature. This collection highlighted that the answers to questions about factors limiting wider application of alternative silvicultural practices vary between and within regions and between and within ownerships. To improve our insight into such highly complex decision-making environments and to better understand the decision factors, we found it useful to organize the assessment into a framework structured around the five silvicultural principles listed above (partial harvest, natural regeneration, small-scale variability in stand structures, mixed species, avoidance of intensive operations). Within these principles, we determined that the challenges to the implementation of these principles could be grouped into five categories and structured our results and discussion accordingly: ecological; economic; logistical and administrative; educational and informational; and cultural and historical. All authors repeatedly provided insights and feedback regarding the number and distinctions between the silvicultural principles and categories. At the same time, we went through repeated iterations in which the authors interacted to expand and organize the list of challenges. Thus, even though this does not comprise a systematic review of all existing literature (which would go beyond the scope of a single publication), we refined the list of challenges until we felt comfortable that all influential topics were covered in the discussion and provided a solid basis to initiate further discussions.

RESULTS AND DISCUSSION

Ecological Challenges

The emphasis on continuous canopy structures (avoidance of clearcuts) typifies forests growing under natural disturbance regimes characterized by small scale disturbances, e.g., pathogen-induced mortality or mechanical damage (Lutz and Halpern 2006). In much of the world, the spatial scale of natural disturbance regimes can be represented by a J-shaped size frequency distribution, i.e., a combination of frequent small-scale and fewer large-scale disturbances (Veblen et al. 1981; Seymour and Hunter 1999; Shorohova et al. 2009; Yasuda et al. 2013). Examples of disturbances that generate these patterns include wildfires, e.g., in the western United States and Australia, or windstorms, e.g., in storm-prone regions of Western Europe (Schelhaas et al. 2003) or in the temperate forests of southern Chile (Veblen 1985). In such regions, a focus on any single silvicultural approach will only reflect a subset of patterns and dynamics found under natural conditions (Seymour and Hunter 1999). For example, region-wide applications of a management regime that focused only on continuous overstory cover would not promote conditions and patterns associated with severe, larger-scale disturbances, e.g., certain early successional habitat features. In such cases, the emphasis on smaller-scale variability in stand structure may reflect natural conditions found in individual stands, but it may not reflect the overall landscape-scale dynamics. Such concerns are currently discussed regarding the lack of early-seral habitat on federal land in the Pacific Northwest of the USA (Swanson et al. 2011; Franklin and Johnson 2012). Since natural disturbances such as fire and windstorms cannot be totally prevented or controlled in most forest landscapes, concern about the lack of selected disturbance effects may be of little practical relevance in the long run. Exceptions are situations where management artificially reduces the probability of occurrence of large-scale or stand-replacing disturbances. Examples of such situations include the application of short rotations that prevent trees from reaching heights common in older forests, and thus may reduce windthrow. Alternatively, density management aims to reduce tree stress to the remaining trees by ensuring higher light, water, and nutrient levels, and thus may limit competition related mortality (either directly due to drought or indirectly, e.g., insect related mortality due to low vigor, often found in high-density unmanaged stands (e.g., Neumann and Minko 1981). Also, fuel management may reduce the extent of large-scale fire disturbances. Hence, practices that change "natural" disturbance regimes must be carefully assessed in terms of their implications on landscape dynamics, and restoration may be required in some cases. For example, controlled burning as a restoration technique can be beneficial for threatened species in boreal forests where there has been extensive long-term fire suppression (Kouki et al. 2012).

Lack of large openings and early successional conditions in close-to-nature silvicultural systems operating with small canopy gaps can be detrimental to maintenance of light-demanding tree species in ecosystems where these species are mixed with more shade-tolerant species (e.g., Neyland et al. 2012; Bauhus et al. 2013). On the other hand, under selected conditions, such as low-density stands or species with open tree canopies, intensive management with repeated entries can facilitate survival and growth of light-demanding species, e.g., small-group uneven-aged management in Mediterranean pine forests (Ciancio et al. 1986; Barbeito et al. 2008; Calama et al.2008). However, more often the higher mortality of light-demanding species in shady conditions and the slower responses of suppressed trees after silvicultural release can lead to an increasing dominance of shade-tolerant species (Lüpke and Hauskeller-Bullerjahn 1999; Kneeshaw et al. 2002; Drössler et al. 2013). These trends lead to concerns, as light-demanding species are often desirable in managed forests because of their fast growth rates, desirable timber properties, and dependent biodiversity components. They are also of broader interest under climate change conditions as they can be more drought tolerant than shade-tolerant species (Ninemets and Valladares2006) and their fast initial growth rates and associated shorter rotations have the effect of reducing long-term risks.

The focus on small-scale variability in stand structure means that trees of different species and sizes grow in close proximity to each other. Such stands may have a lower probability of extensive crown fires, as they do not have continuous high density canopies. However, multi-layered stand structures can produce ideal fire ladders (Schoennagel et al. 2004), a great concern in fire-prone areas. The spread of certain pests may also be facilitated in such intimate structures, e.g., large infected trees have the potential to infect nearby regeneration through higher exposure to inocula (Reich et al. 1991; Shaw et al. 2005). Tree-species diversity tends to restrict this threat, however, by reducing host-density for any species-specific pest or pathogen. In addition, smaller trees of one species may be less vulnerable to specific hosts when hidden by taller trees of another species (Castagneyrol et al.2014). Similar arguments for species diversity at small spatial scales can be made in regards to other perturbations, as species differ in their sensitivity to wind, herbivory, and climatic stresses, such as drought or frost periods (for example, see the mixed species discussion in the *Economic Challenges* section below).

Natural regeneration with the consequent avoidance of the need for financial investments in artificial tree regeneration works well where competitive understory vegetation is sparse. Although dense stand conditions prior to regeneration cutting can reduce the amount of understory vegetation, numerous forest types in the world have understory vegetation that is highly shade tolerant and understory species that can respond very aggressively and competitively to canopy opening, even after partial harvests (e.g. Noguchi and Yoshida 2004; Donoso and Nyland2005; Nilsson and Wardle 2005; Bose et al. 2014; Dodson et al. 2014). In these forests, regeneration of desirable tree species (especially, but not limited to shade-intolerant species) may be poor or delayed for years or even decades (Hibbs and Bower 2001; Axelsson et al. 2014).

In summary, ecological constraints underline the imperative of allowing more flexibility in adherence to the five principles in order to provide more opportunities for broader applications of silvicultural alternatives. Based on local ecological conditions, it may be necessary to deviate or "stretch" the principles. For example, larger canopy openings or more intense weed-control practices than preferred under alternative silvicultural approaches may be necessary in selected settings to ensure regeneration of desired tree species. Finally, acknowledging the full diversity of natural disturbance and development patterns may put the role of alternative silvicultural practices in a larger perspective.

ECONOMIC CHALLENGES

Economic considerations are generally strong drivers of land-use decisions, as already made clear by Thünen's (von Thünen 1842) theory on the spatial distribution of forestry and agriculture. A strong focus on economics may not always apply to forestry, where ecological considerations, tradition, social acceptability, culture, lifestyle, and similar factors may dominate management decisions (Kimmins 1992) and long-term economics of decisions are not always clear. However, it is unlikely that alternative silvicultural strategies will be adopted on large scales if they are not financially compatible with economic sustainability.

Economic considerations include the expected profit, uncertainty, and flexibility of management alternatives. The outcome of such economic considerations strongly depends on the specific settings of the land ownership and the services included into the decision matrix (Amsalu et al. 2014). Any application of alternative silvicultural approaches has to be viewed in the context of other management options to consider the opportunity costs for the final silvicultural decisions appropriately. In many places, especially on larger private holdings in the tropics, subtropics, and temperate regions of the Southern Hemisphere, the context includes the economic expectations from short-rotation plantation forestry, generally with exotic tree species, and often aided by political and economic policies (e.g., Niklitschek 2007; Miller Klubock 2014). Even under optimistic expectations, alternative silvicultural approaches are unlikely to reach the short-term financial profitability of intensively managed short-rotation plantations in tropical regions or some temperate regions (Cubbage et al. 2007; Knoke and Huth 2011). In some settings, however, native tree species may compete with exotic tree species when managed in a short-rotation plantation system (Cubbage et al. 2007; Griess and Knoke 2011).

The economic picture for alternative silvicultural methods is more favorable when viewed in a longer-term context, such as typically used in temperate and boreal biomes. For example, Tahvonen et al. (2010) showed the importance of relying on natural regeneration to reduce regeneration costs as a critical criterion that allowed continuous-cover forestry to emerge as the option with the highest net present value in Finnish spruce forests, under nearly all combinations of assumptions used in the study (Note that the study assumed higher regeneration costs due to artificial regeneration in even-aged management regimes). Many other studies support the findings published by Tahvonen et al. (2010) in principle, for example Chang (1981); Haight (1987); Haight and Monserud (1990); Pukkala et al. (2010) and Hyytiäinen and Haight (2010). Another argument often presented in the discussion of economic outcomes of alternative silvicultural practices is that individual trees can be harvested at the time when they peak in terms of their economic values. This can increase the total net present value substantially in stands with an existing wide range of tree values (as determined by species, size, and wood quality) (Knoke 2012). The advantage of harvesting trees at their peak economic value - versus harvesting all trees in a stand at the same time - may be reduced or lost in stands where trees are very homogenous (also, see the discussion about harvesting costs and staffing needs). Thus, it is important to note that the economic attractiveness of alternative silviculture is conditional on the initial state in the short run (Tahvonen et al. 2010). This makes generalizations about comparisons between conventional and alternative silvicultural approaches not very useful for landowners who consider such options for a specific property. The influence of the initial state can be minimized by using a very long (at best an unlimited) time horizon in comparative economic analyses; a requirement that certainly poses additional challenges.

The economic comparison may also change towards alternative silviculture when these approaches result in an increased resilience to natural hazards, e.g., in mixed-species even-aged (Griess et al. 2012; Neuner et al. 2015) or uneven-aged stands (Hanewinkel et al. 2014). Alternative silvicultural approaches may also be more suitable for forest owners with concerns about economic risks, as the risk is lower in stands with higher resilience to disturbances and when timber harvesting is accomplished with multiple operations over time (Roessiger et al. 2013).

Finally, increased stand-level ecological stability in mixed and uneven-aged forests also enhances economic performance of forest operations through improved flexibility because harvest levels can be more easily adjusted to accommodate changes in market conditions (Knoke and Wurm 2006). Management flexibility may be of special importance in a more unpredictable future, e.g., with climate change (Jacobsen and Thorsen 2003; Wagner et al.2014) or novel pest incidences. Consequently, major issues in economic assessments of silvicultural approaches are the relative emphases on economic risks and management and marketing flexibility.

One argument often used to promote application of single-tree selection and also for alternative silvicultural approaches is the benefit of the production of high value, i.e., large diameter and high-quality trees (e.g., "Dutch cuttings", see Schütz et al. 2012) (often in conjunction with the decreased need for non-commercial thinning operations). As markets have shifted, and in regions where much of the wood-processing industry has re-tooled to efficiently process smaller logs, large logs now may have no premium (per unit of volume), and in some cases may even have price penalties (Fitzgerald et al. 2013). This issue will vary by species and may be alleviated by sawmilling innovation in response to a higher availability of predicted assortments from alternative silviculture. In the meantime, in many regions, the theoretically higher mill recovery from larger diameter logs is often offset by the faster speed when processing homogenous (in terms of size and quality) logs, especially in high-volume sawmills. Concerns also exist about lower log quality for trees that grow in highly variable conditions throughout their lives. This variability can exhibit itself in somewhat adverse branching patterns, including large branch sizes and higher variation in juvenile wood, wood density, and growth rings (Schütz 1999b; Macdonald et al. 2010; Piispanen et al.2014).

Harvesting costs per unit volume for similar sized trees are typically higher in partial harvesting operations, especially on steep slopes and in mixed-species stands when different species have to be sorted and marketed separately (Pausch 2005). These higher costs are partially due to reduced efficiency of logging when only small amounts are harvested in a single operation. Early adopters also may incur higher costs and time delays when hiring and training specialized logging contractors. The necessary investments into infrastructure, including the development and maintenance of road systems, may also be a financial barrier to managing alternative silviculture, especially for landowners with smaller properties. Similarly, compared to conventional approaches, the proper implementation of alternative silvicultural approaches may result in higher personnel costs due to the higher skill level required, as well as more frequent site visits to assess progress, mark trees, and develop and supervise the implementation of often more complicated management prescriptions. Another economic concern with partial harvesting may be the risk of physical damage to the retained growing stock. This issue is of special concern in difficult terrain and when large-crowned species are felled or species with little decay resistance are left behind (Nill et al. 2011).

In summary, the ability to rely on natural regeneration, the limited need for site preparation and weed-control operations, in conjunction with a higher proportion of larger, high quality trees at harvest time, are viewed as major economic advantages of alternative silvicultural approaches. Thus, ecological conditions that influence natural regeneration, such as seed production, herbivory levels, and competitive conditions in the understory are crucial factors that can determine the financial viability of alternative silvicultural approaches. Risk and flexibility considerations may also constitute strong arguments in favor of alternative silvicultural approaches. However, the economies of scale of industrialized short-rotation management may more than balance these advantages of alternative silviculture in many parts of the world in the short-term, where the production of large amounts of homogenous timber is the main aim of forestry.

Finally, ecosystem services that may be provided more efficiently in forests managed under alternative silvicultural approaches, e.g., visual quality, hydrological services, or wildlife habitat provision, are not necessarily adequately monetized or evaluated (e.g., Ribe 1999; Daniel 2001). Whether or not such values compensate for a potential loss in timber production after conversion to alternative silviculture regimes will vary with the specific situation and assumptions, e.g., soil quality and interest rates, respectively (Amsalu et al. 2014). In many situations, the application of alternative silviculture would only be economically viable through measures such as conservation payments (Cubbage et al. 2007), monetary incentives as part of forest certification efforts (Nasi et al. 2011), or direct payments for ecosystem services (Wunder 2006), such as water supply and regulation (Nahuelhual et al. 2007).

Logistical and Administrative Challenges

Efficient partial-harvesting operations may require highly trained loggers and specialized machinery. Both can be in scarce supply, particularly in regions where these practices are only emerging. In some areas, forest managers may be able to benefit from loggers with experience in selective harvesting in other settings, such as exploitation cuttings that selectively removed only the highest value trees. Using inexperienced loggers and unsuitable machinery that is not designed for partial harvest operations increases the probability of inadequate prescription application and harvesting damage to residual trees (Jones and Thomas 2004; Putz et al. 2008), both outcomes that can reinforce a perception that alternative silvicultural approaches are inoperable. The damage to retained trees can be very high also in regions with a longer tradition of uneven-aged forest management (Nill et al. 2011). Moreover, advanced regeneration is often cut to improve the visibility for machine operators during harvest procedures or damaged during skidding or forwarding operations (Newton and Cole 2006). In contrast, these concerns can be less prevalent in parts of the world where logging operations are not highly mechanized. For example, manual logging operations with extraction by horses or oxen can be done with relatively little damage to residual trees and regeneration (Bacardit 2014). However, there are notable exceptions, for example where logs are thrown downhill by forest workers through forests on steep slopes, as has been practiced in some parts of China (J. Bauhus, personal observations). Even in stands with little harvesting damage, sudden exposure after partial harvesting can lead to higher evaporative demands (Bladon et al. 2006) and unexpected mortality of residual tress (Yoshida et al. 2006; Aubry et al. 2009).

Health and safety considerations are always important during and after partial harvesting operations due to concerns about the stability of branches and retained trees (Schaetzl et al. 1989). The presence of dead branches or forked trunks leads to safety concerns, but can be taken into account when selecting residual trees. Past scarring, possibly during earlier harvest operations, can also provide indications of possible trunk rot and thus the likelihood of tree fall (Matlack et al. 1993). Training can also reduce workplace safety risks. For example, the injury rate dropped - rather than increased - when variable retention harvesting replaced clearcutting in coastal British Columbia, because of strong organizational focus on worker safety. In contrast, at a larger scale, forests with continuous cover are preferred because of their safety benefits for avalanche protection (Bebi et al. 2009) and protection from landslides (Dhakal and Sidle 2003; Noguchi et al. 2011).

Changing management practices to alternative silvicultural approaches requires the development of different expertise by existing staff and additional educational and training efforts (see below) or, possibly, restaffing. Applying repeated partial-harvesting operations with decisions made at the individual-tree level may initially require a larger workforce until new training programs take effect to reduce or offset the workload. Planning and monitoring processes and inventory systems may have to be modified to more efficiently consider and accurately track the development of tree and stand attributes for anticipated future treatments (Rice et al. 2014).

Alternative silvicultural benefits from continuity of ownership and supervisory management personnel who can oversee the consistent application of silvicultural prescriptions. For example, these approaches may only be feasible if concessions are agreed on beyond the scope of a single harvest to include repeated harvesting operations, and if concessionaires are held responsible for reforestation and other silvicultural operations. Especially on larger private ownerships and in areas with limited infrastructure, the supervision of logging activities is necessary to avoid unauthorized or accidental removal of higher-valued trees intended for retention. Furthermore, the higher diversity of forest conditions (e.g., tree-species combinations – literally dozens in tropical forests - and various stand structural conditions), and the associated lack of simple, standardized instructions and management protocols make it more difficult to cope with rapid turnover of staff. Even where scientific data and understanding already exist from long-term silvicultural trials, foresters often appear to heavily rely on their observations, intuition, and experience when applying alternative silvicultural approaches. Developing and fostering these skills in new personnel is challenging for forest administrators and land owners.

In summary, staffing challenges will require increased training efforts, e.g., in reduced impact logging (RIL, Tropical Forest Foundation 2008). New machinery and modification of harvesting practices may help overcome challenges when implementing alternative silvicultural practices. New technologies for planning and monitoring (e.g., GPS, GIS, remote sensing tools) will reduce the need for staff, but will require a more highly qualified workforce. Finally, the modification of contracting practices, such as involving concessions, may be necessary in many places to facilitate a transition to alternative silvicultural systems.

Informational and Educational Challenges

Over the last century, many educational programs in forestry schools and continuing professional education programs have focused primarily on conventional forestry practices and have often provided limited exposure to alternative management approaches such as selection systems (Schütz et al. 2012).

One reason for this has been the decline in the availability of experienced silvicultural professionals to impart this knowledge. Practical examples of alternative silviculture that can be used as demonstration projects for students and practitioners are still lacking in many regions. This is true especially in countries with developing economies, following the decline of previous colonial and overseas development forestry practices. Related to the lack of practical field examples is the limited capture of traditional ecological knowledge of some indigenous communities (Berkes et al. 2000), although this is changing in some regions. For example in the Maya region of Mexico, new silvicultural systems for the regeneration of *Swietenia macrophylla* were developed as a result of collaborations among forest scientists and local communities (Menzies 2007).

The amount and type of scientific information available to support alternative silviculture reflect the bottom-up development of these approaches. In contrast to conventional forestry operations, which are based on the agricultural model or efficiency paradigm (Puettmann et al. 2009), alternative silvicultural approaches have not received the same level of attention by scientists. Studies that document the long-term response to alternative silvicultural treatments in terms of the full suite of ecosystem goods and services are lacking. Instead, in many instances, foresters adopted these approaches after observing trends in their own forests or visiting forests on which alternative silviculture had already been successfully applied. Thus, the spread of silvicultural alternatives is related to the availability of examples as well as to scientific progress. The limited awareness and utilization of scientific studies covering relevant aspects, such as the physiological basis for seedling and sapling responses to overstory removals or natural regeneration of species mixtures (Puettmann and Ammer 2006), remain a constraint in many regions, particularly in the tropics. New field installations, simulation exercises (Yasuda et al. 2013), and other research programs are currently beginning to provide valuable results (e.g., Thorpe et al. 2010; Neyland et al.2012). At this time, organizations and forest managers may be (understandably) reluctant to take risks and accept uncertainties associated with trying out unfamiliar management approaches without being able to rely on extensive scientific literature and empirical results from field experiences. This is likely one of the major limitations for a more rapid spread of alternative silviculture.

Just as for forest managers, training opportunities are often lacking (or too expensive) for loggers to obtain information and experience with these new management systems. Successful logging operations require a workforce with a good understanding of the reasons for and basic principles behind different silvicultural treatments. Harvesting operators may also have limited access to innovative machinery and techniques suitable for partial harvesting operations that address small-scale variability in stand structure, are able to deal with a wider variety of species, tree sizes, and qualities, and are designed to protect retained trees (including natural regeneration and site and soil conditions).

Cultural and Historical Challenges

Schütz et al. (2012) wrote that "willingness" (defined as "readiness to do something" by Merriam-Webster 2014) is one of the main factors limiting the wider application of alternative management regimes. In addition to the reasons for the lack of "readiness" that have been discussed already, the role of strong intellectual, political, and administrative leadership is especially important in shaping public and professional opinions (Brukas and Weber2009). The degree to which an organizational leadership champions alternative approaches influences the degree to which employees and contractors embrace and apply the practices. Strong support from upper management encourages employees and contractors to overcome operational challenges, as well as prejudices, enabling successful transition to alternative silviculture (Baker 2011). Other reasons may have a deeper, historical background. For example, the expansion of "scientific forestry" (*sensu* Oosthoek 2007; see also Lowood 1990), reflected in controlled management of homogenous, even-aged, mono-cultural plantations from central Europe at the end of the 19th century, was accepted in other countries because it fitted prominent economic ideas about control and efficiency (Lang and Pye 2001; Oosthoek 2007; Cock 2008; Puettmann et al. 2009; Sears and Pinedo-Vasquez 2011). This approach also helped to elevate forestry as an autonomous discipline at universities and research institutions, which facilitated its professional recognition by foresters and society in general. This historical development has led to several phenomena that are still influential today. The resulting expectation of simple, practical operational guidelines cannot easily be met by alternative silvicultural approaches, especially when they emphasize greater silvicultural flexibility (and outcomes) in a wide diversity of forest structures and species compositions. Moreover, the educational focus on conventional forestry has led to "imprinting" or "knowledge lock-in" in many practicing foresters who may consider only the predominant management model as acceptable. An example of this can be found in the management history of beech and black pine forests in the mountains of southern Italy, where selection felling based on traditional local knowledge and applied by private landowners during the last few centuries has been and often still is considered "irrational" by forest administrations and technicians (Ciancio et al. 2006; Nocentini 2009). Thus, the professional and public perceptions of alternative silvicultural approaches may have a major influence on the development of policies and implementation strategies (Brukas and Weber 2009).

Forest policies and practices can still be biased against continuous-forest-cover systems. For example, in Finland and Sweden, forest authorities and professionals have actively encouraged clearcut operations because selective cutting methods applied in the past led to political concerns that private forest owners were devastating their forests (Siiskonen 2007; Brukas and Weber 2009). In 1948, a group of influential Finnish researchers published a declaration against the destructive impacts of selective logging on Finland's forests. Similarly, influential German forestry professors were successful in discrediting the "Dauerwald" (continuous-cover forests) efforts, mainly by stating concerns about the economic performance and management complexity (Pommerening and Murphy 2004). These arguments may exacerbate an existing lack of incentives or desire to switch to alternative silvicultural approaches, given that they are still brought up in current discussions of new laws and regulations.

The "search for simplicity" in silviculture is often also encouraged by staff shortages and overly complex, computerized desk-work environments, where foresters have to deal with multifarious and new, distracting aspects. The reluctance to try more complicated silvicultural prescriptions is also partly the result of highly visible failures of alternative silviculture experiences in some regions. For example, early partial-harvest trials in old-growth *Pseudotsuga menziesii* stands in the Pacific Northwest were pronounced a failure in the 1950s. As noted later by Curtis (1998), "partial cutting trials came to an abrupt end", leading to "the consequent lack of research into alternatives to clearcutting", which "severely handicaps current efforts to meet changing objectives and public concerns". Similarly, "European" silvicultural systems and practices were initially implemented in Australia (and other places around the globe) without an adequate understanding of the ecological functioning of the local ecosystems (Florence 1996). For a long time, silvicultural alternatives were only applied on selected properties, which were viewed as "unique" and in some cases even belittled, making it harder for a broader set of professionals to associate themselves with these practices.

Decisions about management activities are typically the domain of landowners acting within their local ecological, social, and economic contexts. Thus, it is not enough to educate foresters and contractors; it is also necessary to convince private landowners and politicians who have decision power over public land use and policies that impact private landowners. The attitudes of these decision makers regarding alternative silvicultural regimes may be influenced by a variety of factors, including their knowledge and prioritization of desired ecosystem goods and services, as well as the anticipated revenue constraints due to size of the estate. For example, large public or investor-owned properties can afford to spatially separate the production of various forest products or ecosystem services over large areas. This allows them to homogenize portions of their forest holdings for the efficient production of selected forest products or services. In contrast, landholders who manage small and/or fragmented properties may need to obtain a variety of products from their property. In Finland, landowners with small forest land bases often obtain logs of different sizes for firewood or for construction wood and at the same time may harvest berries, mushrooms, and game (Siiskonen 2007). Similarly, in Latin America, campesinos (rural smallholders) may work under the logic of a diversified production to mitigate potential risks. For example, they collect firewood, produce charcoal, sell animals, and produce apple cider—activities that also generate income that may be complemented with off-farm income (Moorman et al. 2013). Alternative silvicultural regimes are typically more suitable in such settings where integrated land use is important because their emphasis on smaller spatial scales facilitates simultaneous production of various ecosystem goods and services (Hein et al. 2006).

The regional or national context also influences decisions about silvicultural approaches. For example, high human population densities may increase the importance of providing high-amenity recreational settings as an ecosystem service, leading to alternative silvicultural approaches that are more visually appealing (Rydberg and Falck 2000). Similarly, mountainous topographies may lead to an emphasis on protection from landslides, e.g., in Japan (Noguchi et al. 2011) and Italy (see discussion about safety above). On the other hand, a well-established and economic and politically powerful forest industry sector can influence the availability of wood processing facility and thus markets for forest products (Miller Klubock 2014). This may limit or bias market accessibility towards products (tree species, sizes, qualities) derived from industrial forest lands through conventional management, effectively reducing the value of a variety of products that could come from forests managed under alternative regimes (for example, see the above discussion about the impact of log sizes on wood value). A strong forest industry sector may also directly or indirectly influence the focus of educational and research themes according to their management interests (Benner and Sandström 2000).

A long history of forest management also means that private and public ownerships are organized and staffed to accommodate conventional approaches, making any transition more difficult. Furthermore, some of the challenges mentioned above may actually be legally embedded in forest policy and regulations. For example, well-intentioned reforestation regulations may specify acceptable tree species and required seedling densities, effectively limiting the choice of species and discouraging natural regeneration (for example, see the Oregon Forest Practices Act, Oregon Department of Forestry 2014). It should be acknowledged that not only environmental conditions, silvicultural options, and markets determine the implementation of alternative management methods, but traditions and stakeholders' views also affect ongoing political discourses (Arts et al. 2010).

CONCLUSIONS

In summary, the diversity of settings in which alternative silvicultural approaches are of interest reflects the diversity of challenges that limit their wider implementation. The range of arguments presented here suggests that ecological, economic, logistical and administrative, educational and informational, and cultural and historical challenges are all important. The relative importance of any of these challenges varies among and within geographical regions and ownerships. We suggest that these challenges can be addressed at two levels. At a lower level, each situation can be analyzed for the specific major challenges or limitations. Any such assessment of challenges benefits from viewing alternative regimes as a collection of silvicultural principles. Knowing which principle specifically is related to the major challenge in the situation at hand provides detailed insights that will facilitate discussion of whether that specific principle is necessary or can be modified or ignored, while still achieving overall management objectives. For example, foresters often will forgo natural regeneration if they foresee regeneration failures, and they will plant seedlings—even in close-to-nature forest operations. Having such detailed assessments will allow interested groups, for example landowners or politicians, to initiate efforts to implement policies, procedures, and/or incentives to help overcome these challenges. Examples could include new laws or regulations, subsidies, and scientific and educational efforts that are targeted to specific situations. Although this may be successful in individual settings, it may be an inefficient process for tackling overall problem areas.

At a broader level, our overview provides some general insights that suggest opportunities for more coordinated efforts. It appears obvious that in this context, efforts that allow foresters to take advantage of the widest possible array of silvicultural tools to accommodate a wide variety of local ecological, economic, and social conditions are especially useful. This would be facilitated by a more formal shift of alternative silvicultural approaches away from the focus on desired or acceptable stand structures, often defined as specific diameter distributions or other visual targets, such as photos or sketches of desired stand structures (Larsen and Nielsen 2007) (Figure 2). Instead, a more targeted focus on the suite of desired ecosystem goods and services will allow foresters to integrate a broader set of disturbances, stand dynamics, and successional trends into their management practices. Such a focus will allow more flexibility in the future and encourage practices to be "customized" to optimally fit specific ecological, economic, and social settings without the confining restrictions of labels.

For further progress, especially to overcome information deficits and educational, economic, cultural, and historical challenges, the spread of alternative management approaches would benefit greatly from the continued development of a scientific basis. This basis includes investigations of specific detailed ecological, social, economic, and ethical issues. In addition, developing a larger-scale scientific framework that provides more explicit linkages between basic scientific theories and principles and practical applications of alternative silvicultural approaches will also facilitate progress by allowing better coordination—including internationally—of research and educational activities. This will facilitate an understanding of the distinctions between idiosyncrasies and regularities. It also will help provide answers to questions such as whether experiences are unique to specific situations or how far experiences in one forest can be extrapolated into other settings—questions that were already at the heart of the Dauerwald discussions almost 100 years ago. Furthermore, an expanded scientific basis will be very helpful in developing a coherent and efficient educational and—maybe most important—outreach and policy agendas. Finally, great economic challenges remain for the creation of economic modeling approaches that are more biologically realistic and thus capture the ecological complexity of alternative silvicultural management approaches.

While not a formal review of all pertinent literature, the expertise represented by the authors covering a variety of regions and ecological and social settings allowed us to build a comprehensive framework to provide hypotheses and suggest opportunities to advance the implementation of alternative silvicultural regimes. However, we acknowledge that experiences in specific settings may differ or complement the ones provided by us. We hope that our arguments and such differences will provide the basis for further fruitful discussion.

DECLARATIONS

Acknowledgements

This material is based upon work supported by the National Institute of Food and Agriculture, United States Department of Agriculture, McIntire-Stennis Funds under ID number OREZ-FES-850-P to the lead author and by the Edmund Hayes Professorship. PJ Donoso acknowledges the support from FONDECYT Grant No 1110744. Jürgen Bauhus received funding for this work from the European Union Seventh Framework Programme (FP7/2007-2013) under grant agreement no. 265171. The authors would like to acknowledge many colleagues, practitioners, and students for their contributions.

Competing interests

The authors declare that they have no competing interests.

AUTHORS' CONTRIBUTIONS

KP initiated the work and coordinated the contributions from the co-authors. SW, SB, PD, LD, GA, BH, TK, YL, SN, FP, TY, and JB provided input into the analytical structure and contributed to writing and revising the manuscript. All authors read and approved the final manuscript.

REFERENCES

1. Amsalu DW, Jacobsen JB, Lundhede TH (2014) Economic assessment of use values of near-natural forestry compared with rotational forestry in Denmark. Eur J For Res 133(4):611–622
2. Arts B, Appelstrand M, Kleinschmit D, Pülzl H, Vissen-Hamakers I (2010) Discourses, actors and instruments in international forest governance. IUFRO World Ser 28:57–73
3. Attiwill PM (1994) The disturbance of forest ecosystems - the ecological basis for conservative management. Forest Ecol Manag 63:247–300
4. Aubry KA, Halpern CB, Peterson CE (2009) Variable-retention harvests in the Pacific Northwest: a review of short-term findings from the DEMO study. Forest Ecol Manag 258:398–408
5. Axelsson EP, Lundmark T, Högberg P, Nordin A (2014) Belowground competition directs spatial patterns of seedling growth in boreal pine forests in Fennoscandia. Forests 5(9):2106–2121
6. Bacardit P (2014) Evaluación del daño provocado por cortas de selección sobre los árboles residuales en un bosque del Tipo Forestal Siempreverde en Llancahue, provincia de Valdivia. Undergraduate Thesis, Faculty of Forest Sciences and Natural Resources, Universidad Austral de Chile, Valdivia, Chile
7. Baker SC (2011) Seeking a Balance Between Forestry and Biodiversity – The Role of Variable Retention Silviculture. Insights from Western USA and Canada. FWPA Project Report PG D167–0910. FWPA, Melbourne, p 60
8. Baker SC, Read SM (2011) Variable retention silviculture in Tasmania's wet forests: ecological rationale, adaptive management and synthesis of biodiversity benefits. Austral For 74:218–232
9. Barbeito I, Pardos M, Calama R, Cañellas I (2008) Effect of stand structure on Stone pine (*Pinus pinea* L.) regeneration dynamics. Forestry 81:617–629
10. Bauhus J, Schmerbeck J (2010) Silvicultural Options to Enhance and Use Forest Plantation Biodiversity. In:

Bauhus J, van der Meer P, Kanninen M (eds) Ecosystem Goods and Services from Plantation Forests. Earthscan, London, pp 96–139

11. Bauhus J, Puettmann KJ, Kühne C (2013) Close-to-nature forest management in Europe: does it support complexity and adaptability of forest ecosystems? In: Messier C, Puettmann KJ, Coates KD (eds) Managing forests as complex adaptive systems: building resilience to the challenge of global change. Routledge, New York, pp 187–213
12. Bebi P, Kulakowski D, Rixen C (2009) Snow avalanche disturbances in forest ecosystems—state of research and implications for management. Forest Ecol Manag 257(9):1883–1892
13. Benner M, Sandström U (2000) Institutionalizing the triple helix: research funding and norms in the academic system. Res Policy 29:291–301
14. Bergeron Y, Leduc A, Harvey B, Gauthier S (2002) Natural fire regime: a guide for sustainable forest management of the Canadian boreal forest. Silva Fennica 36:81–95
15. Berkes F, Colding J, Folke C (2000) Rediscovery of traditional ecological knowledge as adaptive management. Ecol Appl 10:1251–1262
16. Biolley H (1901) Le jardinage cultural. J Forestier Suisse 52:67–104
17. Bladon KD, Silins U, Landhäusser SM, Lieffers VJ (2006) Differential transpiration by three boreal tree species in response to increased evaporative demand after variable retention harvesting. Ag For Meteorol 138:104–119
18. Bliss JC (2000) Public perceptions of clearcutting. J Forestry 98:4–9
19. Bose AK, Harvey BD, Brais S, Beaudet M, Leduc A (2014) Constraints to partial cutting in the boreal forest of Canada in the context of natural disturbance-based management: a review. Forestry 87(1):11–28
20. Brang P, Spathelf P, Larsen JB, Bauhus J, Boncina A, Chauvin C, Drössler L, Garcia-Guemes C, Heiri C, Kerr G, Lexer MJ, Mason B, Mohren F, Mühlethaler U, Nocentini S, Svoboda M (2014) Suitability of close-to-nature silviculture for adapting temperate European forests to climate change. Forestry 87(4):492–503, doi:10.1093/forestry/cpu018
21. Brukas V, Weber N (2009) Forest management after the economic transition—at the crossroads between German and Scandinavian traditions. For Pol Econ 11:586–592
22. Brunson MW, Steel BS (1994) National public attitudes toward federal rangeland management. Rangelands 16:77–81
23. Butler BJ, Leatherberry EC (2004) America's family forest owners. J Forestry 102:4–14
24. Calama R, Barbeito I, Pardos M, del Rio M, Montero G (2008) Adapting a model for even-aged *Pinus pinea* L. stands to complex multi-aged structures. Forest Ecol Manag 256:1390–1399
25. Cameron A (2002) Importance of early selective thinning in the development of long-term stand stability and improved log quality: a review. Forestry 75(1):25–35
26. Castagneyrol B, Jactel H, Vacher C, Brockerhoff EG, Koricheva J (2014) Effects of plant phylogenetic diversity on herbivory depend on herbivore specialization. J Appl Ecol 51(1):134–141
27. Chang SJ (1981) Determination of the optimal growing stock and cutting cycle for an uneven-aged stand. Forest Sci 27:739–744
28. Ciancio O, Nocentini S (2011) Biodiversity conservation and systemic silviculture: concepts and applications. Plant Biosyst 145:411–418
29. Ciancio O, Cutini A, Mercurio R, Veracini A (1986) Sulla struttura della pineta di pino domestico di Alberese. Annali dell'Istituto Sperimentale della Selvicoltura, Arezzo, p 17
30. Ciancio O, Iovino F, Menguzzato G, Nicolaci A, Nocentini S (2006) Structure and growth of a small group selection forest of Calabrian pine in Southern Italy: a hypothesis for continuous cover forestry based on traditional silviculture. Forest Ecol Manag 224:229–234
31. Cock AR (2008) Tropical forests in the global states system. Int Aff 84(2):315–333
32. Cubbage F, MacDonagh P, Sawinski J Jr, Rubilar R, Donoso PJ, Ferreira A, Hoeflich V, Morales V, Ferreira G, Balmelli G, Jacek Siry M, Báez N, Alvarez J (2007) Timber investment returns for selected plantations and native forests in South America and the Southern United States. New For 33(3):237–255
33. Curtis RO (1998) "Selective cutting" in Douglas-fir - History revisited. J Forestry 96:40–46
34. Daniel TC (2001) Whither scenic beauty? Visual landscape quality assessment in the 21st century. Landsc Urban Planning 54:267–281
35. Dhakal AS, Sidle RC (2003) Long-term modelling of landslides for different forest management practices. Earth Surf Proc Land 28:853–868
36. Dodson EK, Burton JI, Puettmann KJ (2014) Multi-scale controls on natural regeneration dynamics after partial overstory removal in Douglas-fir forests in western Oregon, USA. Forest Sci 60(5):953–961
37. Donoso PJ, Nyland RD (2005) Seeding density according to structure, dominance and understory cover in old-growth forest stands of the evergreen forest type in the coastal range of Chile. 2005. Rev Chil Hist Nat 78(1):51–63
38. Donoso PJ, Promis A (2013) Silvicultura en Bosques Nativos. Avances en la investigación en Chile, Argentina y Nueva Zelandia. Estudios en Silvicultura de Bosques Nativos, Vol. 1. Ed. Marisa Cuneo, Valdivia, Chile. p 253.https://sites.google.com/site/alvaropromis/Home/libro-silvicultura-bosques-nativos.
39. Drever CR, Peterson G, Messier C, Bergeron Y, Flannigan M (2006) Can forest management based on natural disturbances maintain ecological resilience? Can J For Res 36:2285–2299

40. Drössler L, Nilsson U, Lundqvist L (2013) Simulated transformation of even-aged Norway spruce stands to multi-layered forests: an experiment to explore the potential of tree size differentiation. Forestry 87(2):239–248, doi:10.1093/forestry/cpt037

41. Duncker PS, Barreiro SM, Hengeveld GM, Lind T, Mason WL, Ambrozy S, Spiecker H (2012) Classification of forest management approaches: a new conceptual framework and its applicability to European forestry. Ecol Soc 17(4):51,http://dx.doi.org/10.5751/ES-05262-170451

42. Dvorák L, Bachmann P, Mandallaz D (2001) Sturmschäden in ungleichförmigen Beständen. Schweiz Zeitschr Forstw 152:445–452

43. Evans Z (2006) What is ecological forestry?.http://www.forestguild.org/ecological_forestry/Ecological_Forestry_evans_06.pdf. Accessed 10 November 2014

44. FAO (Food and Agriculture Organization of the United Nations) (2010) Global forest resources assessment 2010. Food and Agriculture Organization of the United Nations, Rome

45. Fedrowitz K, Koricheva J, Baker DC, Lindenmayer DB, Palik B, Rosenvald R, Beese W, Franklin JF, Kouki J, Macdonald E (2014) Can retention forestry help conserve biodiversity? A meta-analysis. J Appl Ecol 51(6):1669–1679, doi:10.1111/1365-2664.12289 Central

46. Filotas E, Parrott L, Burton PJ, Chazdon RL, Coates KD, Coll L, Haeussler S, Martin K, Nocentini S, Puettmann KJ, Putz FE, Simard SW, Messier C (2014) Viewing forests through the lens of complex systems science. Ecosphere 5:1,http://dx.doi.org/10.1890/ES13-00182.1

47. Fitzgerald SA, Oester PT, Parker R (2013) Individual tree selection (ITS) in a northeast Oregon mixed conifer forest. Extension Service, Oregon State University, Corvallis, OR, USA

48. Florence RG (1996) Ecology and silviculture of eucalypt forests. CSIRO, Melbourne, p 400

49. Franklin JF, Forman TT (1987) Creating landscape patterns by forest cutting: ecological consequences and principles. Landscape Ecol 1:5–18

50. Franklin JF, Johnson KN (2012) A restoration framework for federal forests in the Pacific Northwest. J Forestry 110:429–439

51. Franklin JF, Mitchell RJ, Palik BJ (2007) Natural disturbance and stand development principles for ecological forestry, General Technical Report NRS-19. United States Department of Agriculture, Forest Service, Northern Research Station, Newtown Square, PA

52. Gayer K (1886) Der gemischte Wald, seine Begrűndung und Pflege, insbesondere durch Horst-und Gruppenwirtschaft. Parey Verlag, Berlin

53. Griess VC, Knoke T (2011) Can native tree species plantations in Panama compete with Teak plantations? An economic estimation. New For 41:13–39

54. Griess VC, Acevedo R, Härtl F, Staupendahl K, Knoke T (2012) Does mixing tree species enhance stand resistance against natural hazards? A case study for spruce. Forest Ecol Manag 267:284–296

55. Gurnaud A (1882) Le controle et le regime forestier. Rev Eaux For 21:1–23

56. Gustafsson L, Kouki J, Sverdrup-Thygeson A (2010) Tree retention as a conservation measure in clear-cut forests of northern Europe: a review of ecological consequences. Scand J For Res 25:295–308

57. Gustafsson L, Baker SC, Bauhus J, Beese WJ, Brodie A, Kouki J, Lindenmayer DB, Lõhmus A, Martínez Pastur G, Messier C, Neyland M, Palik B, Sverdrup-Thygeson A, Volney WJA, Wayne A, Franklin JF (2012) Retention forestry to maintain multifunctional forests: a world perspective. Bioscience 62:633–645

58. Haight R (1987) Evaluating the efficiency of even-aged and uneven-aged stand management. Forest Sci 33:116–134

59. Haight RG, Monserud RA (1990) Optimizing any-aged management of mixed-species stands: II. Effects of decision criteria. Forest Sci 36:125–144

60. Hanewinkel M (2002) Comparative economic investigations of uneven-aged and uneven-aged silvicultural systems: a critical analysis of different methods. Forestry 75:473–481

61. Hanewinkel M, Kuhn T, Bugmann H, Lanz A, Brang P (2014) Vulnerability of uneven-aged forests to storm damage. Forestry 87:525–534, doi:10.1093/forestry/cpu008

62. Harlow RF, Downing R, Van Lear D (1997) Responses of wildlife to clearcutting and associated treatments in the Eastern United States, Technical Paper no. 19. Department of Forest Resources Clemson University, Clemson, SC

63. Hein L, van Koppen K, De Groot RS, van Ierland EC (2006) Spatial scales, stakeholders and the valuation of ecosystem services. Ecol Econ 57(2):209–228

64. Hibbs DE, Bower AL (2001) Riparian forests in the Oregon Coast Range. Forest Ecol Manag 154:201–213

65. Higgs E, Falk DA, Guerrini A, Hall M, Harris J, Hobbs RJ, Jackson ST, Rhemtulla JM, Throop W (2014) The changing role of history in restoration ecology. Front Ecol Environ 12:499–506

66. Hugosson M, Ingemarson F (2004) Objectives and motivations of small-scale forest owners; theoretical modelling and qualitative assessment. Silva Fennica 38:217–231

67. Hyvaerinen E, Kouki J, Martikainen P (2006) Fire and green-tree retention in conservation of red-listed and rare deadwood-dependent beetles in Finnish boreal forests. Conserv Biol 20:1710–1719

68. Hyytiäinen K, Haight RG (2010) Evaluation of forest management systems under risk of wildfire. European J Forest Res 129:909–919

69. Jacobsen MK (2001) History and principles of close to nature forest management: a central European perspective. Naconex 3:56–58

70. Jacobsen JB, Thorsen BJ (2003) A Danish example of optimal thinning strategies in mixed-species forest under changing growth conditions caused by climate change. Forest Ecol Manag 180(1):375–388

71. Jones TA, Thomas SC (2004) The time course of diameter increment responses to selection harvests in *Acer saccharum*. Can J For Res 34:1525–1533

72. Kimmins H (1992) Balancing act: environmental issues in forestry. UBC Press, Vancouver, BC

73. Kneeshaw D, Williams H, Nikinmaa E, Messier C (2002) Patterns of above- and below-ground responses of understory conifer release 6 years after partial cutting. Can J For Res 32:255–265

74. Knoke T (2012) The economics of continuous cover forestry. In: Pukkala T, Gadow Kv (eds) Continuous cover forestry. Managing forest ecosystems Vol. 23, pp 167–193. doi:10.1007/978-94-007-2202-6_5

75. Knoke T, Huth A (2011) Modelling forest growth and finance: Often disregarded tools in tropical land management. In: Günter S, Weber M, Stimm B, Mosandl R (eds) Silviculture in the tropics Tropical Forestry 8, part 3, pp 129–142. doi:10.1007/978-3-642-19986-8_11. Springer Verlag, Heidelberg.

76. Knoke T, Wurm J (2006) Mixed forests and a flexible harvest policy: a problem for conventional risk analysis? Eur J For Res 125(3):303–315

77. Knoke T, Moog M, Plusczyk N (2001) On the effect of volatile stumpage prices on the economic attractiveness of a silvicultural transformation strategy. Forest Pol Econ 2(3):229–240

78. Kohm KA, Franklin JF (1997) Creating a forestry for the 21st century: the science of ecosystem management. Island Press, Washington, DC

79. Kouki J, Hyvarinen E, Lappalainen H, Martikainen P, Simila M (2012) Landscape context affects the success of habitat restoration: large-scale colonization patterns of saproxylic and fire-associated species in boreal forests. Divers Distrib 18:348–355

80. Kuuluvainen T (2009) Forest management and biodiversity conservation based on natural ecosystem dynamics in northern Europe: the complexity challenge. AMBIO 38:309–315

81. Lang C, Pye O (2001) Blinded by science: the invention of scientific forestry and its influence in the Mekong Region. Watershed 6:25–34

82. Larsen JB, Nielsen AB (2007) Nature-based forest management—where are we going?: laborating forest development types in and with practice. Forest Ecol Manag 238:107–117

83. Lindenmayer DB, Franklin JF, Lõhmus A, Baker SC, Bauhus J, Beese W, Brodie A, Kiehl B, Kouki J, Pastur GM, Messier C, Neyland M, Palik B, Sverdrup-Thygeson A, Volney J, Wayne A, Gustafsson L (2012) A major shift to the retention approach for forestry can help resolve some global forest sustainability issues. Conserv Lett 5:421–431

84. Little C, Lara A, McPhee J, Urrutia R (2009) Revealing the impact of forest exotic plantations on water yield in large scale watersheds in South-Central Chile. J Hydrol 374:162–170

85. Long JN (2009) Emulating natural disturbance regimes as a basis for forest management: a North American view. Forest Ecol Manag 257:1868–1873

86. Lowood HE (1990) The calculating forester: quantification, cameral science, and the emergence of scientific forestry management in Germany. In: Frängsmyr T, Heilbron JL, Rider RE (eds) The quantifying spirit in the eighteenth century. University of California Press, Berkeley, http://ark.cdlib.org/ark:/13030/ft6d5nb455/

87. Lu Y, Liu X, Lei X, Wang H, Hong L, Guo H (2013) Development of silvicultural models for multi-functional forest management: the systematic consideration and application in experiment center of tropical forestry, Southwestern China. In: Fehrmann L, Kleinn C (eds) Forests in climate change research and police: the role of forest management and conservation in a complex international setting. Proceedings of the 3rd International DAAD workshop. Cuvillier Verlag Goettingen, Germany.

88. Lüpke B, Hauskeller-Bullerjahn M (1999) Kahlschlagfreier Waldbau: wird die Eiche an den Rand gedrängt? Forst und Holz 54:363–368

89. Lutz JA, Halpern CB (2006) Tree mortality during early forest development: a long-term study of rates, causes, and consequences. Ecol Monogr 76:257–275

90. Macdonald E, Gardiner B, Mason W (2010) The effects of transformation of even-aged stands to continuous cover forestry on conifer log quality and wood properties in the UK. Forestry 83:1–16

91. Manolis JC, Andersen DE, Cuthbert FJ (2000) Patterns in clearcut edge and fragmentation effect studies in northern hardwood-conifer landscapes: retrospective power analysis and Minnesota results. Wildlife Soc B 28:1088–1101

92. Marten GG (2005) Environmental tipping points: a new paradigm for restoring ecological security. J Policy Studies 20:75–87

93. Mason B, Kerr G, Pommerening A, Edwards C, Hale S, Ireland D, Moore R (2003) Continuous cover forestry in British conifer forests. Forest Res Annu Rep Acc 2004:38–53

94. Matlack GR, Gleeson SK, Good RE (1993) Treefall in a mixed oak-pine coastal plain forest: immediate and historical causation. Ecology 82:1559–1566

95. Menzies NK (2007) Our forests, your ecosystem, their timber. Communities, conservation and the state in community-based forest management. Columbia University Press, New York

96. Merriam-Webster (2014) Merriam Webster Dictionary. http://www.merriam-webster.com/. Accessed 1 May 2014

97. Messier C, Puettmann KJ, Coates D (2013) Managing forests as complex adaptive systems: building resilience to the challenge of global change. Routledge, New York

98. Miller Klubock T (2014) La Frontera: forests and ecological conflicts in Chile's frontier territory. Duke University Press, Durham, NC

99. Möller A (1923) Der Dauerwaldgedanke. Erich Degreif Verlag, Oberteuringen

100. Moorman M, Nelson S, Moore S, Donoso P (2013) Stakeholder perspectives on adaptive co-management as a Chilean conservation management strategy. Soc Natur Resour 26:1022–1036

101. Nahuelhual L, Donoso PJ, Lara A, Núñez D, Oyarzún C, Neira E (2007) Valuing ecosystem services of Chilean temperate rainforests. Environ Dev Sustainabil 9:481–499

102. Nasi R, Putz FE, Pacheco P, Wunder S, Anta S (2011) Sustainable forest management and carbon in tropical Latin America: the case for REDD+. Forests 2:200–217

103. Neumann FG, Minko G (1981) The sirex wood wasp in Australian radiata pine plantations. Aust For 44:46–63

104. Neuner S, Albrecht A, Cullmann D, Engels F, Griess VC, Hahn A, Hanewinkel M, Härtl F, Kölling C, Staupendahl K, Knoke T (2015) Survival of Norway spruce remains higher in mixed stands under a dryer and warmer climate. Global Change Biol 21:935–946

105. Newton M, Cole EC (2006) Harvesting impacts on understory regeneration in two-storied Douglas-fir stands. West J Appl For 21:14–18

106. Neyland M, Hickey J, Read SM (2012) A synthesis of outcomes from the Warra Silvicultural Systems Trial, Tasmania, Australia: safety, timber production, economics, biodiversity, silviculture and social acceptability. Aust For 75:147–162

107. Niklitschek ME (2007) Trade liberalization and land use changes: explaining the expansion of afforested land in Chile. Forest Sci 53(3):385–394

108. Nill M, Kohnle U, Sauter U (2011) Rindenschäden mit mutmaßlichem Bezug zur Holzernte im Spiegel der Betriebsinventuren in Baden-Württemberg. Forstarchiv 9(6):216–224

109. Nilsson M-C, Wardle DA (2005) Understory vegetation as a forest ecosystem driver: evidence from the northern Swedish boreal forest. Front Ecol Environ 3:421–428

110. Ninemets Ü, Valladares F (2006) Tolerance to shade, drought, and waterlogging of temperate Northern Hemisphere trees and shrubs. Ecol Monogr 76:521–547

111. Nocentini S (2009) Structure and management of beech (*Fagus sylvatica* L.) forests in Italy. iForest 2:105–113,http://www.sisef.it/iforest/show.php?id=499

112. Noguchi M, Yoshida T (2004) Tree regeneration in partially cut conifer-hardwood mixed forests in northern Japan: roles of establishment substrate and dwarf bamboo. Forest Ecol Manag 190:335–344

113. Noguchi M, Okuda S, Miyamoto K, Itou T, Inagaki Y (2011) Composition, size structure and local variation of naturally regenerated broadleaved tree species in Hinoki cypress plantations: a case study in Shikoku, southwestern Japan. Forestry 84:493–504

114. O'Hara KL, Ramage BS (2013) Silviculture in an uncertain world: utilizing multi-aged management systems to integrate disturbance. Forestry 86:401–410

115. Oosthoek J (2007) The colonial origins of scientific forestry in Britain. Essay in: Environmental History Resources.http://www.eh-resources.org/colonial_forestry.html. Accessed 22 February 2015.

116. Oregon Department of Forestry (2014) Oregon forest practices act.http://www.oregon.gov/ODF/privateforests/pages/fpareforestation.aspx. Accessed 1 May 2014

117. Paquet J, Bélanger L (1997) Public acceptability thresholds of clearcutting to maintain visual quality of boreal balsam fir landscapes. Forest Sci 43:46–55

118. Paquette A, Messier C (2009) The role of plantations in managing the world's forests in the Anthropocene. Front Ecol Environ 8:27–34

119. Pausch R (2005) Ein System-Ansatz zur Darstellung des Zusammenhangs zwischen Waldstruktur, Arbeitsvolumen und Kosten in naturnahen Wäldern Bayerns. Forstliche Forschungsberichte München 199

120. Piispanen R, Heinonen J, Valkonen S, Mäkinen H, Lundqvist S-O, Saranpää P (2014) Wood density of Norway spruce in uneven-aged stands. Can J For Res 44(2):136–144

121. Pommerening A, Murphy S (2004) A review of the history, definitions and methods of continuous cover forestry with special attention to afforestation and restocking. Forestry 77:27–44

122. ProSilva Europe (2014) ProSilva Europe: integrated forest management for resilience and sustainability across 25 countries. http://prosilvaeurope.wordpress.com. Accessed 13 August 2014

123. Puettmann KJ (2011) Silvicultural challenges and options in the context of global change: simple fixes and opportunities for new management approaches. J Forestry 109:321–331

124. Puettmann KJ, Ammer C (2006) Trends in North American and European regeneration research under the ecosystem management paradigm. Eur J For Res 126:1–9

125. Puettmann KJ, Ek AR (1999) Status and trends of silvicultural practices in Minnesota. North J Appl For 16:203–210

126. Puettmann KJ, Coates KD, Messier C (2009) A critique of silviculture: managing for complexity. Island Press, Washington, DC

127. Pukkala T, Lähde E, Laiho S (2010) Optimizing the structure and management of une-ven-sized stands of Finland. Forestry 83:129–142

128. Putz F, Sist P, Fredericksen T, Dykstra D (2008) Reduced-impact logging: challenges and opportunities. Forest Ecol Manag 256:1427–1433

129. Reich RM, Mielke PW Jr, Hawksworth FG (1991) Spatial analysis of ponderosa pine trees infected with dwarf mistletoe. Can J For Res 21:1808–1815

130. Ribe RG (1989) The aesthetics of forestry: what has empirical preference research taught us? Environ Manage 13:55–74

131. Ribe RG (1999) Regeneration harvests versus clearcuts: public views of the acceptability and aesthetics of Northwest Forest Plan harvests. Northw Sci 73:102–117

132. Ribe RG (2005) Aesthetic perceptions of green-tree retention harvests in vista views: the interaction of cut level, retention pattern and harvest shape. Landscape Urban Plan 73:277–293
133. Rice B, Weiskittel A, Wagner R (2014) Efficiency of alternative forest inventory methods in partially harvested stands. Eur J For Res 133:261–272
134. Roessiger J, Griess VC, Härtl F, Clasen C, Knoke T (2013) How economic performance of a stand increases due to decreased failure risk associated with the admixing of species. Ecol Model 255:58–69
135. Rosenvald R, Lohmus A (2008) For what, when, and where is green-tree retention better than clear-cutting? A review of the biodiversity aspects. Forest Ecol Manag 255:1–15
136. Rydberg D, Falck J (2000) Urban forestry in Sweden from a silvicultural perspective: a review. Landscape Urban Plan 47:1–18
137. Schaetzl RJ, Johnson DL, Burns SF, Small TW (1989) Tree uprooting: review of terminology, process, and environmental implications. Can J For Res 19:1–11
138. Schelhaas M, Nabuurs GJ, Schuck A (2003) Natural disturbances in the European forests in the 19th and 20th centuries. Glob Change Biol 9:1620–1633
139. Schoennagel T, Veblen TT, Romme WH (2004) The interaction of fire, fuels, and climate across Rocky Mountain forests. BioScience 54:661–676
140. Schütz J-P (1999a) Naturnaher Waldbau: gestern, heute, morgen. Schw Z Forstwes 150:1–12
141. Schütz J-P (1999b) Principles of functioning of mixtures in forests stands; experience of temperate central European forest conditions. In: Olsthoorn AFM, Bartelink HH, Gardiner JJ, Pretzsch H, Hekhuis HJ, Franc A (eds) Management of mixed-species forest; silviculture and economics, IBN Scientific Contribution 15. Inst. For Forestry and Nature Research, Wageningen, pp 219–234
142. Schütz J-P (2002) Silvicultural tools to develop irregular and diverse forest structures. Forestry 75:329–337
143. Schütz J-P, Pukkala T, Donoso PJ, von Gadow K (2012) Historical emergence and current application of CCF. In: Pukkala T, von Gadow K (eds) Continuous cover forestry. Springer Verlag, Dordrecht, Heidelberg, p 6
144. Sears RR, Pinedo-Vasquez M (2011) Forest policy reform and the organization of logging in Peruvian Amazonia. Dev Change 42:609–631
145. Seymour RS, Hunter ML (1999) Principles of ecological forestry. In: Hunter ML (ed) Maintaining biodiversity in forest ecosystems. Cambridge University Press, Cambridge, pp 499–524
146. Shaw DC, Chen J, Freeman EA, Braun DM (2005) Spatial and population characteristics of dwarf mistletoe infected trees in an old-growth Douglas-fir – western hemlock forest. Can J For Res 35:990–1001
147. Shorohova E, Kuuluvainen T, Kangur A, Jogiste K (2009) Natural stand structures, disturbance regimes and successional dynamics in the Eurasian boreal forests: a review with special reference to Russian studies. Ann For Sci 66:201
148. Siiskonen H (2007) The conflict between traditional and scientific forest management in 20th century Finland. Forest Ecol Manag 249:125–133, doi:10.1016/j.foreco.2007.03.018
149. Steventon JD, MacKenzie KL, Mahon TE (1998) Response of small mammals and birds to partial cutting and clearcutting in northwest British Columbia. Forest Chron 74(5):703–713
150. Swank W, Vose J, Elliott K (2001) Long-term hydrologic and water quality responses following commercial clearcutting of mixed hardwoods on a southern Appalachian catchment. Forest Ecol Manag 143:163–178
151. Swanson ME, Franklin JF, Beschta RL, Crisafulli CM, DellaSala DA, Hutto RL, Lindenmayer DB, Swanson FJ (2011) The forgotten stage of forest succession: early-successional ecosystems on forest sites. Front Ecol Environ 9:117–125
152. Tahvonen O, Pukkala T, Laiho O, Lähde E, Niinimäki S (2010) Optimal management of uneven-aged Norway spruce stands. Forest Ecol Manag 260:106–115
153. Thorpe HC, Astrup R, Trowbridge A, Coates KD (2010) Competition and tree crowns: a neighborhood analysis of three boreal tree species. Forest Ecol Manag 259:1586–1596
154. Tönnes S, Karjalainen E, Löfström I, Neuvonen M (2004) Scenic impacts of retention trees in clear-cutting areas. Scand J For Res 19:348–357
155. Tropical Forest Foundation (2008) Sustaining tropical forests with reduced impact logging. http://www.tff-indonesia.org/index.php/en/r-i-l/ril-criteria-and-indicators. Accessed 14 April 2015
156. Urquhart J, Courtney P (2011) Seeing the owner behind the trees: a typology of small-scale private woodland owners in England. Forest Policy Econ 13:535–544
157. Veblen TT (1985) Forest development in tree-fall gaps in the temperate rain forests of Chile. Natl Geogr Res 1:162–183
158. Veblen TT, Donoso ZC, Schlegel FM, Escobar RB (1981) Forest dynamics in south-central Chile. J Biogeogr 8:211–247
159. von Thünen JH (1842) Der isolierte Staat in Beziehung auf Landwirtschaft und Nationalökonomie, 2nd edn. Leopold, Rostock, Germany
160. Wagner S, Nocentini S, Huth F, Hoogstra-Klein M (2014) Forest management approaches for coping with the uncertainty of climate change: trade-offs in service provisioning and adaptability. Ecol Soc 19(1):32
161. West PW (2014) Growing plantation forests. Springer, Dordrecht, Heidelberg, London, New York
162. Wilson SMG (2013) Adoption of alternative silvicultural systems in Great Britain: a review. Q J Forest 100:279–293
163. Wunder S (2006) Are direct payments for environmental services spelling doom for sustainable forest management in the tropics. Ecol Soc 11:23

164. Yasuda A, Yoshida T, Miya H, Harvey BD (2013) An alternative management regime of selection cutting for sustaining stand structure of mixed forests of northern Japan: a simulation study. J For Res 18:398–406

165. Yoshida T, Noguchi M, Akibayashi Y, Noda M, Kadomatsu M, Sasa K (2006) Twenty years of community dynamics in a mixed conifer broad-leaved forest under a selection system in northern Japan. Can J For Res 36:1363–1375

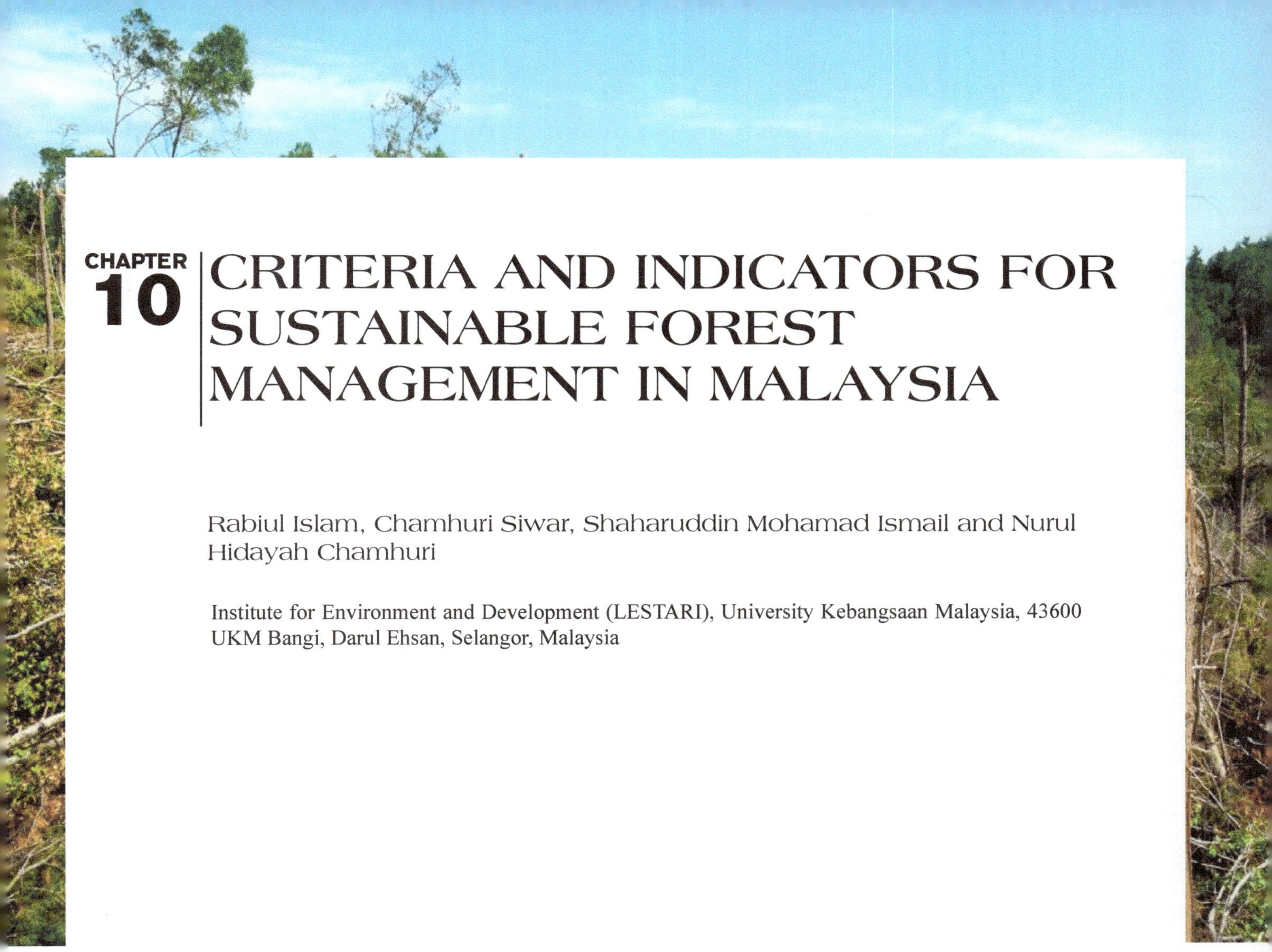

CHAPTER 10 CRITERIA AND INDICATORS FOR SUSTAINABLE FOREST MANAGEMENT IN MALAYSIA

Rabiul Islam, Chamhuri Siwar, Shaharuddin Mohamad Ismail and Nurul Hidayah Chamhuri

Institute for Environment and Development (LESTARI), University Kebangsaan Malaysia, 43600 UKM Bangi, Darul Ehsan, Selangor, Malaysia

ABSTRACT

Problem statement: This study investigated the timber certification of sustainable forest management and the criteria and indicators for sustainable forest management. Approach: Sustainable forest management is an important role in the forestry sector and takes a major player in Malaysia's sustainable development. It is a management regime that integrates and balances social, economic, ecological, cultural and spiritual needs to present and future generations. Results: We analyzed the pillars of sustainable forest management, criteria, indicators, activities and standards of performance for sustainable forest management in Malaysia. Conclusion: The aim of this study was to highlight and clarify the impacts of trade and environment on sustainable forest management in Malaysia.

INTRODUCTION

Sustainable Forest Management (SFM) is the way of management in which growth exceeds timber harvest, now also encompasses economics, environmental and social qualities that contribute to the sustainability of forest dependent communities and ecosystems as well as the forest itself. Malaysia has a plan more environmentally-friendly and responsible business practices. This will help reduce operating costs in the long-run and is a wise investment in the future, safeguarding the natural resources depending on corporations and communities. For example, local corporations in forest industries are joining WWF's Global Forest and Trade Network. They are targeting European and US markets, where consumers are increasingly demanding wood products from sustainably managed forests (WWF-Malaysia, 2008).

According to International Tropical Timber Organization (ITTO, 1992) sustainable forest management is defined as the process of managing permanent forest land to achieve one or more clearly specified objectives of management with regard to the production of a continuous flow of desired forest products and services without undue reduction of its inherent values and future productivity and without undue undesirable effects on the physical and social environment while FAO (1993) defines it as one which ensures that the values derived from forest meet present day needs while at the same time ensuring their continued availability and utilization to long-term development needs. Sustainable Forest Management (SFM) is impossible to

Citation: Rabiul Islam, Chamhuri Siwar, Shaharuddin Mohamad Ismail and Nurul Hidayah Chamhuri, Criteria and Indicators for Sustainable Forest Management in Malaysia, ISSN 1553-345X.

achieve if a country does not have a management system. In this regard, the use of more systematic approach in managing the forests in Peninsular Malaysia began in 1901 when the first forest officer was appointed (Ismail, 1996). Since then, forest management practices in Peninsular Malaysia had been subjected to constant review and refinement so as to ensure their suitability in achieving forest renewal and sustained yield. Sustainable forest management is a part of sustainable development. Rabiul et al. (2010) stated the impacts of trade and environment on sustainable development. They also discussed their paper about the development of market access, trade barriers, trade liberalization, environmental degradation and pollution, sustainable forest management and sustainable development.

Kamaruzaman and Dahlan (2008) state the sustainable forest management practices and environmental protection in Malaysia. In response to the needs to promote SFM, the Forestry Department Peninsular Malaysia has produced the Malaysian Criteria and Indicators (MC & I) that clarify major activities that have to be complied with sustainable basis. To ensure that the forest is better conserved, the annual coupe in Peninsular Malaysia has been scaled down for each Malaysian Plan starting from the 4th Malaysia Plan (1981-1985) of 74,869-36,940 ha year^{-1} under the 9th Malaysia Plan (2006-2010).

Rahim et al. (2009a) describe the sustainable forest management and west Malaysian sawntimber supply analysis. The calculation of the allowable cut was an important element of forest management. It is expressed in term of forest area of wood volume to be harvested. The goal was to obtain a sustainable yield of the best possible yield. Mohd Shahwahid and Awang Noor (2002) revealed that the annual coupe was lower than the official approved annual coupe by 34%. In fact, in Peninsular Malaysia, annual coupes have been steadily declining and this is part of the conservation strategy to ensure sustainable timber production. Hence, the sustainable yield has been one of the basic views of forest management for a considerable length of time. Initially, the only goal or at least the main goal in mind of forester was usually wood production.

Rahim et al. (2009b) represents the sustainable forest management practices and west Malaysian log market. The sustainable management definition and objectives required maintenance of the forest stand regeneration not only for timber production, also for environmental stability, biodiversity conservation, but recreational values preservation and other forest products and forest components conservation. Motivated by ITTO target, National Committee Sustainable Forest Management was established in 1994 under the Ministry of Primary Industries. This committee has formulated a total of 92 activities, based on 5 criteria and 27 indicators to implement the ITTO criteria at the national level (Thang, 2002). SFM has led to greater planning and monitoring of the environment. In term of economic, there are indications to incur incremental costs.

Rabiul and Chamhuri (2010) addressed the trade and environment in the forestry sector towards sustainable forest management. They discussed their paper about the trends in production for different types of forestry wood products. With regards to SFM practices, Schwarzbauer and Rametsteiner (2001) have analyzed the potential impacts of SFM certification on forest products markets using a simulation model of the Western European forest sector. The empirical evidence shows that rather modest changes are to be expected from SFM certification in forest products markets. The market impact of timber supply reduction from certified forest would be more distinct than the impact of chain of custody costs. That means, a decrease of harvesting levels in certified forest will affect forest products markets more than the increase costs related to the installation and maintenance of SFM certification in the forest sector.

The purpose of this study is to obtain the timber certification, criteria and indicators for sustainable forest management and the impacts of trade and environment on sustainable forest management in Malaysia.

MATERIALS AND METHODS

Data Attainment

The study is conducted in University Kebangsaan Malaysia, Bangi since July, 2008 to November, 2009. The data for analysis is perceived from secondary sources in Malaysia. The significant manipulations for acquired data are sustainable forest management, timber certification and criteria and indicators for sustainable forest management. The corresponding outcomes are demonstrated from the analysis of collected data.

Sustainable Forest Management

Sustainable forest management implies the sustainable utilization of forest resources for the benefit of communities and states. The concept not only aims to maintain the value of forest resources, it also has a huge potential for creating employment, income and wealth for the populations and states concerned. The sustainable forest management programme seeks to improve the management of natural tropical forests by increasing the adoption of sustainable forest management practices by forest managers from industries to communities. Good forest management can both reduce the negative impacts of timber harvesting on other forest resources and services and increase yields of desired products and services from a given area of forest.

Criteria and indicators for sustainable forest management in Malaysia

Sustainable forest management is considered as one of the most important contributions which the forestry sector can make to the sustainable development objectives of any nation, particularly those richly endowed with forest. The UNCED, held in Rio de Janeiro in 1992, set the guiding principles for sustainable forest management as a contribution towards sustainable development. MTCC (2001) states that the Malaysian Criteria, Indicators, activities and Standards of Performance for Forest Management Certification (MC & I) will be used for assessing forest management practices in the Permanent Forest Estate (PFE) at Forest Management Unit (FMU) level for the purpose of certification.

Wolfslehner et al. (2005) addresses the application of the analytic network process in multi-criteria analysis of sustainable forest management. The demand to assess forest-management regimes and alternatives in regards to their specific benefits and to sustainability in general has lead to the use of Criteria and Indicators (C & I) (Prabhu et al., 1996; Van Bueren and Blom, 1997). Criteria are the essential elements against which sustainability is assessed to the productive, protective and social roles of forests and forest ecosystems. Each criterion relates to a key element of sustainability and may be described by one or more indicators. Indicators are parameters which can be measured and correspond to a particular criterion (Prabhu et al., 1999). C & Iapproaches appear to be highly capable of measuring aspects of SFM at national, regional and forest management unit levels.

Regional Level Criteria and Indicators

Criteria and indicators for sustainable forest management have become an important issue in many countries around the world. At present, many countries are taking part the sets of criteria in all nine processes that SFM comprises the following thematic areas:

- Extent of forest resources
- Biological diversity
- Forest health and vitality
- Protective functions of forests
- Productive functions of forests
- Socio-economic functions
- Legal policy and institutional framework (FAO, 2005)

National Level Criteria and Indicators

National level criteria and indicators help decision makers, including planners and policy makers, to define what makes up SFM and to establish guidelines and monitor trends in the sustainability of the goods and services being provided by a nation's forests.

Major Criteria and Indicator Processes and Initiatives

- International Tropical Timber Organization (ITTO)
- Pan-European
- Montreal
- Tarapoto
- Dry Zone Africa
- Near East

Forest Management Unit Level Criteria and Indicators

Forest management unit level criteria and indicators have been increasingly developed and implemented to complement national criteria. The forest management unit level indicators depend on local, often site-specific, environmental factors such as forest type and topography, local economic and social considerations and priorities. The criteria at forest management unit level are likely to be identical or very similar to those defined at national level, although they are more flexible. Thus, they must be mutually compatible to help ensure complementarily over the country.

C & I-approaches is that it can be used to collect and report information within a system, which is usually characterized by a lack of knowledge, uncertainties and missing information about impacts, dependencies and feedbacks (Rametsteiner, 2001; Brang et al., 2002). C & I has become a common approach to assess the aspects of SFM. Among these approaches, Multi-Criteria Analysis (MCA) techniques have been adapted to structure and implement the C & I-based assessment of SFM (Varma et al., 2000; Bousson, 2001; Mendoza and Prabhu, 2003). Ecologically and environmentally sound forest conservation and management practices have been developed to ensure that the forest resources are managed for the sustainable timber production, ensuring climatic stability and ecological balance, safeguarding of water supply and environmental quality and the conservation of biological diversity. Today, Malaysia, as a member of the International Tropical Timber Organization (ITTO) which is fully committed to ITTO's objectives, is striving towards the achievement of forest management that is environmentally responsible, socially beneficial and economically viable. ITTO has developed a set of key Criteria and Indicators (C & I) for the sustainable management of tropical forests (ITTO, 2005).

Timber Certification

Baharuddin and Simula (1994) states that timber certification is a process which results in a written statement (a certificate) attesting to the origin of wood raw material and its status the following validation by an independent third party. Certification is designed to allow participants to measure their forest management practices against standards and to demonstrate compliance with those standards. Timber certification typically includes two main components: certification of sustainability of forest management and product certification. Certification of forest management covers forest inventory, management planning, silvi-culture, harvesting, road construction and other related activities, as well as the environmental, economic and social impacts of forest activities. In product certification, round-wood and processed timber products are traced through the successive phases of the supply chain. Certification of forest management thus takes place in the country of origin; product certification covers the supply chain of domestic and export markets.

Kamaruzaman and Dahlan (2008) have stated that Malaysia is a producer member country of the International Tropical Timber Organization (ITTO) which is committed to achieve sustainable forest management of sustainable development. Developing countries follow timber certification in order to protect domestic timber products. Today, timber certification is the slogan in forest harvesting. Malaysia timber certification began in 1994 in stages. The Malaysian Timber Certification Council (MTCC) is an independent organization established in 1998 to develop and operate the Malaysian Timber Certification Scheme (MTCS) in order to provide independent assessments of forest management practices in Malaysia as well as to meet the demand for certified timber products. Timber certification is a market-linked tool to promote and encourage sustainable forest management as well as to provide an assurance to buyers that the timber products they buy come from sustainably managed forests. The Malaysian Criteria, Indicators, activities and Standards of Performance for Forest Management Certification (MC & I) is the standard that will be used for assessing forest management practices in the Permanent Forest Estate (PFE) at Forest Management Unit (FMU) level for the purpose of certification.

Thang (2003) states that timber certification can be defined as a process which entails an independent assessment of a forest management operation, according to specific economic, social, environmental and ecological criteria, indicators, activities and management specifications or standards of performance. Forest Management Unit is defined as an area of forest land that is managed by an organizational entity which decides on and subsequently implements forest activities to ensure the economic, ecological, biological and socio-cultural sustainability of the area. It addressed the issues of forest inventory, management planning, silvi-culture, harvesting, forest road construction and other related forest management operations.

MTCC addresses the eight forest management units in Malaysia, namely the states of Johor, Kedah, Negeri Sembilan, Pahang, Perak, Selangor, Terengganu and Kelantan. They are covering 4.7 m ha of the PRFs in Peninsular Malaysia which are certified using the set of Malaysian Criteria and Indicators (MTCC, 2001). The standard currently being used for assessing Forest Management Units (FMUs) is the Malaysian Criteria, Indicators, Activities and Standards of Performance for Forest Management Certification (MC & I). The MC & I are based on the 1998 ITTO Criteria and Indicators for Sustainable Management of Natural Tropical Forests. It contains the key elements for sustainable forest management covering economic, social, environmental and conservational aspects and incorporates the corresponding standards of performance for Sabah, Sarawak and Peninsular Malaysia identified during the regional and national level consultations held in 1999.

RESULTS AND DISCUSSION

Sustainable Forest Management (SFM) is the process of managing forest land to achieve one or more clearly specified objectives of management without undue reduction of its inherent values and future productivity or undesirable effects on the economic, social and environment (ITTO, 1992) and integrates and balances social, economic, ecological, cultural and spiritual needs to present and future generations (United Nations, 1992). There are three pillars of SFM include the Fig. 1.

Economic: The capacity of the forests to attract investment and support economically viable forest uses in the present and the future is undiminished. The forest is not used beyond its long-term capacity for production of wood and non-wood forest products.

Social: There are a variety of pillars include social such as:

- The rights of indigenous peoples and local communities are respected and protected
- Forest workers are healthy, safe and their rights are protected (e.g., freedom of association, right to bargain, child labor, forced labor, equal remuneration and non-discrimination)
- Local communities, including indigenous peoples, benefit economically from forest management
- Sites of religious, spiritual, archaeological, historic as well as of aesthetic and recreational value are preserved

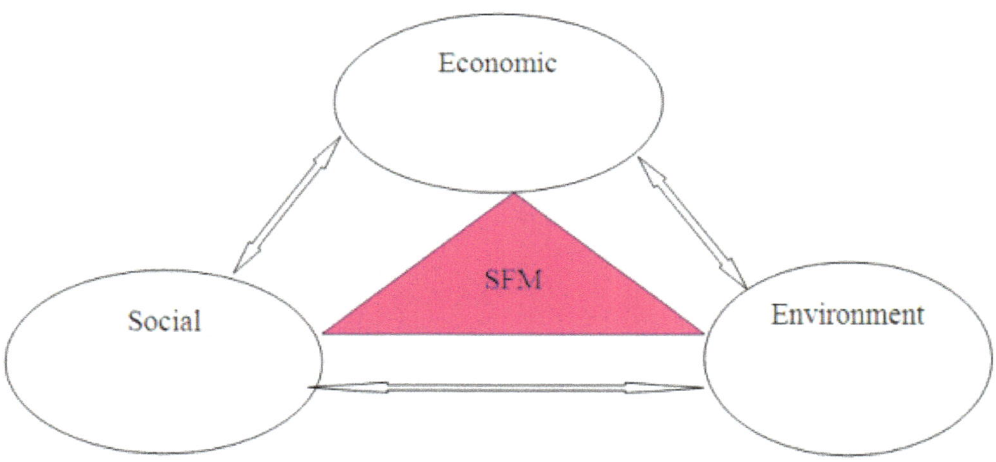

Figure 1: Pillars of sustainable forest management

Table 1: Criteria, indicator, activities and standards of performance for sustainable forest management at the national level in the MC & I

Criteria	Indicators	Activities	Standards of performance+
Enabling condition for sustainable forest management	9	31	20
Forest resource security	5	22	17
Forest ecosystem health and condition	5	20	13
Flow of forest produce	12	37	31
Biological diversity	8	25	22
Soil and water	9	26	29
Economics, social and cultural aspects	16	39	38
Total 7	64	200	170

+: For peninsular Malaysia only

Table 2: Criteria, indicators, activities and standards of performance for sustainable forest management at the forest management unit level in the MC & I

Criteria	Indicators	Activities	Standards of performance+
Enabling condition for sustainable forest management	8	26	17
Forest resource security	5	22	17
Forest ecosystem health and condition	3	12	9
Flow of forest produce	12	37	31
Biological diversity	7	21	18
Soil and water	9	26	29
Economics, social and cultural aspects	12	27	29
Total 7	56	171	150

+: For peninsular Malaysia only

Table 3: Relationship and number of criteria, indicators, activities and standards of performance formulated at different management levels in the MC & I

Management levels	Criteria	Indicators	Activities	Standards of performance+
National level (SFM) #	7	64	200	170
FMU# level (SFM)	7	56	171*	150**
FMU level (forest management certification)	7	53	162++	142++

+: For Peninsular Malaysia only; (SFM) #: Sustainable Forest Management; FMU: Forest Management Unit; *: 86% are identical to those formulated at the National level; **: 88% are identical to those formulated at the National level; ++: A subset of activities/standards of performance taken from those formulated at the FMU level (SFM)

Environmental: Forest use protects biodiversity (ecosystems, species, genes and ecological processes) and the capacity to maintain ecosystem processes and services such as watershed protection, pollination, protection against mudslides, aesthetic beauty and carbon storage. The economic pillar is a suitable mix of wood products and non-wood products (plants and animals), that does not diminish the productive capacity of the forest. Social pillars include respect for labor and indigenous rights, the health and safety of forest workers, sharing of economic benefits and protection of sites of spiritual or historic value. Environmental pillars can include soil protection, biodiversity, maintenance of air and water quality and aesthetics. The appropriate balance of these pillars will vary among regions and contexts.

MTCC harmonized through meetings with various stakeholders. The MTCC (2001) contain key elements for SFM covering economic, social and environmental aspects. The MC & I contains 7 criteria, 64 indicators, 200 activities and 170 standards of performance were formulated at the national level (Table 1); while 7 criteria, 56 indicators, 171 activities and 150 standards of performance were formulated for assessing sustainable forest management at the forest management unit (Table 2). In Malaysia, the criteria, indicators, activities and management specifications at the national level would be used for reporting progress towards achieving sustainable forest management and in particular, the ITTO Year 2000 Objective (Table 3). Those formulated at the forest management unit level would be used by the Forestry Departments to monitor and assess progress in sustainable forest management at the state level including forest certification by independent third party assessors.

CONCLUSION

Malaysia is fully aware of the need to manage the forest effectively from single use of sustained supply of timber to multiple-use of forestry including maintaining of the environmental stability. Malaysia, like many countries around the world, is working hard to find ways to use its forests for community benefit within the constraints imposed by ecological processes and ensuring that future options are not foreclosed. Over the last several years, C & I have emerged as one of the most powerful forest management tools designed to promote SFM. Criteria and indicators developed for application will be reviewed and refined periodically to reflect new concepts of sustainable forest management based on the evolving knowledge about the functioning of forest ecosystem, anthropogenic intervention on the forests whether planned or unplanned and the changing needs of society for forest goods and services. The level of management will be refined once the current silvicultural management systems are further developed for application at a lower management level, perhaps at the forest district level, forest reserve level or even at the compartment level. Malaysia continues to strongly support international efforts to promote and ensure sustainability in forest management. The implementation of timber certification will continue to be actively pursued to ensure market access of Malaysian timber product, especially in environmentally sensitive markets. Timber certification is very essential in the following reasons:

- To restrict the importation of tropical timber products due to illegal logging and the associated trade in illegally sourced timber products
- The measures to implement public timber procurement policies
- The Forest Law Enforcement, Governance and Trade (FLEGT) action plan which is being implemented by the EU, Voluntary Partnership Agreements (VPAs)

ACKNOWLEDGEMENT

Financial assistance provided by the Research University Grant (GUP), Institute for Environment and Development, University Kebangsaan Malaysia is gratefully acknowledged.

REFERENCES

1. Baharuddin, H.G. and M. Simula, 1994. Certification Schemes of All Types of Timber. ITTO.
2. Bousson, E., 2001. Development of a Multicriteria Decision Support System Adapted to Multiple-Use Forest Management: Application to Forest Management

at the Management Unit Level in Southern Belgium. In: Criteria and Indicators for Sustainable Forest Management at the Forest Management Unit Level, Franc, A., O. Laroussinie and T. Karjalainen (Eds.). European Forest Institute, Joensuu, Finland, pp: 14.

3. Brang, P., B. Courbaud, A. Fischer, I. Kissling-Naf, D. Pettenella et al., 2002. Developing indicators for the sustainable management of mountain forests using a modeling approach. For. Policy Econ., 4: 113-123.

4. FAO., 1993. Forest Resources assessment 1990: Tropical Countries. FAO Forestry Paper 112, Rome, Italy.

5. Ismail, A., 1996. Sustainable forest management in Malaysia-the way forward. Proceeding of the Thailand Environment Institute Annual Conference, July 9-9, Bangkok, Thailand, pp: 26.

6. ITTO., 1992. Criteria for the Measurement of Sustainable Tropical Forest Management. ITTO., Yokohama, Japan, pp: 5.

7. ITTO., 2005. Revised ITTO Criteria and Indicators for Sustainable Management of Tropical Forests Including Reporting Format. ITTO., ISBN: 4- 902045-20-6, pp: 40.

8. Kamaruzaman, J. and H. Dahlan, 2008. Sustainable Forest management practices and environmental protection in Malaysia. WSEAS Trans. Environ. Dev., 3: 191-199.

9. Mendoza, G.A. and R. Prabhu, 2003. Qualitative multicriteria approaches to assessing indicators of sustainable forest resource management. For. Ecol. Manage., 174: 329-343.

10. Prabhu R., C. Colfer, P. Venkateswarlu, L.C. Tan, R. Soekmadi and E. Wollenberg, 1996. Testing Criteria and Indicators for the Sustainable Management of Forests. Phase I. CIFOR Special Publication, Jakarta

11. Prabhu, R., C.J.P. Colfer and R.G. Dudley, 1999. Guidelines for Developing, Testing and Selecting Criteria and Indicators for Sustainable Forest Management. Center for International Forestry Research, Jakarta.

12. Rabiul, I. and S. Chamhuri, 2010. Trade and Environment in the Forestry Sector: Towards Sustainable Forest Management. Asian J. Sci. Res., 3: 1-17.

13. Rabiul, I., S. Chamhuri and M.I. Shaharuddin, 2010. Impacts of trade and environment on sustainable development. Am. J. Environ. Sci., 6: 11-19.

14. Rahim, A., A. Samad and M.S.H. Othman, 2009a. Sustainable forest management and west Malaysian Sawn timber supply analysis. J. Sustain. Dev., 2: 58-64.

15. Rametsteiner, E., 2001. SFM Indicators as Tools in Political and Economic Context: Actual and Potential Roles. In: Criteria and Indicators for Sustainable Forest Management, Raison, R.J., A.G. Brown and D.W. Flinn (Eds.). CABI Publishing, Wallingford, UK, pp: 24.

16. Schwarzbauer, P. and E. Rametsteiner, 2001. The impact of SFM-certification on forest product markets in Western Europe-an analysis using a forest sector simulation model. For. Policy Econ., 2: 241-256.

17. Thang, H.C., 2002. Towards achieving sustainable forest management in Peninsular Malaysia. Proceeding of the National Seminar on Practicing Sustainable Forest Management: Lessons Learned and Future Challenges, Aug. 20-22, Kota Kinabalu, Sabah, Malaysia.

18. Thang, H.C., 2003. Development and Application of Criteria and Indicators for Sustainable Forest Management in Malaysia. Proceeding of the Attachment Training of Bhutanese Foresters in Malaysia, Oct. 11-31, Kuala Lumpur, Malaysia.

19. Van Bueren, E.M.L. and E.M. Blom, 1997. Hierarchical Framework for the Formulation of Sustainable Forest Management Standards, Principles, Criteria, Indicators. The Tropenbos Foundation, pp: 82.

20. Varma, V.K., I. Ferguson and I. Wild, 2000. Decision support system for sustainable forest management. For. Ecol. Manage., 128: 49-55.

21. Wolfslehner, B., H. Vacik and M.J. Lexer, 2005. Application of the analytic network process in multi-criteria analysis of sustainable forest management. For. Ecol. Manage., 207: 157-170. DOI: 10.1016/j.foreco.2004.10.025

22. WWF-Malaysia, 2008. Annual Review. World Wide Fund for Nature, Malaysia.

CHAPTER 11

PERCEIVING SUSTAINABLE FOREST SPACES: GOVERNANCE ASPECTS OF PRIVATE AND COMPANY OWNED FORESTS IN NORTH-KARELIA, FINLAND

Moritz Albrecht

Department of Geographical and Historical Studies, University of Eastern Finland

The integration of improved environmental or sustainable aspects in forest management is often affiliated with the rise of market-driven governance systems, such as forest certification. In terms of forest resource peripheries, like NorthKarelia, Finland, these are often attributed to environmental business and consumer demands from the green Central European markets. While acknowledging these aspects related to the supply chains of wood-based products, this study evaluates the actual perceptions about environmental forest governance and its spaces in the resource peripheries themselves. It displays the perceived changes and practices in forestry by comparing private and corporate ownership and their governance networks. This is accomplished by a qualitative, interview based case study of North Karelian and Finnish forestry actors. Transnational forest governance is hereby treated as a relational space, with forest certification systems as possible technologies used to achieve improved, sustainable forest management. Utilizing the North-Karelian forestry sector, the varying positionalities of actors and institutions within such a relational space shape the knowledge networks, perceptions and decision-making. The study evaluates how these local-global positionalities of actors and individuals shape their understanding, and guide the direction of sustainable forest management in Finland while it (re-)produces opposing regimes of practice. With the discourse on forest certification being twofold, a more complex picture emerges if aspects of evenversus uneven-aged forest management in Finland are integrated. Shaped by the actor's positionalities and related knowledge networks, perceptions regarding the quality of forest management practices and technologies used to achieve sustainability differ and thereby shape governance processes. The green markets are not perceived as the main driving force and a strong governmental influence, particularly related to private ownership aspects, is noted in the Finnish case. Forest certification systems, and other political technologies for sustainable forest management, are embedded in or strongly restricted by these aspects.

Keywords:

Finland, forest governance, qualitative case-study, sustainable spaces, ownership positionalities, certification

INTRODUCTION

Regarding sustainable forest management (SFM), Finnish forestry has publicly provided a twofold picture on the international scale during the last decade. On the one hand, a positive image of sustainable managed forestry, almost entirely

Citation: Albrecht, Moritz (2012). Perceiving sustainable forest spaces: governance aspects of private and company owned forests in North-Karelia, Finland. Fennia 190: 1, pp. 3–18. ISSN 1798-5617.

covered by forest certification appears and is promoted by large forest and wood-based companies, as well as forest owner associations and the state forest service Metsähallitus (e.g. Metsähallitus 2009; FFIF 2011a; FMA 2011a). On the other hand, large environmental non-governmental organizations (ENGO) blame Finnish forestry for following unsustainable management practices and promoting the destruction of Europe's last remaining old growth forests (e.g. Harkki 2004; Harkki 2008; Greenpeace 2009). Even though international protests about the destruction of old growth forests in Finnish Lapland have ceased after joint agreements (FFIF 2010), debates about the sufficient means and management tools or practices to achieve SFM in Finnish forestry continue (e.g. Luonto-Liitto et al. 2009; PEFC 2011).

A continuous integration of environmental aspects into forest management, harvesting practices and related laws is taking place in Finnish forestry (e.g. Mielikäinen & Hynynen 2003; Parviainen & Västilä 2011). Increased integration of environmental and sustainable aspects into forest management are often attributed to market driven governance systems, such as forest certification, based on customer demands and ENGO protest campaigns in the Central European markets (Cashore et al. 2004; Albrecht 2010a). However, it remains unclear how implementing actors outside the academic or expert communities perceive changes in environmental criteria as well as their drivers. For Finland, multiple studies are concerned with forest owners' opinions about conservation or environmental management practices (e.g. Karppinen 1998; Horne 2004; Hänninen & Kurttila 2004), yet, taking into account that up to 95% of commercial forestry is carried-out by contracting companies, these are the main implementing actors for management practices in Finland (Koneyrittäjät 2009). These actors' perceptions are of primary importance when evaluating aspects of environmental forest governance. As perceptions and rationalities guide actors' activities (Merlingen 2003), it is important to understand the varying actors' values regarding what has been achieved and what is regarded as sufficient for SFM, and how they define such sustainable forest spaces. While local forestry stakeholders' perceptions on uneven-aged forest management are described in a Swedish casestudy by Axelsson and Angelstam (2011), their account primary treats the knowledge of the stakeholders on technical aspects of forest management. Suitable for certain comparison, it lacks the relational and spatial aspects evaluated by this study.

On account on these aspects presented above the paper utilizes the following research questions as a guideline to provide an improved account of the processes at work in transnational forest governance:

- How do the perceptions of actors, involved or linked to forestry, differ in terms of their attitudes/definitions towards SFM?
- What is the impact of the actors' positionality in regard to the (re-)production of their rationalities on, and regimes of practice for SFM?
- How do such perceptions of their sustainable spaces of forestry guide their decision making and acceptance of various technologies for SFM?
- How do these varying positionalities and perceptions shape the (re-)production of Finnish forestry in terms of SFM and transnational forest governance spaces in general?

These questions should not be understood in a normative fashion to determine general patterns in Finnish or transnational forestry but aim to highlight the multiplicity of processes which shape forest governance and the utilization of related technologies such as certification. Transnational forest certification systems are prominently utilized to define SFM, whereby two systems play a major role in global forestry. First, there is the strongly ENGO backed Forest Stewardship Council (FSC) and second, the Programme for the Endorsement of Forest Certification (PEFC), which is mostly supported by forest owner associations and forestry institutions (e.g. Albrecht 2010b). In Finland, with a PEFC coverage of 95% in commercial forests (FMA 2011b) compared to a just recently approved and marginally implemented FSC the debate can be viewed as the defence of PEFC as a tool for SFM by promoting its achievements. Generally, also in Finland, FSC entails stricter protective criteria and requires to set 5% of the certified forest aside for conservation of biodiversity (see also PEFC 2009; FSC 2011). Nevertheless, since national standards and implementation differs, the paper refrains from ranking the certification systems on their practical achievements for SFM. This has been studied elsewhere with varying results (e.g. Harkki 2004; UPM 2005; Schlyter et al. 2009; Indufor 2010). Additionally, governmental means such as the Forestry Act of 1996 (MMM 1996) or guidelines by varying forestry related institutions play a decisive role in regard to the SFM debate in Finnish forestry. Hereafter, environmental transnational forest governance is regarded as a relational space, including the core market – resource periphery relations of wood product commodity chains (Albrecht 2010a, 2010b, 2012). The local case of the North-Karelian forest sector aims to display how actors in the resource peripheries (Fig. 1) perceive SFM based on their locally embedded relations. As they are integrated to (re-)produce the relational space they are entangled with, this paper argues that this local positionality based on ownership patterns and other relations challenges purely market-driven modes and influences resource governance.

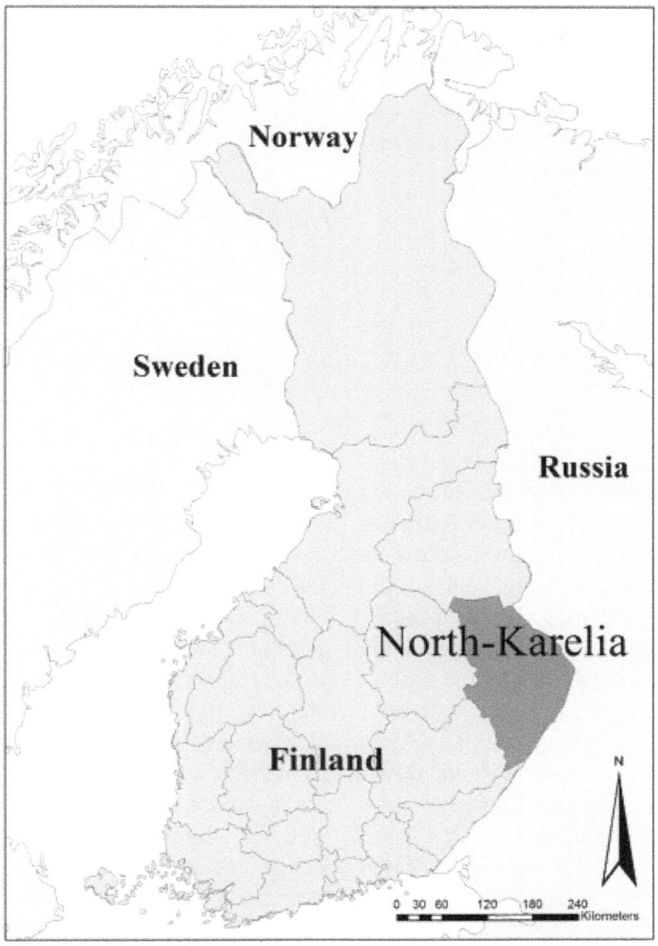

Figure 1. Finnish administrative regions and location of case study area of North-Karelia in a North European context.

FINNISH AND NORTH-KARELIAN FORESTRY SYSTEMS

Finnish as well as North-Karelian forestry are distinguished by their high private, non-commercial ownership of 52% (Forest Centre PK 2010; METLA 2010). Regarding productive forests the share of private owners in Finland is as much as 60%, while the share of final felling in private forests is 77% of Finnish forestry (METLA 2010). Company ownership of Finnish forests amounts to 9% whilst 35% of forests are state owned, of which only 26% are productive forests. In terms of company-owned forests, North-Karelia deviates from the Finnish mean with 23% (Forest Centre PK 2010). This is due to the prominent role of Tornator Oy (Tornator) in North-Karelia, an affiliated company of the global forest product company Stora Enso and responsible for the management of Stora Enso's former forest areas. State ownership is 20%. In relation to management operations and wood harvesting, despite the large share of private ownership, forestry operations are carried-out by contracting companies in 95% of cases (Koneyrittäjät 2009). Apart from the age structure of private forest owners, with 56% being over 60 in age, this owes to the specific forest planning and management system for private forests (METLA 2010). Finland has 13 forestry districts and regional Forestry Centres, North-Karelia being one. Aside preparing the regional forestry programme the Forestry Centres have several tasks concerning private owned forests: preparation of 10 year management plans for private forest owners and their holdings, and distribution of information, support for forest management, subsidies and education for forest owners and professionals. Additionally, by approving forest use declarations prior to loggings and carrying-out follow-up evaluations in chosen harvesting sites, Forest Centres are the responsible institutions for supervising the implementation of the Finnish Forest Act (Forest Centre 2011). Hence, they strongly guide and influence forest management in Finland.

Other important actors are the Forest Management Associations (FMA) and the Forest Owners Union (FOU). The FMA has legislative rights to collect a forest management fee, private forest owners are automatically members of the FMA (FMA 2011a). The FMA supports forest owners in terms of planning, education and timber sale. Harvesting and management plans are developed and tenders for logging rights are offered by power of attorney. North-Karelia's FMA buys all plans made by the Forest Centre and integrates them into its forestry planning support. Harvesting activities in private forests are carried-out twofold. For thinning and intermediate felling, the FMA marks stands and hires contractors, while for final felling this task is primarily done by the companies after acquisition of logging rights. The FOU, on the other hand, is mainly concerned with broader organizational aspects and supports cooperation among local FMAs. Additionally, the FOU holds and administers the regional PEFC certificate for its forestry district. In North-Karelia, compared to most other forestry districts, the FMA and the FOU are one entity separated only on paper.

Table 1. List of acronyms with English and original (Finnish) names.

Acronym	English name	Finnish name
ENGO	Environmental Non-Governmental Organization	
FANC	Finnish Association for Nature Conservation	Suomen luonnonsuojeluliitto
FFIF	Finnish Forest Industry Federation	Metsäteolisuus ry
FMA	Forest Management Association	Metsänhoitoyhdistys
FOU	Forest Owner Union	Metsänomistanjien liitto
FSC	Forest Stewardship Council	
NL	Nature League	Luonto-liitto
PEFC	Programme for the Endorsement of Forest Certification	
Tapio	Forest Development Center Tapio	Tapio
SFM	Sustainable Forest Management	
WWF	World Wide Fund for Nature (Finland)	

Company forests are less embedded into this structured system. While being obliged to prepare forest use declarations to Forest Centres the same as private forest owners, follow-up checks are carried out by the Forest Development Centre Tapio (Tapio) in most cases. Management planning is carried-out by their own staff while the marking of logging sites is mostly performed by the respective buyers' personnel. In the case of Tornator, up to 90% of logging rights are sold to Stora Enso which hires contractors to carry-out logging activities. With respect to state owned forests, most responsibilities and duties rest with the State Forestry Service Metsähallitus. The large majority of protected areas are situated in state-owned forest (METLA 2010). Private and corporate owned forests entail numerous entities involved in forest management, who are influencing forest governance with their perceptions, knowledge and the resulting practices. Since these entities are affected by their relations among each other and towards external actors or aspects of environmental forest management, the next section will present the underlying theoretical framework of this study based on relational space and knowledge networks.

PERCEIVING SUSTAINABLE SPACES OF FOREST GOVERNANCE

Local actors and entities are part of a wider space of transnational forest governance. Hereafter, space is regarded as relational (e.g. Massey 2005; Murdoch 2006), just as transnational environmental forest governance is perceived as a relational space (e.g. Albrecht 2010a, 2010b, 2012). Interlinked actors involved within this space, whether on a global or on a local scale, (re-)produce and affect the performativity of this space, due to the varying relations they are entangled with. Relations in this regard include social and biophysical components (Massey 2005). For example, knowledge or cooperation networks, varying forestry practices, or ecological aspects of forests are relations which enable or restrict entities in their activities or choices (e.g. Albrecht 2010b). As relational spaces are open and (re-)produced by a multiplicity of relations and actors (Massey 2005), Murdoch (2006) points out its consensual and contested character. This notion is highly visible in the debate about SFM practices, if on a global scale, or in localized examples, like the Finnish debate discussed by this paper.

A common problem in debates on SFM and its practices lies in the definition of the criteria necessary for one to achieve sustainability. As stressed by Hudson (2005) in his account on sustainable economic practices, flows and spaces, the delineation of sustainable spaces is critically dependent upon the definition of sustainability in the realm itself. Taking into account the social co-construction of nature (Castree & Braun 2001), the varying perceptions of actors, and entities and their knowledge networks on what is claimed to be the truth, respectively, the necessary means to achieve SFM are steering governance processes and individual behaviour (Rutherford 2007; Dean 2010). Thus, evaluating perceptions of actors and entities directly linked to and involved in forest management provides information on the overall performance of transnational and local forest governance. It further enables one to critically access claims about non-state market-driven processes of environmental forest governance via forest certification (e.g. Cashore et al. 2004, 2007).

The two forest certification systems, FSC and PEFC, are utilized as instruments for SFM by multiple actors, to varying degrees. Thus, these systems can be seen as governmental technologies promoted by competing regimes of practice (Baldwin 2003; Dean 2010). According to Dean (2010), regimes of practice produce and distribute knowledge based on their internal, yet multiple rationalities and thereby stress varying problematizations related to the space to be governed. The lack of conservation in forest management stressed by ENGOs might be regarded as such. Regimes of practice with their knowledge networks can be a relatively stable set of relations which struggle for supremacy, possibly marginalizing competing regimes (Murdoch 2006; Dean 2010). However, due to the open character of space, these sets are prone to change and marginalization might be overcome (Massey 2005). Yet, if maintained and kept more or less together, a set of relations with its dominant regime of practice may appear increasingly natural (Sheppard 2002). The institutional setting of FMAs, FOUs and Forestry Centres is the dominant regime of practice in the Finnish forest sector. Yet, the perceptions and rationalities of actors, for instance on environmental aspects of forest management, influence those regimes and guide their behaviour (Dean 2010).

Actors and entities in practical forest management guide their decisions about what they perceive as truth and what suits their rationality in terms of SFM requirements. These perceptions are influenced by the actors' positionality within the respective space. Positionality as proposed by Sheppard (2002: 318) might be used "...to describe how different entities are positioned with respect to each other in space/time". Thereby, the actors' positionality and the resulting rationalities are influenced by external influences. In terms of SFM, this means that rationalities concerned with the issue are not merely (re-)produced based on knowledge or perceptions directly related to the subject but include various aspects. For instance, economic reasons, personal values or institutional structures shape perceptions and thereby decisionmaking and management. Additionally, perceptions about the drivers for environmental achievement are possible indicators on how certain regimes of practice, based on their positionality and problematizations, promote various governmental technologies such as certification, forestry guidelines or law for their aims. Evaluating perceptions of environmental achievements, changes and conflicts, as well as investigating the opinions about their driving forces, enables one to integrate a local case into the wider processes of transnational environmental forest governance (e.g. Albrecht 2010a, 2010b, 2012).

A QUALITATIVE CASE-STUDY OF NORTH KARELIAN PRIVATE FORESTRY

This study has been conducted as a qualitative case-study of the private and corporate owned forest sector in North-Karelia, Finland. In-depth, open-ended interviews were conducted with managers and owners of three forestry contracting companies1 in North-Karelia and with the directors of the North-Karelian Forestry Centre, the FMA and the FOU. Empirical data on corporate owned forests was retrieved from the forest company Tornator, as it represents the largest private/ corporate ownership in the region. Group interviews were conducted with the companies' Forestry and Resource Manager, the Corporate Responsibility Superintendent and a forestry team manager. Additionally, two visits to logging sites, one thinning and one final cut, with contractors and an extended visit to Tornator's forests were conducted in January 2011. These forest visits provided additional information from forest workers. Further, an ENGO representative for the regional forestry council was interviewed.

Several institutions on the national scale have been integrated. Interviews were conducted with officials from ENGOs; the Finnish Association for Nature Conservation (FANC), the Nature League (NL) and the World Wide Fund for Nature Finland (WWF). Further, a certification and auditing expert from Tapio and the Head of forestry affairs of the Finnish Forest Industry Federation (FFIF) were interviewed. Altogether, 17 persons have been interviewed in direct relation to this study. To support the qualitative data and to provide, as stressed by Yin (2003), a multiplicity of data sources, position papers of the respective, as well as other institutions, were evaluated and interview data from previous studies was utilized (Albrecht 2010a, 2010b, 2012). Owing to the study's main focus on the actors' perceptions of environmental related management changes and their drivers, in-depth expert interviews are most suited to provide valuable data (Yin 2003; Silverman 2006) and provide the bulk of information. Being a possible point of criticism, the choice to not include private forest owners' opinions in the study is due to the fact that the focus is on forest management practice and the previously mentioned low number of forest owners being involved in active management of their forests (Koneyrittäjät 2009). The forest organizations and contractors were chosen as the focus due to their deemed importance in relation to the topic (Cloke et al. 2004) and according to the administrational structure underlying Finnish forestry.

Interviewees were asked to present their views on the state of SFM. In addition to providing an answer on the whether or not SFM is practiced in Finnish forestry, their views on specific achievements and problems, as well as the related drivers for those issues were evaluated. Due to the in-depth, open-ended character of the interviews and the heterogeneity of involved entities no standardized set of questions was utilized but the interviewees were requested to tell about aspects of SFM they deemed important. Thereby, their perceptions, requirements and rationalities on their sustainable spaces in forest management are displayed with a reduced danger to be pulled into a certain direction by the interviewers' preconceptualized framework about the topic (Silverman 2006; Dean 2010). The following section describes these various perceived achievements and problems of environmental forest management and its related driving forces or technologies. Hence, impacts and aspects related to forest certification will be scrutinized, as are the Finnish Forest Act and further means to achieve SFM or improve environmental performance in forestry.

PERCEPTIONS OF SUSTAINABILITY AND ITS DRIVERS IN FINNISH FOREST MANAGEMENT

Keeping in mind the twofold representation of Finnish forestry from the outset, it is important to be aware of some aspects concerning environmental issues and practices in Finland to understand and evaluate the actors' perceptions. In Finland, about 95% of commercial forests are PEFC certified by regional group certification (FMA 2011b). Being a member of FMA, the private forest owners become automatically part of the regional group certificate. This system is criticized, as it lacks audits prior to certification and may include actors unaware of their certification (e.g. Harkki 2004; Greenpeace 2011). Regional forest management institutions or regional contracting companies apply for participation in this regional group certificate. In early 2011, after a discussion that lasted almost 10 years (Albrecht 2010b), a Finnish FSC standard was approved and first areas of the global forestry company UPM-Kymmene have been certified (UPM 2011). Demanded by ENGOs, this FSC standard should improve Finnish forestry towards SFM.

The most prominent national debate concerning practical forest management concerns the issue of even-aged versus uneven-aged forest management2 . The former has been well-established practice for more than three decades and promoted by forest law, while the latter, being commonly practiced until the 1960s (Siiskonen 2007), has recently gained support as environmental and further non-economical management aspects increase. While a forest professional and academic discourse on the feasibility of the two approaches takes place (e.g. Tahvonen 2007; Kuuluvainen 2009; Laiho et al. 2011; Pukkala et al. 2011), the inclusion of uneven-aged forest management into forest law is being discussed by politics and separates the minds of institutions and actors related to forestry. A similar process is taking place in Sweden (e.g. Axelsson & Angelstam 2011). Thus many of the perceptions presented below concerning environmental forest governance in Finland circulate around these governmental technologies being used to promote SFM.

Forestry Institutions' Perceptions

According to the Forest Centre and FMA/FOU representatives, the current forest management system, backed by the Finnish Forest Act, Tapio's recommendations for SFM and PEFC group certificates, is generally perceived as sufficient to achieve SFM in Finnish and North-Karelian forestry. Environmental aspects addressed by law and the additional requirements by PEFC were described to be well integrated into the management system with only minor breaches appearing.

Hence, it was specifically pointed out that new approaches, for instance stricter certification criteria through FSC or uneven-aged forest management as demanded by most ENGOs, are deemed unnecessary. Accordingly, a recent proposal by a working group of the Ministry of Agriculture and Forestry on liberalizing the possibilities of forest management, thus including uneven-aged forest management, is seen as critical. Nevertheless, it was mentioned that law as well as certification are minimum requirements for SFM from an environmental protective aspect, while the criteria could always be tightened, however related to massive cost increases on the forest owner and management side. The Finnish Forest Act, in particular its 10th section concerning the protection of valuable habitats is seen as a major driving force for guaranteeing SFM practices, while PEFC is seen as an extra topping, adding some criteria. For instance, valuable habitats, additional to the legal requirements found in PEFC criteria were mentioned in this regard. However, certification is also regarded as a cost-raising factor driven mainly by company demand. In the case of FSC, it was pointed out that companies only demand it to please the ENGOs. Both certification systems, particularly FSC, were said to require unnecessary bureaucratic efforts, while stricter environmental criteria could be imposed without them if wished so by involved actors. Market-driven influences aside certification demands by companies are based on economical aspects, as wood prices, and not in relation to direct changes in environmental management practices, as stated by some interviewees.

Private forest owners are not regarded as drivers for improved environmental protection in forestry; yet, hostile attitudes towards environmental aspects and protection created by ENGO campaigns and the weakly planned implementation of Natura 2000 areas in the past are deteriorating. According to the interviewees, this development is mainly due to the METSO protection program with its voluntary protection approach (see Table 2). In relation to private forest owners, it was further stressed that their knowledge of environmental criteria or certification varies strongly. This owes much to the well-structured management system of Forest Centres, FMA/FOU and contracting companies, while most forest owners rely upon that these management aspects are taken care of by professionals, who do the planning and harvesting of their forests. Acceptance of environmental regulations when directly explained is high.

As to their own impact on the environmental performance of forest management, it was stated that environmental aspects, based on law, PEFC and Tapio's SFM recommendations are integrated into several processes. For instance, environmental issues are handled aside from economical aspects, in the Regional Forestry Programme by the Forest Centre, while FMA follows all environmental criteria deriving from law and certification in their forestry planning and harvesting. Further, the Forest Centre controls the environmental quality of regional forestry practices by recording breaches of law, as well as the implementation of environmental criteria like retention trees and valuable habitat demarcation. Additionally, both institutions provide and organize education and information about environmental related forestry practices to forest professionals and forest owners.

The Contractors' View

Information retrieved from the interviews with managers and owners of the three North-Karelian contracting companies displayed a cohesive picture about recent forestry practices and their applicability for SFM with the associations' perceptions voiced above. Nevertheless, aspects of environmental forest management were stressed from the implementation side, while several problems or discourses concerned with the same were mentioned. This relates to the aspect that contracting companies work for organizations in private, state or company owned forests, yet are not representing those ownerships but merely providing services based on the customer demands. Still, they are most concerned with the practicality of environmental criteria.

Table 2. METSO 2008–2016, overview (MMM 2010).

METSO Forestry Biodiversity Programme for Southern Finland 2008–2016	• compensation based protection scheme for private and state owned forest lands (voluntary forest owner agreements) • First step: 2008-2012, 180 million € funds • ~14.000 ha protected in 2008–2009 (3661 ha in private forests) • ~6.400 ha restored habitats (300 ha in private forests)
Targets	• improved protection/management of valuable habitats • improve network of protected areas in Southern Finland • halt decline of forest based species
Main critics	• insufficient funding • mainly active forest owners participate

According to the contractors, change concerning the integration of environmental aspects into forest management has appeared on two levels. First, practical issues, such as improved protection of valuable habitats, water protection through buffer zones and protection of waterways, as well as protection of retention trees, have improved the environmental quality of forest management. Second, parallel to the appearance of these practices, attitudes have shifted and the above-mentioned protection measures are described as becoming common practice while their purpose is being slowly understood among forestry professionals. One of the interviewees stressed that there remains a need to explain, especially to machine drivers, the importance of even small protective areas or habitats. Nevertheless, the current environmental criteria are regarded as sufficient for SFM, one contractor stating that some habitats such as "trickles on hill slopes" do not need to be protected due to their abundance in Finnish forests. This statement must be understood in reference to requirements laid down by the section 10 (2/1) of the Forest Act (MMM 1996) and PEFC criterion 10 (PEFC 2009).

Describing the law as a basic and obligatory framework for environmental criteria, most specific changes are attributed to PEFC certification. Hence retention trees and required buffer zones on lake shores, both not demanded by law, were mentioned as examples (cf. Forest Act 1996). The environmental quality of the former, consisting of a required minimum of 5–10 trees/hectare left standing after harvesting, were found to improve with increasing machine-driver understanding of their importance (cf. PEFC 2009). Still, in some cases, retention trees are logged in the post harvesting period by misinformed private owners for fire wood. A contractor specialized in insular logging pointed out that the latter, while contributing to water protection measures in general, has positive impacts on landscape protection values when moving on Finnish waterways. Regarded as a critical aspect of Finnish forestry is the missing management agenda of many private forest owners, which leads to neglected and bushy forest stands, a view commonly shared by Swedish foresters (Axelsson & Angelstam 2011).

It was further mentioned that the protection of valuable habitats related to law and PEFC have led to a fragmentation of logging sites making the job of contractors more challenging as more and more aspects and restrictions have to be taken into account. This results in smaller logging sites and constantly requires environmental education for machine drivers and planning professionals. To ensure proficient knowledge for machine drivers most large customers were said to demand an environmental management degree of them. This degree, provided by Tapio in cooperation with Forest Centres requires 4–6 days of courses and an additional amount of self-study to prepare forest professionals to recognize valuable habitats and to take environmental aspects in forest management and harvesting into account (Tapio 2011). Contractors stated that most of their employees, if not all, possess this degree. In addition, the large companies, like Stora Enso or UPM-Kymmene, provide environmental education to keep their contractors up-to-date on environmental regulation or changes in law. Attitudes among contractors towards these trainings varied to some degree. On the one hand, it was mentioned that training should be made mandatory in general and that a lack of possibilities for training exists, while on the other hand, the amount of training was regarded as too high. The additional costs (only educational costs are covered by the large companies) are criticized due to the loss of valuable working time if employees have to take classes instead of being in the forest logging.

Information in general was presented as being of the utmost importance to guarantee the performance of SFM practices in forestry. Since harvesting is done preferably in winter times with frozen soils, Finland's often extensive snow cover restricts recognition of valuable habitats, such as small springs, rivulets or bird-nesting sites. Machine drivers have to rely on their harvesting plans. Harvesting plans are mostly prepared in summer by the companies hiring the contractors' work, based on management plans for the respective forest sites. Thereby, quality of mapping was said to differ. For instance, when harvesting for Stora Enso in a forest owned by Tornator, valuable areas were described as well charted and provided as a GPS map to contractors, partially even including premarked retention tree groups. Thus, it is easy for machine drivers to recognize the environmentally important areas by using their built-in GPS devices, while in private forests, even though for the same company, such digital maps are often absent, which increases the risk of mistakes. Apart from its benefits, the continuous shift to digital maps and GPS is seen as critical, since marking in the forests ceases, and the abilities of machine drivers to use new systems were described as still limited. Generally, it was pointed out that harvesting guidelines provided by large companies take environmental issues of law and certification strongly into account, while at least for FMA this was questioned to some degree.

As for the drivers, no specific group was mentioned in relation to the increasing regulation and environmental criteria. Assumptions about environmentalists, some institutional authority and customer demands were provided as drivers for certification. The need for additional certification systems, specifically with stricter criteria was strongly denied and a critical attitude was expressed concerning uneven-aged forest management. However, ENGO demands are partially seen as rightful, yet not a large problem for contractors and in most cases regarded as solvable. It appeared from the interviews that contractors and their association take a more neutral stance towards the larger political debates and leave those to politics, large forestry and wood-based product companies and forest owner associations while concentrating on the practical issues of SFM.

Company perceptions: Tornator Oy

With forest holdings of 595,000 hectares in Finland, Tornator is the third largest forest owner in Finland while its focus on Eastern Finland makes it the most important non-state owner in NorthKarelia. Managing the forests formerly owned by Stora Enso which still holds 41% of Tornator shares, between 80–85% of cutting rights from its forests are sold to Stora Enso (Tornator 2011). Forest maintenance and planning is largely carriedout by the own forestry professionals while harvesting activities, specifically final felling, is carried-out by external contractors, mostly by Karel Wood, Stora Enso's main contractor in NorthKarelia. The companies own forests are completely PEFC certified. Planning and maintenance services are further offered to private owners in the vicinity of Tornator's forest areas. In addition to commercial management, Tornator has recently set aside some 1,250 hectares of its forest as part of several state promoted protection programmes, terminating negotiations with the Ministry of Environment lasting almost two decades. Additionally, two areas have been included within the Metso programme since 2009, the latest being a 74 hectare storm damaged area in North-Karelia (Tornator 2011). The Metso programme has been a focus of Tornator's own environmental responsibility programme in 2009 and its forest superintendents have been trained to identify its habitats (Tornator 2010). As burning over criteria have been tightened in PEFC, trials to burn over retention tree areas in each forestry team have been promoted as part of METSO habitat management (Tornator 2011).

Tornator representatives regard the major changes of integrating environmental aspects into forest management in a similar way to the contracting companies. Primary, specifically PEFC certification is said to have united the efforts of different actors to work on and integrate environmental aspects into forestry. It was stressed that forestry organizations, contractors and private owners are now collected within a single framework to handle environmental aspects. Second, in relation to practical changes which promote and improve SFM, retention trees, protection of valuable habitats, water protection and nature management of commercial forests (e.g. Metso) were mentioned. Forest and related environmental laws are considered to provide means for most of these practical changes, while specifically retention trees are credited to PEFC. Another effect brought about by the different environmental regulations is a concentration on economically valuable forest areas away from less productive stands, thus investment in, for instance, peat land has ceased.

SFM is stressed as an important aspect in Tornators' own forest management activities. Hence, requirements by Forest Law, PEFC and Tapio's recommendations for SFM are always integrated into forestry activities. For that purpose, own contractors for forest maintenance and Stora Enso's logging contractors are obliged by agreement to participate in training programs organized by the two companies, often taught by the Forest Centre or Tapio specialists. Contractors are provided with GPS maps entailing valuable habitats and other areas restricted for harvesting practices. Information on retention trees might be provided and respective areas should already be delineated at first thinnings. Nevertheless, this relatively new practice was described as being in need of improvement. Despite the digital support, the above-mentioned trainings should enable machine drivers to decide upon and recognize valuable areas also by themselves.

It follows that Tornator perceives PEFC as widely sufficient, while, for instance, FSC is recently perceived as too strict and based on complicated criteria, and thus it is regarded critically. Yet, implementation would be possible and requests from Stora Enso were received concerning this issue. Nevertheless, increased costs due to logging losses and additional costs in administration are expected and require evaluation, while being a matter of debate between Stora Enso and Tornator concerning plans for FSC certification. FSC demand by Stora Enso is perceived to be customer driven. One aspect highlighted as endangering SFM in Finnish forestry is a decreasing knowledge and activity among a growing number of forest owners. Consisting mainly of young owners living in long distance to their forests, these owners were found to neglect their holdings, which might lead to insufficiently managed forests. Being eventually positive for biodiversity in the short term, this development reduces forest productivity and value for forest owners; both considered aspects of SFM as well.

Despite its acclaimed achievements in SFM in its forests (e.g. Tornator 2010, 2011) Tornator, aside from UPM-Kymmene, is blamed by ENGOs for neglecting the protection of valuable habitats based on Metso criteria. The NL in cooperation with its parent organization FANC and with Greenpeace Finland mapped large areas of Tornator and UPM-Kymmene owned forests in 2008 and 2009 delineating 77 exemplifying areas, valuable for protection, 20 of which are situated in Tornator's forests (Luonto-Liitto et al. 2009). Not being pleased by these ENGO demands, it was agreed to evaluate the areas marked using one's own discretion. No concrete details were provided in the interviews. However, it was mentioned that some areas do meet Metso criteria while others, a marked sapling stand was described as an example, lack any protective values and are not considered for possible protection. It was further mentioned that Stora Enso, as buyer of most logging rights, is closely following the happenings and requests full information if respective stands are marked for cutting. Interestingly, opposite to the often mentioned "carrot and stick" approach (e.g. Cashore et al. 2004) FSC was not regarded as a means to avoid such campaigns but was more expected to increase the level of demands brought forward by ENGOs.

ENGO protests and market campaigns are recognized as drivers for environmental changes and seen as a push factor for FSC. Parallel to this, ENGO attention is described as shifting from Lapland to southern Finland, where forest owning companies appear as major targets for such campaigns. On the other hand, rethinking in forest policy is regarded as an important driver as well, since certain environmental measures have positive economical impacts in today's forestry-based business environment.

NGO perceptions: countering claims of SFM

Perceptions of Finnish ENGO on achievements and the state of SFM in Finland counter most attitudes expressed by the three actor groups above. First, none of the ENGO representatives regard Finnish forestry to be managed in a sustainable manner. On the contrary, most issue a quite low performance for Finnish forestry in terms of SFM and environmental protection. Within their criticism, the interviewees mentioned improvements and at least some positive development. It is interesting to note that they pointed out the very similar improvements mentioned by the previous actor groups; however they regard them as insufficient to reach SFM according to their definition. Additionally, these improvements, like retention trees, buffer zones and valuable habitats, are commonly attributed to law or the SFM recommendations by Tapio then to PEFC, despite being part of PEFC criteria.

In general, PEFC is rejected as a system providing SFM in Finland, due to its weak criteria, and lax implementation and control. Criteria exceeding law, like retention trees or lakeside buffer zones were deemed a nice thought but too weak to bring about significant changes. Only the representative of WWF Finland attributed improvements and certain positive aspects to PEFC as a system to refer immediately to the high amount of loopholes, and inferiority to FSC. Despite its annual regional audits, PEFC was described more as a toothless tiger, as several breaches of criteria dis covered by NGOs remained with no consequences for the certificate holders. Apart from these aspects, ENGOs criticized the fact that the minor improvements achieved with retention trees, buffer zones and the protection of some valuable habitats are promoted by forestry associations and partially by industry as the solution to the problem of insufficient SFM.

According to the representatives of ENGOs, Finland's forestry is in need of restructuring on a larger scale then with beauty corrections such as five retention trees per hectare. Therefore, the debate of even-aged versus uneven-aged forest management was more prominent in their talks. Opposing the positions of forest associations and some contractors, ENGOs promote uneven-aged forest management as the better alternative. In addition to environmental aspects such as improved biodiversity, economic gains through savings in forest regeneration and planting are presented to back their claim. In this regard, the political sphere was described as moving towards more possibilities for forest owners, however in a slow manner with FMAs opposing changes to the recent system.

The campaign which aimed at Tornator and UPM-Kymmene differed widely to the public protest campaigns related to old-growth forest destruction in Lapland. Instead of large publicity or protests to blame the respective companies, proposals based on fieldwork findings by NL concerning valuable habitats were submitted to the companies. While the data was published in Finnish online, no large publications (cf. Greenpeace 2009) have been prepared to target an international audience as compared to the Lapland forest conflict. Mapping of valuable habitats was conducted based on Metso criteria and on the improved connection of protected networks (LuontoLiitto et al. 2009). The response of UPM-Kymmene was described as cooperative; Tornator, on the other hand, was described as more reserved on the issue. It was stressed though that for Tornator, with its main revenues deriving from forest management and wood sale, decisions on excluding areas from exploitation weight heavier than for a transnational, integrated forest-product company like UPM-Kymmene. Yet, to increase pressure, proposals were sent to Stora Enso, the main customer of Tornator

As company owned forests were described as thoroughly managed, yet not sustainable in most cases, the situation for private forests was described as strongly variable. Criticized by the regional forestry institutions and contractors, the forest owners' lack of a management agenda was coupled with positive effects for biodiversity (e.g. no thinnings). Concerning certification, Finnish ENGOs expect the major companies to certify their forests with FSC in the future, UPM taking the first-mover role. Private owners, especially due to the negative attitudes of their associations are seen as less likely to switch to FSC in the coming years. Surprisingly, and contrary to the public chorus of international ENGOs, like WWF or Friends of the Earth, FSC is not regarded as the sole solution to achieve SFM in Finland. Although it would improve forest management in commercial forests, it was described as unsustainable with regards to forest energy harvesting, while increased political efforts were stressed in order to improve the network of protected areas. Thus while Finnish ENGOs' perceptions on Finnish forestry varies from that of most other forestry related actors, it also differs from the international, public NGO chorus.

The above-mentioned examples display the various rationalities, based on the differing positionalities of actors, which aim at stabilizing or including their regimes of practice in local and transnational forest governance. Political technologies (e.g. Baldwin 2003; Dean 2010) are thereby promoted to different ends and attributed with varying effects. The following critically discusses the performance of these various perceptions and their possible effects on the processes of transnational environmental forest governance.

DISCUSSION

The twofold public promotion of Finnish forestry practices is (re-)produced throughout the interviews as all actors, apart from the Finnish ENGOs, regard Finnish forestry as sustainable on their account. Yet, keeping in mind that those perceptions about SFM are closely tied to the actors' rationalities and influenced by various regimes of practice (Dean 2010), a more diverse picture unfolds. Rationalities of actors are co-produced by their positionality (Sheppard 2002), which makes actor groups or regimes of practice to focus on problematizations from their perspective (Dean 2010). Hence, perceptions expressed by the interviewees relate to a heterogeneous playing field despite an apparently unified arena. So, the question is not merely if there is SFM in Finland but which aspects of forestry are included in the rationalities of involved entities to decide upon that matter and how varying positionalities of corporate and private forest owners affect the processes of forest governance.

The actors in this study are largely concerned with commercial forestry and its sustainability. Thus, they access forestry with regard to commercially utilized forests and aspects of their management (see also Axelsson & Angelstam 2011). ENGOs, on the other hand, see forests from a more universal, environmental perspective. Hence, one must assume a variety of underlying definitions concerning the criteria and values of SFM on which the delineation of sustainable spaces rests (Hudson 2005). While an in depth evaluation of those definitions is not the aim of this paper, the utilization and promotion of different political technologies promoted on the basis of those is accessed. PEFC certification, Forest Law and Tapio's SFM recommendations are the most prominent technologies mentioned in the Finnish debate. Yet, they are attributed with various achievements towards different ends.

Supported by the dominant regime of practice consisting of Forest Centres, FMA and FOU the role of PEFC is the one of the defending champion, while FSC is yet to rise to become a contestant in the same weight class in Finland. The unifying role accredited to PEFC in environmental issues in the interviews is an important aspect of its strength. Since it is backed-up by the monopoly like structured institutions of Finnish forestry it has a large capacity to promote itself. This was supported by an ENGO forestry expert who stressed the onesided, pro PEFC information provided throughout this monopoly system to forest owners and professionals. However, based on positive third party audits, as well as on checkups carried-out by its supportive institutions, PEFC has been able to improve forestry and, further, to strengthen forest law enforcement, as pointed-out by a forest industry representative (see also Indufor 2010). While ENGOs oppose this claim, some of the advertised improvements are accepted, yet attributed to different drivers. Achievements such as retention trees, lake-side buffer zones or additional habitats are affiliated with Tapio's SFM recommendation rather than to PEFC, despite PEFC criteria being theoretically binding rather than Tapio's recommendations. This shows how even insufficient deemed achievements are attributed to another political technology rather than to run the danger of creating an eventual support/agreement with an opposed regime of practice. Generally, ENGOs describe the achievements mentioned throughout the actor groups as green-washing of forestry practice by forestry institutions and companies, while distracting from the larger problem such as decreasing biodiversity due to a lack of protected areas.

With respect to these wider problematizations stressed by ENGOs, it has to be taken into account that this plays only a marginal role within the contractors' or forest owning companies' perceptions of SFM. Exemplified by a Tapio forestry expert statement that not all species can be expected to cope in commercially used forests, wider aspects such as large protected areas, or high biodiversity rankings are not necessarily regarded as part of their business. At least not without sufficient compensation paid (e.g. METSO). This stresses the importance of positionality effects related to actors' rationalities and their resulting practices (Albrecht 2010b), and is akin to findings in Swedish forestry (Axelsson & Angelstam 2011). Thus, as the debate about SFM includes multiple forest images, the common attempt by ENGOs to kill two birds with one stone, for instance with FSC, buries risks of rejection. These spheres are separated in the perceptions of contractors, institutions and Tornator. Thus, while PEFC, forest law and Tapio's SFM recommendation provide sufficient means for SFM in commercial forests, increasing a protected area network seems to not be affiliated within their direct range of tasks. While, compared to the global FSC equals SFM chorus by many ENGOs, Finnish ENGOs, on the contrary, are aware of that issue and try to avoid accusations of confusing SFM with forest conservation, an accusation mentioned in an interview with a large Finnish forestry company.

With rationalities related to SFM being influenced by deemed external relations as well, for instance distrust in ENGOs or private ownership aspects (Albrecht 2012), PEFC, while deemed sufficient, is not promoted as the solution to all problems, even by its supporters. Problems like varying quality of retention trees due to changing forest professional knowledge and interest, or their post harvesting removal by forest owners for their firewood are acknowledged. The latter is a common problem in Sweden, despite its FSC and PEFC forests as well (Hysing & Olsson 2005). However, the achievements in commercial forestry, the felt fragmentation of logging sites due to habitat and buffer zone protection, and the high amount of environmental related education requirements for forest workers are relations that shape rationalities and provide most forest professionals with the perception that they are doing their share to improve forestry. As even PEFC certification is regarded a burden, requiring costs, education and changed practices, the attitude towards increased restrictions and regulations, be it FSC or any other involuntary measures, are prone to be rejected for the time being. This also seems to relate to perceptions of Finnish SFM practices in global comparison. Interviewees from FMA, FOU, FFIF and Tapio regarded them as top level, while, for instance, forestry in Russia was not seen as being superior despite its FSC certification.

Additionally, the demand for certification by the large buying companies is seen as only being for their reputational sake, and not necessarily based on altruistic attitudes to save the environment. The fact that UPM refrained from FSC certification of all its forests, due to low market demand, and disconnection to its own processing facilities, supports these claims (Woodmark 2011). It was criticized by contractors and by Tornator that many costs remain with them while the companies which demand certification restrain from paying higher prices for those sustainable products. This aspect additionally shapes a positionality that creates rationalities which rejects further regulations. Thereby, PEFC, in concert with Finnish forest law and Tapio's SFM recommendation based on the positionalities of most Finnish forest actors, continues to (re-)produce itself as the dominant regime of practice (e.g. Dean 2010). This, in turn, has strong implications for transnational forest governance. As the majority of Finnish forestry products are exported (FFIF 2011b), PEFC products continue to dominate the certified product market by numbers. Yet, as stated by Massey (2005) on account of the relative instability and openness of space, such domination is constantly challenged. In Finland, the debate of even- vs. uneven-aged forest management is just such an unfolding problematization which affects the various regimes of practice and the space they aim to govern (Dean 2010). Thereby, even-aged forest management, which, following Sheppard (2002), is concurrently promoted as the "natural" regime of practice by Forest Centres, FMA and FOU is in jeopardy.

The opening up of forestry practices, thus including uneven-aged forest management on a large scale, is also related to the dismantling of the monopoly structure of the Forest Centres and FMAs. Thus, forest owners not only should be enabled to choose their mode of harvesting but have free choice from where to acquire consultation and management services. As such a change is currently politically debated, it is highly welcomed by ENGOs and is further supported by the FFIF and, to some degree, the FOUs, based on the right of forest owners to do as they wish with their forest property. Further, the increased distance of forest owners from their holdings diminishes the influence and informational flow deriving from regional Forest Centres or FMAs, thus detaching the owners from those still dominant regimes of practice. While such a shift might provide more possibilities for FSC and other alternative practices to be integrated, it by no means equals an abrupt end to the dominant regime consisting of the amalgamation of forest law, PEFC and Tapio's recommendations. Additionally, influences on forest management and utilization are not merely seen as positive for SFM or biodiversity.

The lack of a management agenda by private forest owners and the resulting increase in unmanaged, bushy forests criticized by contractors was to some degree affiliated by ENGOs with the onesided management possibilities of the current system. As these areas often entail increased biodiversity, due to their natural character, opening up new management possibilities might lead to a reintegration of such areas into commercial forestry, thereby resulting in a possible loss of biodiversity, as stressed by a Tapio expert. Further, a possible decrease of dead wood was described as a possible effect, while missing expertise regarding large scale uneven-aged forestry was stressed to be the main hindering factor (see also Axelsson & Angelstam 2011). Nevertheless, while being probably less productive, uneven-aged forestry is seen as less costly since reforestation and clear-cut site management costs lapse. The fact that forest regeneration is Tornator's largest single expense fosters that claim (Tornator 2011). Nevertheless, uneven-aged forestry is seen as more likely to be integrated into private forest holdings by the interviewees compared to corporate holdings, due to their focus on wood production. Yet, implications on the certification systems as political technologies for SFM performance (Baldwin 2003; Albrecht 2010b) in Finland remain tied to the positionalities of the implicating actors in their local settings.

Being recently promoted due to business demand through large companies like UPM-Kymmene and Stora Enso, both being supportive for PEFC in Finland at the same time (Albrecht 2012), the awaited opening-up of forestry management services adds possibilities for FSC to gain its share of the previously dominated PEFC market. This especially accounts for corporate holdings, as they are prone to ENGO campaigns and market pressure, while implementation of complex standards is facilitated due to their large-scale and professional management. The recent certification of UPM-Kymmene forest areas is a good example of this. Yet, the rationalities of implementing forest actors are (re-)produced by their positionalities (Sheppard 2002; Murdoch 2006). Thus, despite the new possibilities opening up, the current system of Forest law/PEFC seems better suited to target the sustainable spaces perceived by most implementing actors. Hence, the strong and positive role affiliated by most interviewees to the state driven METSO programme counters the all-in-one market driven SFM alternatives, to some degree. It separates the struggle for SFM in commercial forests, according to most actors sufficiently managed by Forest law/PEFC, and improves protection measures through state funds on a voluntary basis. The METSO programme, based on its wide recognition, apart from the criticism of insufficient funding by ENGOs and FFIF, has managed to be a supported political technology that fits the rationalities of most actors belonging to opposing regimes of practice. Thereby, from a local perspective, it strongly influences market driven aspects such as certification demand and further highlights the importance of local steering mechanisms for transnational environmental forest governance.

CONCLUSIONS

The debate on sufficient measures for SFM will continue in Finland, with an expected increase in its complexity, due to the integration of wood energy aspects. The system's current restructuring opens up possibilities for previously excluded alternative political technologies to be integrated into Finnish forestry since positionalities change and with them the rationalities of involved entities. It is, however, questionable if these changes will bring the awaited improvements hoped for by ENGOs or if the current dominant regime of practice entails the resilience to maintain its position. While FSC is expected to increase on corporate owned land, due to business demand, the positionality of forest owners in Finland is preventing such a sudden change and there can be doubt if it will rise to become the dominant system in Finland. On the other hand, uneven-aged forest management is more likely to be integrated into privately owned forests. Thus, it remains to be seen if respective forest owners choose FSC in concert or if Forest Centres, FMAs and FOUs provide uneven-aged aside from even-aged forestry with a continuous reliance on PEFC as the more pragmatic and forest owner trusted system. Thus, will they adjust their system to remain the dominant regime of practice?

Following the research aims presented in the introduction, this paper has displayed how local perceptions based on the positionality of actors (re-)produce the various rationalities of actors on SFM. Further, they display these rationalities' influence on decision-making and practices for SFM and transnational forest governance per se. While the positionality of corporate forest owners, tied to market demand and vulnerable to ENGO campaigns, is more strongly affected by market driven aspects of forest governance (Cashore et al. 2004; Albrecht 2010b), private forest owners are strongly influenced by their locally embedded relations which (re-)produce their positionality. These include governmental instruments (e.g. METSO), knowledge networks or practicalities which all play their role in influencing and shaping transnational forest governance practices, and the utilization and promotion of its political technologies as forest certification schemes. Thus, local regimes of practice in resource peripheries with their affiliated knowledge networks have the strength to resist and shape market-driven domination in forest governance. Based on the theoretical approach of this paper, it is likely that these findings can be generalized to other areas of natural resource management (e.g. sustainable bioenergy), as long as private ownership exists alongside corporate and state owned premises. However, to strengthen this claim further research is required.

NOTES

[1] One of the companies, Anaika Wood had just changed its focus from forestry contracting to bioenergy production, nevertheless it was active in contracting during recent years.

[2] Even-aged forest management is commonly practiced by a rotational process of thinnings, clear-cuts and reforestation. Uneven-managed forest management refers to practicing continuous growth forestry (uneven aged tree cover) with selective loggings and mostly natural regeneration (e.g. Axelsson & Angelstam 2011).

ACKNOWLEDGEMENTS

The research has been funded as part of the Academy of Finland project (no. 14770) "Transnationalization of Forest Governance" and by the University of Eastern Finland (project no. 931429). Special thanks go to Eija Kärkkäinen for her valuable help with the Finnish interviews and translations. Further I like to thank the North Karelian forestry companies, Institutions and Finnish NGOs who have participated in this research.

REFERENCES

1. Albrecht M 2010a. Sustainable forest management through forest certification in Russia's Barents region: Processes in the relational space of forest certification. In Fryer P, Brown-Leonardi C, & Soppela P (eds). Encountering the changing Barents: Research challenges and opportunities, 24–34. Arctic-Centre, Rovaniemi.
2. Albrecht M 2010b. Company positionality and its effects on transnational forest governance: Two companies and

their approaches to forest certification. Geographische Zeitschrift, 98: 2, 116–132.

3. Axelsson R & Angelstam PK 2011. Uneven-aged forest management in boreal Sweden: local forestry stakeholders' perceptions of different sustainability dimensions. Forestry, 84: 5, 567–579

4. Baldwin A 2003. The nature of the boreal forest: Governmentality and forest-nature. Space & Culture, 6: 4, 415–428.

5. Cashore B, Auld G & Newsom D 2004. Governing through markets: forest certification and the emergency of non-state authority. Yale University Press, London.

6. Cashore B, Egan E, Auld G & Newsom D 2007. Revising theories of nonstate market-driven (NSMD) governance: Lessons from the Finnish forest certification experience. Global Environmental Politics 7: 1, 1–44.

7. Cloke P, Cook I, Crang P, Goodwin M, Painter J & Philo C (eds) 2004. Practising human geography. SAGE, London.

8. FSC 2011. FSC Standard for Finland. Finnish FSC Association, Helsinki.

9. Greenpeace 2009. Kein Kahlschlag für Kopierpapier. Greenpeace, Hamburg.

10. Greenpeace 2011. On the ground: the controversies of PEFC and SFI. Greenpeace.

11. Harkki S 2004. Certifying extinction? An assessment of the revised standards of the Finnish Forest Certification System. Greenpeace, Helsinki.

12. Harkki S 2008. Out of Control: high Conservation Value Forest logging und FSC controlled wood in Finland. Greenpeace, Helsinki.

13. Horne P 2004. Forest owners' acceptance of incentive based policy Instruments in biodiversity conservation – A Choice experiment based approach. Australian Agricultural and Resource Economics Society Conference, Melbourne.

14. Hysing E & Olsson J 2005. Sustainability through good advice? Assessing the governance of Swedish forest biodiversity. Environmental Politics, 14: 4, 510–526.

15. Hänninen H & Kurttila M 2004. Metsänomistajien tiedot luonnon monimuotoisuutta vaalivan metsänhoidon velvoitteista ja suosituksista. Metsätieteen Aikakauskirja 3, 285–301.

16. Indufor 2010. Comparison report on the changes in the revised PEFC FI 1002 standard and their impact on forest management. Indufor, Helsinki.

17. Karppinen H 1998. Values and objectives of non-industrial private forest owners in Finland. Silva Fennica, 32: 1, 43–59.

18. Laiho O, Lähde E & Pukkala T 2011. Uneven- vs even-aged management in Finnish boreal forests. Forestry, 84: 5, 547–556.

19. Merlingen M 2003. Governmentality: Towards a Foucauldian framework for the study of IGOs. Cooperation and Conflict: Journal of the Nordic International Studies Association 38: 4, 361–384.

20. Mielikäinen K & Hynynen J 2003. Silvicultural management in maintaining biodiversity and resistance of forests in Europe-boreal zone: case Finland. Journal of Environmental Management 67, 47–54.

21. Parviainen J & Västilä S 2011. State of Finland's Forests 2011. Ministry of Agriculture and Forestry, Helsinki.

22. PEFC 2009. Criteria for group certification; Level of Forest Centre or a Forest Management Association. PEFC Finland, Helsinki.

23. Pukkala T, Lähde E, Laiho O, Salo K & Hotanen J 2011. A multifunctional comparison of even-aged and uneven-aged forest management in a boreal region. Canadian Journal of Forest Research 41: 4, 851–862.

24. Rutherford S 2007. Green governmentality: insights and opportunities in the study of nature's rule. Progress in Human Geography 31: 3, 291–307.

25. Schlyter P, Stjernquist I & Bäckstrand K 2009. Not seeing the forest for the trees? The environmental effectiveness of forest certification in Sweden. Forest Policy and Economics 11: 5–6, 375–382.

26. Siiskonen H 2007. The conflict between traditional and scientific forest management in 20th century Finland. Forest Ecology and Management 249, 125–133.

27. Silverman D 2006. Interpreting qualitative data: Methods for analysing talk, text and interaction. 3rd ed. SAGE, London.

28. Tahvonen O 2007. Optimal choice between evenand uneven- forest management systems. Working Papers of the Finnish Forest Research Institute, 60.

29. UPM 2005. Parallel field testing of forest certification standards. UPM Forestry and Wood Sourcing Environmental Forestry Affairs, Valkeakoski.

30. Yin RK 2003. Case study research: Design and methods. 3rd ed. SAGE, Thousand Oaks.

CHAPTER 12

AN INDEX OF FOREST MANAGEMENT INTENSITY BASED ON ASSESSMENT OF HARVESTED TREE VOLUME, TREE SPECIES COMPOSITION AND DEAD WOOD ORIGIN

Tiemo Kahl, Jürgen Bauhus

Chair of Silviculture, Faculty of Environment and Natural Resources, University of Freiburg, Germany

ABSTRACT

Forest management intensity often affects biodiversity, ecosystem processes and ecosystem services. To assess the influence of past management intensity on current ecosystem properties, management intensity must be quantified in a meaningful and reproducible approach. Here we developed the simple yet effective Forest Management Intensity index (ForMI), which is based only on inventory data of the living stand, stumps and dead wood. The ForMI is the sum of three components taking into account: 1. the proportion of harvested tree volume (Iharv), 2. the proportion of tree species that are not part of the natural forest community (Inonat) and 3. the proportion of dead wood showing signs of saw cuts (Idwcut). Each component ranges between 0 (no sign of management) and 1 (intensive management). Our analysis suggests that the ForMI can be used to assess management intensity in Central European forests for the last 30 to 40 years, depending on decay rates of stumps and dead wood. Our approach was tested using data of 148 forest plots of 1 ha in size in Germany. We found a significant distinction between plots that were previously described as managed and unmanaged as well as between plots comprising trees species of the natural forest community and those with additional, introduced coniferous tree species. We conclude that the index is applicable to a wide range of forest management types, but should not be misinterpreted as an index for old-growth structure.

INTRODUCTION

Forest management intensity is suggested to be a main driver of biodiversity, ecosystem processes and ecosystem services (Fischer et al. 2010, Gustafsson et al. 2012). To assess the influence of past management intensity on current ecosystem properties, management intensity must be quantified in a meaningful and reproducible approach. Here we are concentrating on influences of silvicultural forest management at the stand level, which represents impacts at the ecosystem level. So far, there is no straightforward and agreed approach for its quantification. The main challenge in developing such an index of forest management intensity is to incorporate the different elements of forest management that affect stand structure and composition and to quantify these. Recently, approaches have been suggested that require theoretically derived variables, which are not directly related to a particular

Citation: Kahl T, Bauhus J (2014) An index of forest management intensity based on assessment of harvested tree volume, tree species composition and dead wood origin. Nature Conservation 7: 15–27. doi: 10.3897/natureconservation.7.7281.

forest stand in question (Luyssaert et al. 2011, Schall and Ammer 2013). Schall and Ammer (2013) provided an extensive review on concepts and approaches for assessing land-use and specifically forest management intensity to provide the scientific basis for their approach. We believe that the two previous approaches produced indicators of forest management intensity that are not entirely transparent, not easy to quantify, and not strictly or exclusively related to human pressure on forest ecosystems; seeNiemeijer and de Groot (2008) for a discussion of properties of useful environmental indicators. Here we propose an alternative approach, the index of Forest Management Intensity (ForMI), which is derived from stand inventory data of living trees, stumps and dead wood. The ForMI is the sum of three components taking into account: 1. the proportion of harvested tree volume (*Iharv*), 2. the proportion of tree species that are not part of the natural forest composition (*Inonat*) and 3. the proportion of dead wood showing signs of saw cuts (*Idwcut*). In the following we provide the rational for constructing the index from these three components.

One core activity of forest management consists of the harvest of living trees. There are a variety of silvicultural approaches leading to even-aged forest stands characterized by distinct age-classes or uneven-aged forest stands resulting from selection cutting. Despite the large differences between those silvicultural approaches the result is always the removal of wood. Unfortunately, the records about the amount of wood volume removed are often not specific at the stand level but are documented for compartments or management units, and not all assortments may be traced; e.g. firewood may not be recorded. However, tree removals are traceable by tree stumps left after cutting. To gain a relative measure of the amount of harvested tree volume, the difference between the observed standing volume to the maximum volume based on the self-thinning relationship for a particular age and tree species (Spellmann et al. 1999) may be used (Luyssaert et al. 2011, Schall and Ammer 2013). One major drawback of this approach is the reliance on a theoretical maximum stand carrying capacity, which is influenced by site conditions and tree species. Since these relationships are mainly established for mono-specific and even-aged stands, mixed and uneven-aged stands are rather poorly described (Pretzsch 2009). In addition, these approaches neglect disturbances, which may have affected a particular stand and influenced the difference between a potential and the actual wood volume. The indices of Schall and Ammer (2013) and Luyssaert et al. (2011) are therefore not a direct measure of the harvesting intensity for a particular forest stand and they do not distinguish between human pressure and natural reductions in standing tree volume. Thus we propose to quantify the proportion of wood that has been removed over a period of time that is still indicated by the stumps. This is a direct expression of human pressure, as is often used in environmental indicator sets (Niemeijer and de Groot 2008), on the particular forest ecosystem in question.

In addition to harvesting, the composition of tree species is one of the main factors influenced by forest management. Although preindustrial forest management and land use change may have altered the natural tree species composition already, there is no doubt that the shift in tree species composition following the early industrial period towards secondary forests comprising coniferous species (*Picea abies*, *Pinus sylvestris*) in Central Europe has had a profound influence on forest ecosystems (Bohn et al. 2004, Ellenberg and Leuschner 2010). This element of the index therefore considers the influence on ecosystem properties and processes triggered by human-induced changes of tree species composition. Introduced tree species may change the habitat for herbivores and their predators (e.g. Gossner and Ammer 2006), the disturbance dynamics because they are more susceptible to certain disturbance factors (e.g. Albrecht et al. 2012), or ecosystem nutrient cycling and soil properties (e.g. Binkley and Giardina 1998) to name a few of such alterations. Here we refer to "natural" or "non-natural" tree species as those, which are, or are not, part of the potential natural vegetation (Tüxen 1956) that would occur as an expression of soil and climate at the observed site

Since in actively managed forests, wood is regularly removed, both the quantity and quality of dead wood are substantially altered when compared to unmanaged forests (Jonsson et al. 2005). At the same time, dead wood is of critical importance for many ecosystem functions (Harmon et al. 1986), of which the most relevant are C sequestration (e.g. Litton et al. 2007, Kahl et al. 2012), nutrient cycling (Brunner and Kimmins 2003, Kuehne et al. 2008), and provision of habitat for wood-dwelling organisms (Lonsdale et al. 2008, Rondeux and Sanchez 2010). For example, in Germany, 56% of all forest Coleoptera are dependent on dead wood (Köhler 2000). Consequently, the amount of standing and lying dead wood has been adopted as one of the Pan-European Indicators of sustainable forest management (MCPFE 2002). However, since the quantity of dead wood is highly variable in managed as well as in unmanaged forests of Central Europe (own data), it is unlikely to be a good indicator for forest management intensity. While in long-term unmanaged forests, large quantities of dead wood will accumulate (Christensen et al. 2005), this may not occur in the short-term, in particular when harvesting ceases at a time of stand development, when little dead wood is created through processes, such as self-thinning or disturbances (Harmon 2009). For example, an intensive harvesting operation may leave large quantities of dead wood if the utilization standards are low. Taking this into account, and in order to keep dead wood quantity as an independent variable for further statistical analysis, we selected a variable that is largely independent of dead wood quantity but describes its quality by means of its origin. Forest management generally tries to avoid the natural death of trees since this would reduce the quality of the products or the amount of harvestable wood volume. This in turn leads to a decrease of dead wood originating from natural mortality of trees. At the same time in managed forests, more dead wood originates from the cutting of trees and therefore shows saw cuts. Hence we used the proportion of dead wood with signs of saw cuts as the third component of the proposed index.

In the following, we present the definition of the proposed ForMI and show an application to a set of 148 forest plots as well as a comparison with two other indices that have been calculated for the same plots.

METHODS

Definition of ForMI

The first component of the index *Iharv* describes the proportion of harvested tree volume. Our approach is based on the assumption that the sum of the living standing tree biomass (1), harvested tree biomass removed from the stand (including the logging waste in the stand) (2), and the dead wood biomass of trees that died naturally (no chain saw cuts) (3) can be used as a proxy for the potential cumulative merchantable volume (Pretzsch 2005) at a given stand age. This approach makes the assumption of a theoretical maximum stand carrying capacity redundant and is independent of site quality, previous disturbances and tree species composition. Cut stumps are used to derive estimates of the harvested volume.

The main drawback of this approach is the uncertainty about the time until the stumps and dead wood are decayed. The decay rate of stumps depends mainly on tree species, site conditions and size (Shorohova et al. 2012). For southern Finland stump turn-over times of 61, 56 and 43 years for pine (*Pinus sylvestris*), spruce (*Picea abies*) and birch (*Betula* spec.), respectively, were found (Shorohova et al. 2012). In the case of European beech (*Fagus sylvatica*), which is the dominant deciduous tree species in Central Europe, we assume that the stump turnover time is similar to the turnover time of logs, about 54 years (Kahl et al. 2012). These values should be considered as maxima for the longevity of stumps and a time of 30 to 40 years seems realistic for the probability to find remains of cut stumps. Cut stumps of Central European tree species, such as oak (*Quercus* spec.), which are known to be more decay resistant, would add a bias to this stump-based approach. Stump size (diameter) seems to have only little influence on stump turnover time. No influence of size on stump turnover time was found for *Picea abies* in Sweden (Melin et al. 2009). The same was true for stumps of *Picea abies* and *Pinus sylvestris* in a study in Finland, while the turnover time of *Betulus* spec. stumps in the same study slightly increased with size (Shorohova et al. 2012). In addition, contradictory results have been reported for the influence of diameter on the decay rate of wood. For example, decay rates for dead wood of *Picea abies* increased (Næsset 1999) or decreased with increasing diameter (Brown et al. 1996). Regarding these studies we believe that stump size is far less important for stump turnover time than may be assumed. Since stump turnover time is the most critical point in this context, we suggest restricting this approach to regional scale comparisons of forest management intensity to minimize the influence of climate on decomposition rates of stumps.

Although stump turnover time maybe highly variable, this approach has the advantage that the influence of tree harvest recognisable as stumps virtually decays with time. Harvesting events that occurred recently can be fully considered, whereas harvesting events that date further back will have increasingly less influence on this measure, which presumably also reflects a decreasing legacy influence. The harvested tree volume determined in this way is certainly less accurate than could be derived from repeated inventory or detailed, stand-based harvesting records, however, in most cases these do not exist. So the sum of harvested volume, dead wood volume and the living volume provides the proxy for the potential cumulative merchantable volume of the last 30–40 years.

The *Iharv* is calculated as the ratio of harvested volume to the sum of standing, harvested and dead wood volume. A value of 1 represents a clear-cut site, where tree regeneration has not yet reached the minimum tree diameter limit used for inventory purpose and no dead wood occurs, and a ratio of 0 means that no tree harvest occurred within the last 30–40 years. The index *Iharv*can only be applied, where conventional forest management using chain saws or a harvester leaves stumps. In general it can be said that all parameters (e.g. tree species, size, climate) that increase stump turnover time will also increase the component *Iharv*.

The component *Inonat* is a measure for the volume proportion of non-natural tree species. It is estimated as the proportion of harvested, living and dead wood (without saw cuts) volume of non-natural tree species to the sum of harvested, living and dead wood (without saw cuts) volume of all tree species. Again here, the legacy of former stand conditions (e.g. a cohort of introduced spruce trees that were harvested within a time frame of 30–40 years prior to assessment of forest management intensity in a forest composed now entirely of tree species belonging to the natural forest vegetation) is taken into account. A value of 1 means that the whole stand consists of non-natural tree species, whereas a value of 0 is a stand composed of tree species belonging to the natural forest vegetation only.

The component *Idwcut* represents the proportion of dead wood volume with saw cuts to the total amount of dead wood volume. A value of 1 means that all dead wood originated from management activity (wood residue) or that trees that died by natural causes were cut and used, while a value of 0 means that all dead wood is a result of natural tree death. In case where no dead wood occurs but clear signs of management are visible e.g. due to fuel sampling or in plantation forests with whole tree harvesting we recommend to adjust the *Idwcut* to a value of 1.

The forest management index ForMI is the sum of all three components *Iharv*, *Inonat* and *Idwcut*. Although the relative importance of each component as an indicator of forest management intensity is debatable, we decided to weigh each component by a value of 1. For example, a ForMI of 3 would apply to a clear-cut stand of non-natural tree species, where no dead wood of natural origin occurs. A value of 0 would apply to a stand, where no stumps were found, only natural tree species and only dead wood without saw cuts occur. The ForMI describes strictly past management intensity and does not aim to describe forest structure or level of oldgrowthness (c.f. Bauhus et al. 2009). This is due to the fact that stand age and absolute stand volume do not play a role. A value of 0 could equally apply to a 20-years-old stand in an early successional stage after a stand-replacing fire or as primary successional vegetation on a river bank, but it also could represent a 500-years-old primeval forest.

Application of ForMI

To test our approach, we used inventory data of the 150 forest experimental plots of the German Biodiversity Exploratories (Fischer et al. 2010). The 1 ha (100 m × 100 m) sized plots are equally distributed in three exploratories representing the regions: Schwäbische Alb (Alb), Hainich-Dün (H-D) and Schorfheide-Chorin (S-C). The predominant silvicultural systems and forest management types in each region were considered. These produce even-aged coniferous stands (*Picea abies*, *Pinus sylvestris*), even-aged beech (*Fagus sylvatica*) stands, uneven-aged beech stands, and even-aged oak (*Quercus* spec.) stands. In addition, there are extensively managed or unmanaged beech-dominated forests. The even-age stands are later referred to as age-class forest. The inventory data of the living stands are based on a plot inventory (500 m^2) carried out in 2008 and 2009 (Fischer et al. 2010, Hessenmöller et al. 2011). Dead wood inventories were conducted in 2012 using a stratified sampling design with all standing and downed dead wood >25 cm being recorded on the 1 ha plot (volume calculations based on Meyer 1999) and downed dead wood >7 cm and <25 cm being estimated using the line intersect method (Warren and Olsen 1964, Van Wagner 1982) with two perpendicular transects of together 270 m length. Information on saw cuts and dead wood volume was then used to calculate the *Idwcut*. Stumps > 7 cm diameter were estimated on 2 m wide strips along the same transects, which served as central line of the strips. The area covered by these strip-transects was 522 m^2. Volume of harvested trees was calculated based on allometric functions (Muukonen 2007). Data of 148 plots were available for this analysis. The new index ForMI was compared with the existing land use and disturbance intensity (LUDI) approach by Luyssaert et al. (2011) on all available 148 plots and with the Silvicultural Management Index (SMI) by Schall and Ammer (2013) on a subset of 30 plots. All analyses were performed in R 3.0.2 (www.r-project.org).

RESULTS AND DISCUSSION

The ForMI provided a clear distinction between managed and unmanaged forests (Wilcox Test $p<0.001$) (Fig. 1). The main difference between managed and unmanaged forests was caused by the low number of stumps and the low amount of dead wood with saw cuts in unmanaged forests. Four out of 21 unmanaged stands had a ForMI of 0, which suggests that they have been unmanaged for at least 30-40 years. This result is congruent with other information about these unmanaged forest plots (Fischer et al. 2010). The largest contribution to the high ForMI value of coniferous forests was attributable to the *Inonat* component, which is close to 1 in mono-specific coniferous forests. The three variables of the ForMI presented here were calculated based on the volume of living, harvested and dead trees. For the calculation of *Iharv* and *Inonat* also the use of basal area was tested for 29 selected plots. Because basal area of dead wood could not be estimated, *Iharv* was calculated based on harvested and standing tree volume only. A comparison based on a linear model between the ForMI based on volume calculations as predictor variable and the basal area approach as predicted variable shows that both yield comparable result ($a=0.97$, $b=0.13$, $p<0.001$, $R^2=0.97$).

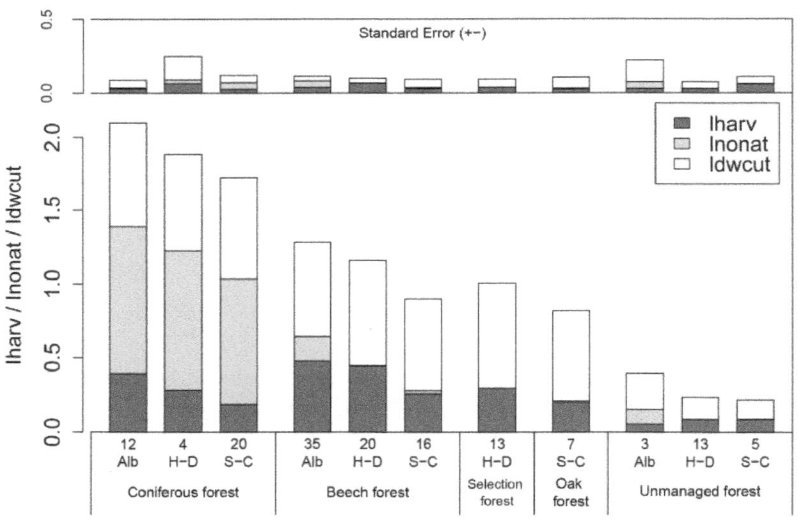

Figure 1. Distribution of the forest management intensity index ForMI and its 3 components across 5 different forest management types in the Biodiversity exploratories Schwäbische Alb (Alb), Hainich-Dün (H-D) and Schorfheide-Chorin (S-C). Numbers below the bars indicate the number of plots per group. The second bar plot above shows the standard error for each component. In the exploratories Alb and H-D, coniferous forest comprises even-aged *Picea abies* stands, and in the S-C exploratory *Pinus sylvestris* age class forest. Beech forest in all exploratories refers to even-aged *Fagus sylvatica* stands. Selection forest in the H-D exploratory is uneven-aged selection cutting stands dominated by *Fagus sylvatica*. Oak forest in the S-C exploratory is even-aged *Quercus* spec. forest. Unmanaged forest in the Alb exploratory represents extensively managed *Fagus sylvatica* forest, in the H-D and S-C it has been unmanaged for 10–60 years.

The ForMI as well as its three components cover the whole range of values across the 148 experimental plots (Fig. 2). The distribution of *Iharv* is right-skewed indicating that only few plots experienced rather intensive recent harvest. The distribution of *Inonat* shows a bimodal shape with two maxima, one at 0 (only natural tree species), reflecting the dominance of beech plots among the exploratories, and one at 1 (only non-natural tree species) and only few stands, where these two tree species groups occur together. *Idwcut* shows a left-skewed distribution indicating that in most stands dead wood has been created by harvesting activities. The distribution of ForMI is slightly right-skewed and peaks at a value of 1. No relationship between*Iharv* and *Inonat* was found while there were significant positive, but not strong relationships between *Idwcut* and *Iharv* and between *Idwcut* and *Inonat* (Fig. 3)

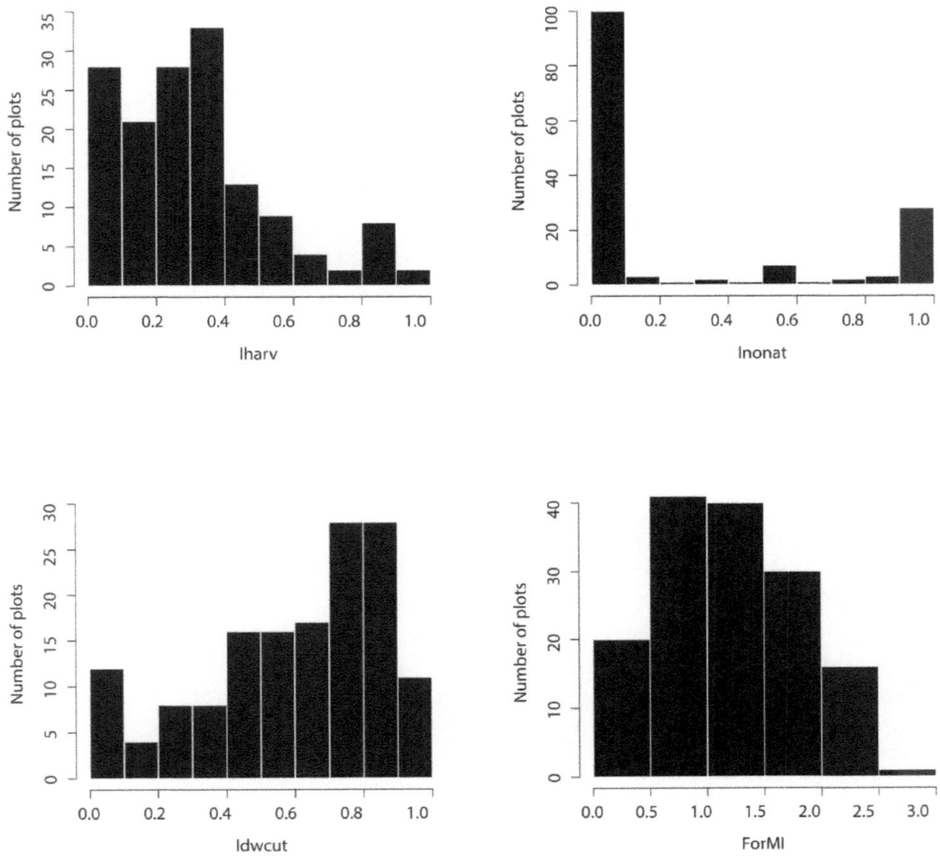

Figure 2. Distribution of the forest management intensity components: Iharv - proportion of harvested volume; Inonat – proportion of non-natural tree species; Idwcut – proportion of dead wood volume with saw cuts; and of the combined ForMI – forest management intensity index. Total number of forest plots is 148.

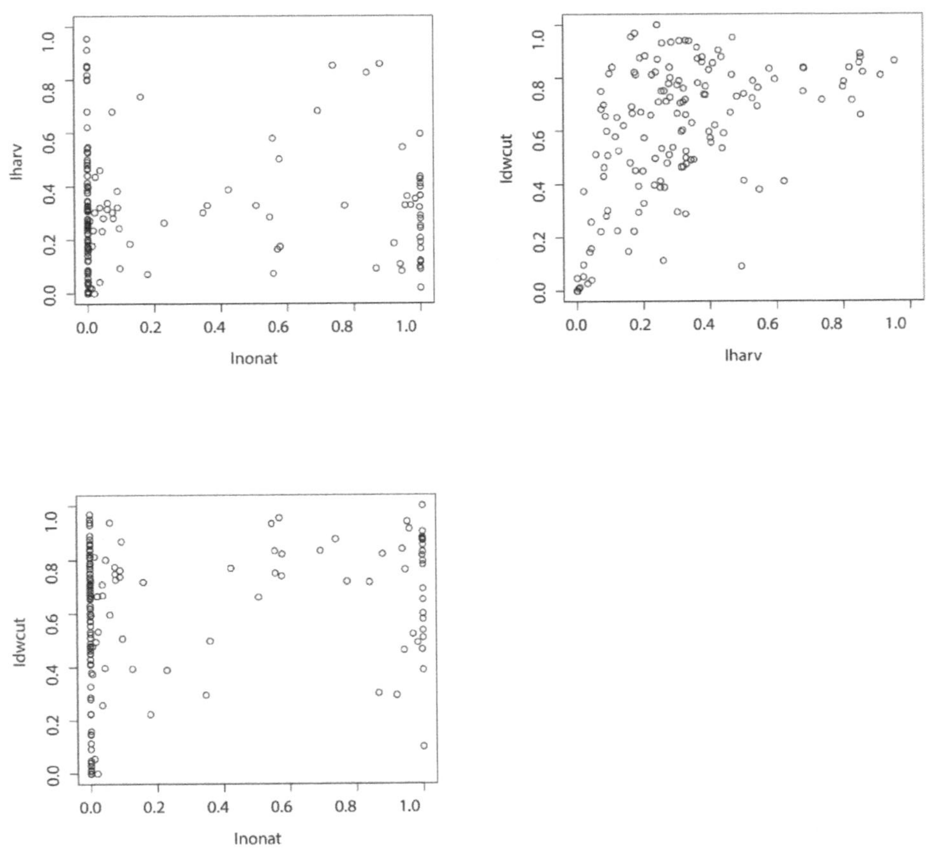

Figure 3. Relationships between the three components of the forest management intensity: Iharv, Inonat and Idwcut. Spearman rank correlation between Iharv-Inonat ($p=0.64$, $r_s=0.04$), Idwcut-Iharv ($p<0.001$, $r_s=0.5$) and Idwcut-Inonat ($p=0.01$, $r_s=0.21$).

Schall and Ammer (2013) proposed the Silvicultural Management Index (SMI), which is composed of a tree species-specific risk and a density component. In their rationale, tree species selection affects among other things the susceptibility of forest stands to natural disturbances. With increasing susceptibility, management intensity, e.g. through thinning or harvesting, must be more frequent to reduce the risk of stand loss. In Central Europe the susceptibility to disturbance tends to be higher for secondary coniferous than for natural hardwood forests. Consequently, the risk component of Schall and Ammer (2013) is to some extent comparable to the *Inonat*, and their density component to the *Iharv*. The SMI approach was applied to a subset of 30 plots of the Biodiversity exploratories. A comparison based on a linear model between the SMI and the ForMI shows that both indices yield comparable results ($p<0.001$, $R^2=0.72$). A comparison of the ForMI with the land use and disturbance intensity (LUDI) approach byLuyssaert et al. (2011), which is based on the same 148 plots used by us, showed that these indices do not converge and obviously capture different attributes of forests. The LUDI is based on the difference between the potential and the actual biomass storage and does not consider the intensification of forest management that is expressed by the change in tree species composition. Across all forest management types there was only a low although significant linear correlation ($p<0.001$, $R^2=0.07$). For plots dominated by non-natural coniferous tree species (*Piceaabies*, *Pinus sylvestris*, n=36), no significant correlation was found (p=0.28) while those plots dominated by hardwood species, such as *Fagus sylvatica* and *Quercus* sp. (n=112), had a significant correlation ($p<0.001$, $R^2=0.41$).

The advantage of the ForMi is its simple yet effective approach, which allows the calculation based only on volume or basal area inventory data of the living stand, stumps, and dead wood. No additional assumptions and models with regard to maximum stand carrying capacity or risk potential have to be employed. The ForMI approach proved to be applicable to all observed silvicultural management systems and allows their direct comparison. Although we have not examined yet how different silvicultural management systems, such as clear-cutting or coppice with standards, or large-scale disturbances, such as storm damage or fire, may affect ForMI, we believe that this index is applicable across the whole range of forest management types, except when trees are harvested including their root stock. The disadvantage of the ForMI is that the assessment of harvesting intensity depends on the visibility and longevity of stumps, and that it does not take into account the management frequency and is therefore not useful to describe situations where management occurs in long time intervals that exceed the longevity of stumps, as may be the case in tropical lowland rainforests.

ACKNOWLEDGEMENT

We thank the managers of the three exploratories, Swen Renner, Sonja Gockel, Kerstin Wiesner, and Martin Gorke for their work in maintaining the plot and project infrastructure; Simone Pfeiffer and Christiane Fischer giving support through the central office, Michael Owonibi for managing the central data base, and Markus Fischer, Eduard Linsenmair, Dominik Hessenmöller, Jens Nieschulze, Daniel Prati, Ingo Schöning, François Buscot, Ernst-Detlef Schulze, Wolfgang W. Weisser and the late Elisabeth Kalko for their role in setting up the Biodiversity Exploratories project. The work has been funded by the DFG Priority Program 1374 „Infrastructure-Biodiversity-Exploratories" (BA 2821/9-2). Field work permits were issued by the responsible state environmental offices of Baden-Württemberg, Thüringen, and Brandenburg (according to § 72 BbgNatSchG). We thank Peter Mühlhäuser, Sandor Haller and Martin Nitsche for conducting the dead wood inventory and Peter Schall and one anonymous reviewer from Peerage of Science for their critical and helpful comments on a previous version of the manuscript.

REFERENCES

1. Albrecht AT, Hanewinkel M, Bauhus J, Kohnle U (2012) How does silviculture affect storm damage in forests of south-western Germany? Results from empirical modeling based on long-term observations. European Journal of Forest Research 131: 229–247. doi: 10.1007/s10342-010-0432-x
2. Bauhus J, Puettmann K, Messier C (2009) Silviculture for old-growth attributes. Forest Ecology and Management 258: 525–537. doi: 10.1016/j.foreco.2009.01.053
3. Binkley D, Giardina C (1998) Why do tree species affect soils? The warp and woof of tree-soil interactions. Biogeochemistry 42: 89–106. doi: 10.1023/A:1005948126251
4. Bohn U, Neuhäusl R, Gollub G, Hettwer C, Neuhäuslová Z, Raus T, Schlüter H, Weber H (2004) Map of the Natural Vegetation of Europe, Scale 1:2500000. (Münster).
5. Brown S, Mo J, McPherson JK, Bell DT (1996) Decomposition of woody debris in Western Australian forests. Canadian Journal of Forest Research 26(6): 954–966. doi: 10.1139/x26-105
6. Brunner A, Kimmins JP (2003) Nitrogen fixation in coarse woody debris of *Thuja plicata* and*Tsuga heterophylla* forests on northern Vancouver island. Canadian Journal of Forest Research 33: 1670–1682. doi: 10.1139/x03-085
7. Christensen M, Hahn K, Mountford EP, Odor P, Standovar T, Rozenbergar D, Diaci J, Wijdeven S, Meyer P, Winter S, Vrska T (2005) Dead wood in European beech (*Fagus sylvatica*) forest reserves. Forest Ecology and Management 210: 267–282. doi: 10.1016/j.foreco.2005.02.032
8. Ellenberg H, Leuschner C (2010) Vegetation Mitteleuropas mit den Alpen; In ökologischer, dynamischer und historischer Sicht. (Stuttgart).
9. Fischer M, Bossdorf O, Gockel S, Hänsel F, Hemp A, Hessenmöller D, Korte G, Nieschulze J, Pfeiffer S, Prati D, Renner S, Schöning I, Schumacher U, Wells K, Buscot F, Kalko EKV, Linsenmair KE, Schulze ED, Weisser WW (2010) Implementing large-scale and long-term functional biodiversity research: The Biodiversity Exploratories. Basic and Applied Ecology 11(6): 473–485. doi: 10.1016/j.baae.2010.07.009
10. Gossner M, Ammer U (2006) The effects of Douglas-fir on tree-specific arthropod communities in mixed species stands with European beech and Norway spruce. European Journal of Forest Research 125: 221–235. doi: 10.1007/s10342-006-0113-y

11. Gustafsson L, Baker SC, Bauhus J, Beese WJ, Brodie A, Kouki J, Lindenmayer DB, Lõhmus A, Martínez Pastur G, Messier C, Neyland M, Palik B, Sverdrup-Thygeson A, Volney WJA, Wayne A, Franklin JF (2012) Retention forestry to maintain multifunctional forests: a world perspective. Bioscience 62: 633–645. doi: 10.1525/bio.2012.62.7.6

12. Harmon ME, Franklin JF, Swanson FJ, Sollins P, Gregory SV, Lattin JD, Anderson NH, Cline SP, Aumen NG, Sedell JR, Lienkaemper GW, Cromack K, Cummins KW (1986) Ecology of Coarse Woody Debris in Temperate Ecosystems. Advances in Ecological Research 15: 133–302. doi: 10.1016/S0065-2504(08)60121-X

13. Harmon M (2009) Woody Detritus Mass and its Contribution to Carbon Dynamics of Old-Growth Forests: the Temporal Context. In: Wirth C, Gleixner G, Heimann M (Eds) Old-growth forests: Function, fate and value. Ecological Studies 207 (Berlin Heidelberg): 159–190. doi: 10.1007/978-3-540-92706-8_8

14. Hessenmöller D, Nieschulze J, von Lüpke N, Schulze ED (2011) Identification of forest management types from ground-based and remotely sensed variables and the effects of forest management on forest structure and composition. Forstarchiv 82: 171–183.

15. Jonsson BG, Kruys N, Ranius T (2005) Ecology of species living on dead wood-lessons for dead wood management. Silva Fennica 39(2): 289–309.

16. Kahl T, Mund M, Bauhus J, Schulze ED (2012) Dissolved organic carbon from European beech coarse woody debris - patterns of input to and retention by surface soil. Ecoscience 19(4): 1–10. doi: 10.2980/19-4-3501

17. Köhler F (2000) Saproxylic beetles in nature forests of the northern Rhineland. Comparative studies on the saproxylic beetles of Germany and contributions to German nature forest research. Schrr LÖBF/LAfAO NRW 18 (Recklinghausen), 1–351.

18. Kuehne C, Donath C, Müller-Using SI, Bartsch N (2008) Nutrient fluxes via leaching from coarse woody debris in a *Fagus sylvatica* forest in the Solling Mountains, Germany. Canadian Journal of Forest Research 38: 2405–2413. doi: 10.1139/X08-088

19. Litton CM, Raich JW, Ryan MG (2007) Carbon allocation in forest ecosystems. Global Change Biology 13: 2089–2109. doi: 10.1111/j.1365-2486.2007.01420.x

20. Lonsdale D, Pautasso M, Holdenrieder O (2008) Wood-decaying fungi in the forest: conservation needs and management options. European Journal of Forest Research 127: 1–22. doi: 10.1007/s10342-007-0182-6

21. Luyssaert S, Hessenmöller D, von Lüpke N, Kaiser S, Schulze ED (2011) Quantifying land use and disturbance intensity in forestry, based on the self-thinning relationship. Ecological Applications 21(8): 3272–3284. doi: 10.1890/10-2395.1

22. Melin Y, Petersson H, Nordfjell T (2009) Decomposition of stump and root systems of Norway spruce in Sweden-A modelling approach. Forest Ecology and Management 257(5): 1445–1451. doi: 10.1890/10-2395.1

23. Meyer P (1999) Totholzuntersuchungen in nordwestdeutschen Naturwäldern: Methodik und erste Ergebnisse. Forstwissenschaftliches Centralblatt 118: 167–180. doi: 10.1007/BF02768985

24. MCPFE: Ministerial Conference on the Protection of Forests in Europe, 2002, Improved Pan-European indicators of sustainable forest management. http://www.foresteurope.org/docs/reporting/Vienna_Improved_Indicators.pdf

25. Muukkonen P (2007) Generalized allometric volume and biomass equations for some tree species in Europe. European Journal of Forest Research 126(2): 157–166. doi: 10.1007/s10342-007-0168-4

26. Næsset E (1999) Decomposition rate constants of Picea abies logs in southeastern Norway. Canadian Journal of Forest Research 29(3): 372–381. doi: 10.1139/x99-005

27. Niemeijer D, de Groot RS (2008) A conceptual framework for slecting environmental indicator sets. Ecological Indicators 8: 14–25. doi: 10.1016/j.ecolind.2006.11.012

28. Pretzsch H (2005) Stand density and growth of Norway spruce (*Picea abies* (L.) Karst.) and European beech (Fagus sylvatica L.): evidence from long-term experimental plots. European Journal of Forest Research 124(3): 193–205. doi: 10.1007/s10342-005-0068-4

29. Pretzsch H (2009) Forest dynamics, growth and yield - from measurement to model. Springer, New York.

30. Rondeux J, Sanchez C (2010) Review of indicators and field methods for monitoring biodiversity within national forest inventories. Core variable: Deadwood. Environmental Monitoring Assessment 164: 617–630. doi: 10.1007/s10661-009-0917-6

31. Schall P, Ammer C (2013) How to quantify forest management intensity in Central European forests. European Journal of Forest Research 132(2): 379–396. doi: 10.1007/s10342-013-0681-6

32. Shorohova E, Ignatyeva O, Kapitsa O, Kauhanen H, Kuznetsov A, Vanha-Majamaa I (2012) Stump decomposition rates after clear-felling with and without prescribed burning in southern and northern boreal forests in Finland. Forest Ecology and Management 263: 74–84. doi: 10.1016/j.foreco.2011.09.006

33. Spellmann H, Nagel J, Böckmann T (1999) Summarische Nutzungsplanung auf der Basis von Betriebsinventuren. Allgemeine Forst- und Jagdzeitung 170(7): 122–128.

34. Tüxen R (1956) Die heutige potentielle natürliche Vegetation als Gegenstand der Vegetationskartierung. Angewandte Pflanzensoziologie 13: 5–42.

35. Van Wagner CE (1982) Practical aspects of the line intersect method. Canadian Forestry Service, Information Report PI-X-12.

36. Warren WG, Olsen PF (1964) A line intersect technique for assessing logging waste. Forest Science 10: 267–276.

CHAPTER 13

IMPACT OF FOREST MANAGEMENT ON SPECIES RICHNESS: GLOBAL META-ANALYSIS AND ECONOMIC TRADE-OFFS

Abhishek Chaudhary[1], Zuzana Burivalova[2], Lian Pin Koh[3] and Stefanie Hellweg[1]

[1] Institute of Environmental Engineering, ETH Zurich, 8093 Zurich, Switzerland

[2] Woodrow Wilson School of Public and International Affairs, Princeton University, 08540 Princeton NJ, USA

[3] Environment Institute, and School of Biological Sciences, University of Adelaide, 5005 Adelaide, Australia

ABSTRACT

Forests managed for timber have an important role to play in conserving global biodiversity. We evaluated the most common timber production systems worldwide in terms of their impact on local species richness by conducting a categorical meta-analysis. We reviewed 287 published studies containing 1008 comparisons of species richness in managed and unmanaged forests and derived management, taxon, and continent specific effect sizes. We show that in terms of local species richness loss, forest management types can be ranked, from best to worse, as follows: selection and retention systems, reduced impact logging, conventional selective logging, clear-cutting, agroforestry, timber plantations, fuelwood plantations. Next, we calculated the economic profitability in terms of the net present value of timber harvesting from 10 hypothetical wood-producing Forest Management Units (FMU) from around the globe. The ranking of management types is altered when the species loss per unit profit generated from the FMU is considered. This is due to differences in yield, timber species prices, rotation cycle length and production costs. We thus conclude that it would be erroneous to dismiss or prioritize timber production regimes, based solely on their ranking of alpha diversity impacts.

INTRODUCTION

Each year, wood valued over US$100 billion is removed from forests globally, mainly accounted for by industrial roundwood, and also including fuelwood[1]. Around 10 million people are employed in the forestry sector and the livelihoods of many more depend on forests[1]. At the same time, it is now clear that forests managed for timber have to play an important role in conserving global biodiversity, as strictly protected areas, where all extractive activities are prohibited, are simply insufficient to achieve global biodiversity conservation goals[2,3]. Preserving forest biodiversity without harming economic interests is a big challenge for nations with forests. Local biodiversity loss due to timber extraction activities can disrupt the long-term resilience of forests, which may in turn cascade into an impoverished delivery of ecosystems services, ultimately affecting also human well-being[4,5].

With increasingly globalized trade, there is a growing choice of timber products from various tropical and temperate tree species[6]. Apart from aesthetic, financial and functional

Citation: Chaudhary, A. et al. Impact of Forest Management on Species Richness: Global Meta-Analysis and Economic Trade-Offs. Sci. Rep. 6, 23954; doi: 10.1038/srep23954 (2016).

preferences, a small but growing group of customers is also increasingly aware of, and interested in the environmental impact associated with wood products from around the globe[7]. The increasing globalization of timber trade also means that the environmental impacts often occur far from the place of consumption[8]. Informing consumers about the biodiversity impacts "hidden" in the wood products could be an important step toward achieving sustainable consumption patterns.

The most commonly mentioned impacts of the timber industry is forest degradation[9,10]. Extracting timber or other products changes the tree age structure, composition of tree species and vertical stratification, thereby affecting local temperature, light, moisture, soil, and litter conditions. This results in changes or complete removal of microhabitats (such as dead wood, cavities, root plates or mature trees) that host forest biodiversity[11]. The magnitude of impact on species diversity depends on several factors. First, forestry is carried out in different climatic zones with different native biodiversity, food web structures, and ecosystem properties. Therefore forestry impacts have a large geographical variability. Second, within the same region, different taxonomic groups may respond in different ways to forestry operations due to variation in, for example, body size, mobility, and diet[12,13]. Third, some types of management may have a larger influence on forest species than others, due to differences in habitat structure and continuity, or microclimatic conditions after the harvest. Additionally, some management regimes might have stronger secondary impacts on biodiversity, such as through increased rates of hunting or fire occurrence.

At an extreme end of the management intensity range is clear-cutting, which results in temporary deforestation of a previously forested area. Reduced Impact Logging (RIL) is an example of harvesting technique at the other end of the timber production management spectrum. Under RIL, carefully selected individual trees are extracted from the forest by trained workers, according to a harvest plan and guidelines designed to minimize the deleterious impact of tree felling, hauling and yarding on the remaining forest[14]. A quantitative understanding of the impacts of different forest management regimes on biodiversity in different regions of the world is crucial to any efforts to reconcile biodiversity conservation and economic interests.

MANAGEMENT TYPES

We chose the most common timber producing forest management regimes to assess the response of species richness to them. Some management types that we considered could be further divided into finer categories and others merged into broader ones, or considered as techniques within another management category. We have striven to achieve a good balance between data availability and information loss from excessive generalization.

Clear-cut (Temperate and Boreal)

Clear-cutting is historically the most common example of even-aged silviculture practice in temperate and boreal biomes[15]. It is technically easy to execute, as the entire stand overstorey is removed in one harvest. Clear-cutting has been criticized for simplifying forest structure and reducing biological diversity, leading to homogeneous forests[16] (but see Greenberg *et al.* for exceptions, where clear-cutting is found to mimic high intensity natural disturbance regimes)[17]. Many countries are now abandoning this practice. The reduction of clear-cut areas is also a part of regulations and standards under many forest certification schemes[18]. We only consider clear-cutting of natural managed forests, rather than of plantations, and without any additional interventions.

Retention (Temperate and Boreal)

In recent decades, silvicultural practices that combine timber harvesting and biodiversity preservation have been promoted to mitigate the impacts of clear-cuts[19]. This has led to other variations of even-aged silviculture, in which individuals (dispersed retention) or groups of trees (aggregated retention) are left on-site to maintain structural diversity (such as patch-cut or green tree retention systems), supply seeds for the next crop (seed tree retention) or to protect the regenerating understorey (shelterwood system)[18,19]. Ideally, this category will be subdivided in the future according to the amount and configuration of trees retained, to assess the impact of individual retention practices.

Selection Systems (Temperate and Boreal)

Selection system is a silvicultural program aimed to maintain uneven-aged stands and is applied as an alternative to clear-cutting[20]. It is designed to remove individual mature trees (single-tree selection), groups of mature trees (group-selection), or a combination of the two to create small openings scattered throughout the stand. This results in heterogeneous stand structures, which are assumed to be less damaging to forest biodiversity than traditional clear-cuts[20]. Selection systems place unique emphasis on maintaining species and structural diversity and regard such non-commodity values as a necessary foundation to achieve sustainable provision of timber and other ecosystem services[21].

Conventional Selective Logging (Tropical)

Selective logging involves the removal of the largest, highest quality trees from a tropical forest stand, leaving the remaining vegetation standing[22]. The term selective logging encompasses a very broad range of interventions, varying in, for example, the intensity of extraction (ranging from <5 m^3ha^{-1} to almost 200 m^3ha^{-1}), the use of bulldozers or cables to extract timber, legality, and ground disturbance[23]. On the one hand, high intensity selective logging can disturb as much as 30–40% of the area (through extensive skid trails and log landings), and damage 40–70% of remaining unlogged trees[24]. On the other hand, in areas with low density of commercially valuable trees, such operations can have a relatively low impact on the remaining forest stand[25].

Note that we distinguish *selection systems* in temperate and boreal regions from *selective logging*, primarily practiced in the tropics, because we argue that selection systems are effectively an attempt towards sustainable forest management, whereas selective logging is typically the default option of timber harvest in the species-diverse tropics, as only certain timber species are of commercial interest.

Reduced Impact Selective Logging (Tropical)

Reduced Impact Logging (RIL) techniques have been proposed in an attempt to manage tropical forests for timber in a more sustainable way by reducing the damage to soil, future crop trees, residual stands, and workers14. RIL includes preparation of detailed harvest plans, worker education, training and supervision, demarcation of log extraction paths, directional felling to reduce collateral damage to surrounding trees and canopy, and protection of riparian areas14. RIL has been conceived to achieve a sustainable level of harvest, but also to improve sustainability in terms of biodiversity, carbon retention and ecosystem services9. Even though RIL techniques can be practiced anywhere and within other management regimes including plantations, here we only consider studies reporting species richness in natural tropical forests managed using reduced impact selective logging.

Timber Plantations

Timber plantations are becoming increasingly popular in some regions as an alternative to extracting timber from natural forests26. Whereas some timber species are relatively easily grown in plantations, others have a very low survival rate in plantations (e.g. mahogany, rosewood). There are likely important differences, in terms of biodiversity, between monoculture and mixed timber plantations, plantations of native *vs.* exotic species, and in the way timber is harvested (clear-cutting, selective logging). However, there is currently insufficient number of studies to distinguish between these effects, and little is known about the dynamics of tropical timber growth in plantations.

Plantations for Fuelwood and Pulp

A large proportion of plantations are monocultures of fast growing, low-density wood species, such as *Pinus spp.* or *Eucalyptus spp.*, used for fuel, or the pulp and paper industry, notably *Acacia mangium*. These plantations also often consist of exotic species, and we consider them as a separate category, because they are typically harvested on a much shorter time cycle than timber plantations. We include this and the following non-timber producing management categories to provide a comparison with the above timber producing management types.

Non-Timber Plantations

Industrial clearing of tropical forests for non-timber plantations (most importantly palm oil but also plantain, bamboo, or rubber, which is planted also for timber) has been a key driver of biodiversity loss in tropical countries27. For example, at least 55% of the expansion of oil palm between 1990 and 2005 in Malaysia and Indonesia occurred at the expense of rainforest28. Monoculture rubber plantations are now the most rapidly expanding tree crop in South-East Asia, due to an increasing global demand. Current area of rubber plantations is equivalent to 57% of oil palm area globally, with more than 2 million hectares established during last decade29. Often, tropical forest is first logged to extract commercially valuable species, and subsequently converted to these non-timber plantations.

Agroforestry

Agroforestry maintains a structural diversity that imitates the native forest better than conventional pastures, row crops, and monoculture plantations30. In agroforestry systems, perennial tree crops such as coconut, rubber, coffee or cacao replace the original forest understorey but some canopy trees are left for shade. Timber from agroforestry systems is rarely sold on international markets, therefore we do not include this management type in our case studies, but include it in the meta-analysis to contextualize other management types. There are many different ways in which agroforestry can be implemented (such as agrosilvopastoral systems, row planting, or companion cropping), and this may also have different effects on biodiversity31. Due to limited data availability, we group all agroforestry types together.

Slash & Burn

Also referred to as swidden or shifting agriculture, slash-and-burn describes an agricultural system in which forest is periodically cleared and burnt to create fields for crops such as rice, cassava, maize, and papaya. After several years of production, the fields are left fallow, and forest typically regenerates to some extent. Whereas slash-and-burn management rarely contributes timber to international markets, it does provide wood for local subsistence32.

OBJECTIVE & SCOPE

To understand the variable impacts of forestry management types on local species richness of different taxa, we conduct a global categorical meta-analysis, using log response ratio (logarithm of the ratio of species richness in managed to unmanaged forest) as a measure of effect size33. Existing meta-analyses on the subject are either restricted to a region (e.g. Europe11, the tropics34), taxon (e.g. plants35), management type (e.g. selective logging23), or consider forestry as one generic land use type without accounting for differences in management types36. We first derive management, taxa-, and continent-specific effect sizes.

To evaluate trade-offs between economic profitability and biodiversity conservation, we consider 10 hypothetical wood-producing Forest Management Units (FMUs) from around the globe, each 5000 ha in size. We calculate the economic profitability of timber harvesting in terms of the net present value of each FMU, over an operation period of 75 years. Our hypothetical FMUs are based on published studies that report the data on management type, harvesting intensity, rotation length, timber species and production costs. The NPV for each FMU was calculated by discounting (10% interest rate) the revenues as well as the production costs associated with future rotation cycles. Net profit over the operation period of 75 years was calculated by multiplying NPV with the area of each FMU (i.e. 5000 ha) in million$.

Next we use the *local* species richness loss estimates derived from our meta-analysis to calculate the *regional* species loss through countryside species-area relationship37. By comparing the regional species loss with net profit in the hypothetical FMUs, we illustrate biodiversity-economic trade-offs connected with different forestry management types.

RESULTS

We compiled 1008 species richness comparisons of managed and unmanaged forests from a total of 287 studies, and used these in an unweighted38 meta-analysis. The overall effect size showed that forest management leads to a 29% reduction in species richness (95% CI=26 to 32%).

Management Type

We grouped the compiled dataset into 10 management categories and calculated the effect size for each group. Three out of six timber producing management regimes - Selection and Retention systems and Reduced Impact Logging - did not alter species richness (Fig. 1). Out of the other three timber producing management types, timber plantations had the highest impact (on average 40% reduction in local species richness compared to natural forests), followed by clear-cutting (22% reduction), and conventional selective logging (13% reduction, Fig. 1).

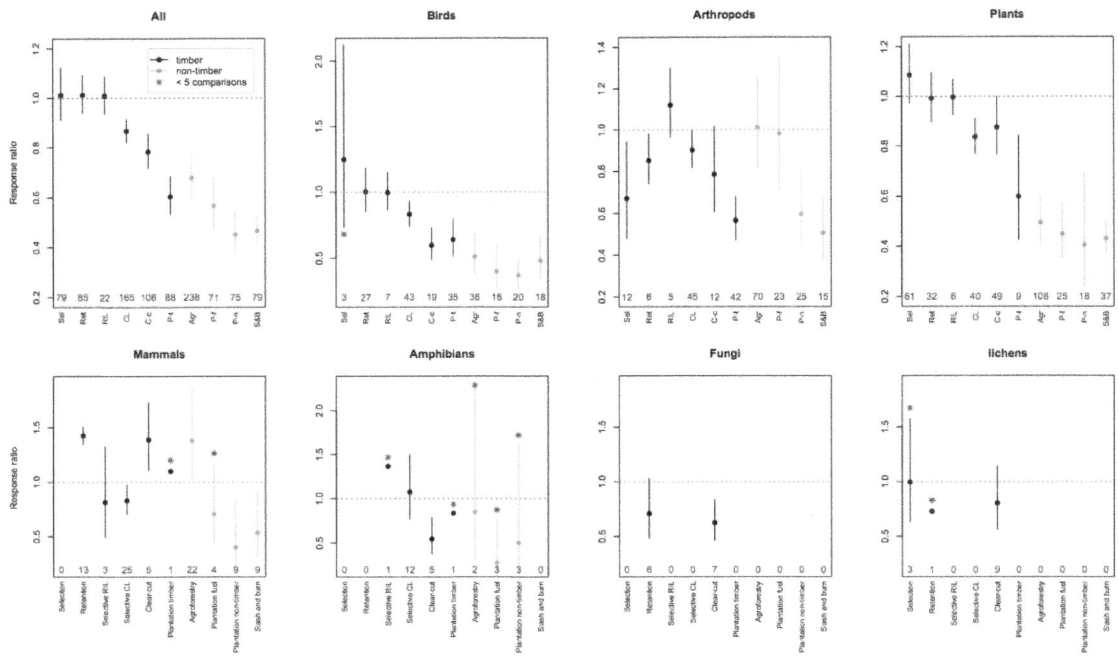

Figure 1: Changes in species richness of different taxonomic groups in response to ten most common forest management regimes. Points represent mean response ratios and lines 95% confidence intervals. The top left panel shows responses for all taxa combined. The management regimes under which timber production is the main goal are showed in black, and non-timber regimes in blue. Horizontal red line signifies no change in species richness. An asterisk designates response ratios based on five or fewer comparisons. Numbers close to the x-axis are the numbers of comparisons.

Out of the non-timber producing management regimes that we considered, agroforestry was least detrimental, retaining on average 68% of species richness found in nearby natural forests. Fuel wood plantations had a very similar impact on species richness as timber plantations (i.e. ~40% reduction). Non-timber tree plantations, referring mostly to oil palm and rubber, and slash & burn reduced the species richness by ~54% which is highest impact among all the 10 management systems considered.

Management type ×Taxa

We calculated the effect size for each combination of management type and taxonomic group (Fig. 1). For most management types, the mean effect sizes differed across taxa, which means that taxa respond differently to a particular management type ($P<0.001$, ANOVA test. The non-timber producing forest management types reduced bird richness more seriously than when richness of all taxa was considered together (Fig. 1). In contrast to the case when all taxa were considered together, clear-cuts (<10 years old) were worse in terms of bird species richness than timber plantations. Only two timber-producing management categories, retention and timber plantations, caused a decrease in species numbers for arthropods (Fig. 1). Both clear-cutting and conventional selective logging decreased arthropod species richness somewhat, but not significantly. Equally, non-timber producing management regimes had less severe impacts on arthropods. Agroforestry and fuel plantations did not decrease arthropod species richness. All timber producing regimes maintained high plant species richness, except timber plantations, which performed similarly to agroforests.

There was a paucity of studies on mammals, amphibians, fungi and lichens for most management types. Mammalian richness was decreased by conventional selective logging, slash & burn, and non-timber plantations. Retention systems and agroforests did not lower the mammalian species richness. Amphibian richness was on average not affected by conventional logging and the spread of responses was very wide. Fungal diversity was lower under both retention forestry and clear-cutting, and lichen diversity did not change under any management regime for which data was available.

Management type ×Region

Next, we grouped management types according to region (continent). Not all forest management types are practiced in all regions, nevertheless, some management regimes showed inter-regional variation. For agroforestry, clear-cut, selective logging, timber plantations, and plantations for fuelwood, the mean effect sizes differed among continents. Timber plantations in Africa maintained on average higher species richness than those on other continents. Other timber producing management regimes did not show a vast difference in impacts between continents. For example, clear-cutting in North America had a very similar effect on species richness as in Europe. Fuel plantations performed similarly on all continents with tropical forests (Africa, Asia, and South America). Non-timber plantations had lower species richness in Africa, Asia, and Central America, when compared with non-timber plantations of South America. Equally, agroforestry in South America caused smaller species losses than in Africa, Asia, and Central America.

Management type ×Region ×Taxa

We further sub-divided the above management type ×region groups according to the taxon. The majority of effect sizes for this three-way combination could not be calculated or were non-significant owing to missing or low number of comparisons available. However, for most combinations of management type and continent, we found differences among the mean effect size of taxa ($P<0.001$, ANOVA test), implying that a taxonomic group of a given continent responds differently to the different management types.

Biodiversity-economic trade-offs

The 10 hypothetical 5000 ha FMUs differed substantially in the fraction of regional species lost, profit generated, and the ratio thereof (Fig. 2). In terms of net present value (NPV), the clear-cutting of *Picea abies* in Sweden with high timber yields (300 m^3/ha) came first, generating 14659 \$/ha, followed by selective logging in Malaysia, which generated 10855 \$/ha. The FMU in Gabon extracting *Aucoumea klaineiana* (oukoumé) with reduced impact logging techniques had the lowest NPV (521\$/ha), mostly due to low harvest per ha (4.23 m^3/ha in first cut).

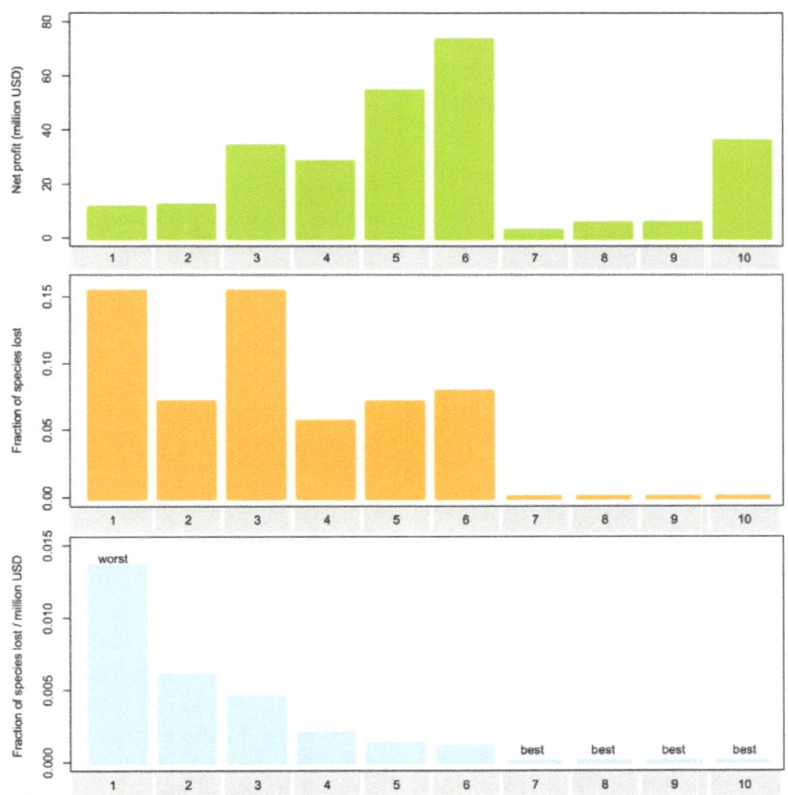

Country	Malaysia	Indonesia	Malaysia	Sweden	Malaysia	Sweden	Gabon	Gabon	Brazil	Canada
Management	Plantation fuel	Selective CL	Plantation timber	Retention	Selective CL	Clear cut	Selective RIL	Selective CL	Selective CL	Selection
Product	*Acacia mangium* (Acacia) logs	various species	*Tectona grandis* (Teak) logs	*Picea abies* (Norway spruce) logs	various species	*Picea abies* (Norway spruce) logs	*Aucoumea klaineiana* (okoumé) logs	*Aucoumea klaineiana* (okoumé) logs	*Swietenia macrophy-lla* (maho-gany) sawnwood	*Thuja plicata* (Western red cedar) sawnwood
Price (USD/m^3)	35	122	140	55	various	55	138	116	220	107
Harvest 1st cut (m^3/ha)	90	54	150	120	117.34	300	4.23	10.68	5	80
Production cost (USD/m^3)	1443*	80.02	1517*	8.07	60.36	6.2	20.9	17.66	15.66	17.5

Figure 2: Biodiversity-economic trade offs for ten hypothetical forest management units (FMU), described in the table at the bottom. The top panel shows Net Present Value per hectare, assuming 75 years of management, and 10% discount rate. The central panel shows fraction of regional species lost when half of the original 10,000 ha of natural forest is subjected to forest management. The bottom panel shows the fraction of species lost per net profit, combining information from the two upper panels.

When the FMUs were ranked according to fraction of species lost per unit profit, *Acacia mangium* fuel wood plantations in Malaysia performed the worst, causing 15% species loss, whilst generating 11.3 million$ profit over 75 years of operation. This is due to the low price of*Acacia* logs (35$/m^3) as well as relatively high production costs. Selective conventional logging in Indonesia's forests caused a lower fraction of species lost (7.1%) but still performed the second worst, owing to high production costs (80.02$/m^3) associated with timber harvesting. The Indonesian FMU performed worse than a similar, selectively logged FMU in Malaysia's dipterocarp forest, even though both resulted in a similar fraction of species lost. The difference was due to higher yields and lower production costs in the Malaysian FMU.

It is interesting to see that although the *Tectona grandis* (teak) timber plantations in Malaysia resulted in the highest fraction of species lost (15.4%) among all FMUs considered, it still performed better than the Malaysian fuel plantation and Indonesian selective logging FMU, because of high yields (150 m^3/ha) and high timber price (140$/m^3). Similarly, although the clear cutting of *Picea abies* in Sweden caused a higher fraction of species loss (7.6%) than a similar FMU in Sweden, operating with a much more biodiversity benign shelterwood retention (5.5% species loss), it generated the highest net income of all FMUs (73.3 million$), and thus resulted in a lower species loss per million$ than the retention system.

No species loss occurred in the case studies corresponding to Reduced Impact Logging of *Acoumea klaineiana* in Gabon and selection system cutting of *Thuja plicata* in Canada (Fig. 1). When adjusted for logging intensity, conventional selective logging of *Acoumea klaineiana* in Gabon and *Swietenia macrophylla* in Brazil also caused almost no species loss.

DISCUSSION

All types of management change some properties of the forest, such as tree age structure, microclimate, or soil conditions. It is unrealistic to expect any type of forestry to have no impact on forest biodiversity. We demonstrated that these impacts vary substantially between forest management types. Through case studies, we illustrated that producing the same economic worth of wood in different ways may result in substantially different biodiversity loss.

We found that in terms of relative species loss at plot level, ways to produce timber can be ranked from best to worse as follows: selection and retention systems in temperate and boreal regions, reduced impact selective logging (RIL) followed by conventional selective logging in tropical regions, clear-cutting in temperate regions, and timber plantations (Fig. 1). Management regimes not focusing on timber production are in general more harmful to species richness than timber producing regimes. A notable exception is agroforestry, which is associated with lower species loss than timber plantations.

Management regimes that mimic small-scale disturbance such as selection harvest, retention, and RIL, result in increased environmental heterogeneity compared to clear cutting, timber plantations or high intensity conventional logging18·39. Selection and retention systems appear to be similarly good at maintaining high species richness in managed forests. The retention trees, areas, and unharvested portions of selections systems provide late successional structures that are more likely to maintain high diversity. Retained trees also enhance habitat connectivity over the landscape and supply refuges for species after harvest40. However, the effect of each of these management types can vary with the proportion of trees retained, and the time since harvest15.

Conventional selective logging reduces species richness more than RIL (Fig. 1). However, in the case of trees, this may be due to the lower logging intensities typically associated with RIL41. In the case of fauna, which is also strongly influenced by logging intensity, RIL has nevertheless been found to have a smaller impact on biodiversity even after taking logging intensity into account42·43.

Interestingly, we found species richness differences between timber, fuel wood, and non-timber plantations (mainly oil-palm and rubber). Whereas plant species richness was similar, fuel wood plantations were poorer in bird species than timber plantations. Birds may prefer to nest in timber plantations, as fuel wood plantations are typically harvested more frequently, perhaps not allowing the trees to grow to sufficient sizes suitable for nesting.

Slash-and-burn agriculture in the tropics was found to have higher impacts on species richness than clear-cutting in temperate regions, even though both systems are based on removing all trees from small forest patches. This could be due to the additional effect of burning or the temporary presence of agricultural crops, which could affect the soil and slow down regeneration44. Additionally, tropical forests have a higher proportion of forest specialist species, that may be more sensitive to the removal of forest patches, habitat change, and edge effects45. However, to truly understand the effect of slash-and-burn agriculture, biodiversity would have to be studied at a broader scale. Certain swidden agricultural systems can provide landscape level biodiversity benefits and be important components of a 'conservation-agriculture matrix'32.

Finally, we found agroforestry to be positioned between timber and non-timber producing regimes in terms of species richness (Fig. 1). Biodiversity within agroforestry systems is enhanced when there is high diversity of native trees in the canopy, low chemical inputs are used, and the understory clearing is not intense46.

The variability in the response of different taxonomic groups to forest management is likely both due to variations in biophysical and life history traits, and due to local context of forestry interventions43·47. For example, primates with slow reproduction and low ecological flexibility show a more marked decline after forest disturbance47. Bird populations are negatively affected by reduction in tree cavities for nesting, absence of leaf-litter microfauna, and foraging sites in the logged forest48. Equally, the amount of dead wood, large logs and snags of different decay stages post-logging determine the survival of many taxa11. We expect that insectivores do better in fuel wood plantations, due to the high abundance of arthropods (Fig. 1) but not frugivorous who need fleshy fruit trees for survival.

Plants showed mixed response to logging. Many studies reported an increase in species richness following logging, which may be a result of early successional colonizers with a negative effect on the late successional shade tolerant species35. Similarly, bryophytes might respond differently to forest management than vascular plants[11], however, more data is needed to tease out such differences. Mammal diversity appeared to benefit from forest disturbance (Fig. 1). This disparity might be because of high tolerance of certain mammal species to degraded forests[49], particularly among bats and small forest floor mammals. Indeed, out of a total of 91 comparisons of mammals compiled in this study, 82 were for either small mammals or bats.

When we evaluated the timber producing management regimes in terms of species loss per unit profit from wood harvesting in our selection of case studies, their rankings changed. For example, while the meta-analysis revealed that conventional selective logging is less harmful to species richness than timber plantations, the Indonesian FMU operating with conventional selective logging resulted in a higher fraction of species lost per million$ profit generated than a *Tectona grandis* timber plantation FMU from Malaysia.

Clearly, our case studies cannot be used to generalize outcomes for the selected timber producing management regimes and countries. We made simplifying assumptions on area surrounding the forestry concessions (matrix) and the spatial extent of timber species within FMUs, all of which have high variability both in space and time. Data for timber prices, yield, rotation cycle, and production costs are highly site-specific and thus caution must be taken while interpreting the results[50,51]. For selective logging regimes, the net present value would be higher if the profits from subsequent conversion to another land use type, such as oil palm or rubber plantation were also included. We also did not include the potential revenues from non-timber forest products. Further, we assumed that all FMUs are replacing natural forests, while in practice forests previously managed through selective logging might have been switched to, for example, RIL.

As our case-studies are hypothetical, we evaluated the outcomes as a fraction of species lost divided by net income generated. However, from an absolute species loss perspective, a 50% temporary extirpation of species in a Swedish concession, where original species richness is relatively meagre and endemic species almost non-existent, may be considered less serious than even a 5% species loss in a hyper-diverse tropical rain forest. Therefore, a more realistic expression of the biodiversity-economic trade-off would be absolute species lost/million$.

We initially compiled some 350 studies from the literature review but over 150 studies, containing >300 pairwise comparisons of species richness had to be excluded because they did not report summary statistics necessary for a meta-analysis, such as standard deviation or sample size. In the selected dataset we found major data gaps owing to which, effect sizes for several three-way combination of management type, taxa and region could not be calculated.

Our meta-analysis did not take into account the surrounding habitat types, as these were mostly not always reported by the studies[45]. Despite their likely importance, we could not decouple the primary impacts of forest management on biodiversity (i.e. due to tree removal), from the secondary impacts (e.g. higher rates of bushmeat hunting, livestock grazing, illegal logging and fires through increased accessibility of the forest), as the majority of studies provide no such information. However those studies that explicitly stated the presence of hunting, grazing etc. were not included in the analysis.

We used local species richness (alpha diversity) as a proxy for biodiversity changes due to different types of forest management. An increase in species richness, such as we found in selection and retention systems, and under reduced impact logging, does not necessarily signify enhanced biodiversity value: rare forest interior species could still be lost under these management types, and replaced with generalist, common, or even invasive species[15,25]. However, a reduction in species richness, such as in other systems, can be safely interpreted as a negative impact. To complement our results, future global meta-analysis should use indicators that compare compositional changes in the community, e.g., Sørensen's similarity index[52] that may be more sensitive to land use impacts than relative species richness.

It would be erroneous to dismiss or prioritize timber production regimes, based solely on their ranking of alpha diversity impacts. First, one has to consider that the impacts on biodiversity at a larger scale and within the context of heterogeneous landscapes may be different than plot-scale effects. For example, our results show that timber and fuel wood plantations reduce local species richness more than any other timber-producing regimes, however, the comparisons we made are all with natural forest. Indeed, plantations, when established in previously degraded and intensively managed areas, may enhance local and regional biodiversity and also contribute to alleviation of pressure on the remaining natural forests. To calculate such impacts at regional scale, we used the countryside species-area relationship (SAR), using data on different species' affinity to various forest use types that we derived through our meta-analysis[53].

Second, timber production systems should be judged also in terms of their economic productivity. The combined quantitative assessment of economic and biodiversity consequences of land use change that we carried out can also help inform the debate on whether it is better to convert large extents of natural forests to regimes relatively benign to biodiversity (e.g. RIL) or to convert smaller extents of the forests to intensive production systems, such as timber plantations (i.e. land sharing vs. land sparing[54]). Similar analyses are also needed to help design appropriate payments for biodiversity schemes[55] and are increasingly relevant for decision makers in countries planning to expand productive areas into natural landscape in near future.

MATERIALS AND METHODS

Data Collection

We searched for research articles using the *ISI Web of Science* database. From this list, we retained those studies that reported species richness comparisons between a managed forest and an unmanaged forest (i.e. forests that had little or no human disturbance) of an equivalent size or equivalent sampling effort in the same region. For each comparison, we recorded the location, management type, taxon and, when available, standard deviation and sample size. Studies that sampled in forests with a disturbance additional to logging, such as hunting, livestock grazing, mining, burning, etc., were excluded. For clear-cut and slash-and-burn management types, we only retained studies that measured species richness within 10 years of the intervention, as time has a large influence on the recovery of biodiversity44.

Data Treatment and Effect Size Calculations

For each comparison, we tabulated the observed mean species richness (X) for both managed (subscript E) and unmanaged groups (subscript C). We calculated the response ratio ($R=X_E/X_C$) for each comparison and took its natural logarithm $L=\ln R=\ln(X_E/X_C)$, as a measure of effect size for the meta-analysis owing to its favorable statistical properties33.

Traditional meta-analyses weight the individual effect sizes by the inverse of the effect size variance, as studies with higher replication and lower variance are considered more precise, and thus weighted more heavily33. However it has been shown that ecological field studies comparing biodiversity in logged and unlogged forests are rife with pseudoreplication (see Ramage *et al.*56 and Halme *et al.*57). Using pseudoreplicated studies in a meta-analysis typically inflates the effect sizes, because the artificially low variation among replicates is due to interdependent samples of the same species pool57. Moreover, weighted meta-analysis requires standard deviation and sample size for each comparison, which is not reported by many field studies. Recent studies have advocated using unweighted meta-analysis to avoid under/over estimation of effect sizes and to include data from more studies38. For these reasons, we performed unweighted, fixed effects meta-analysis.

We carried out a categorical meta-analysis to test for the effect of three factors - region (continent), taxon, and management type, on the magnitude of change in species richness between unmanaged and managed forests. We calculated the mean effect size L_j and its standard error (S_j) for the entire data set, for each subgroup $j$of the above three variables, for the two-level combination of these three variables (management type × region and management type × taxon), and finally for their three-way combination (management type × region × taxon)33.

The confidence interval (CI) for L_j is given by:

$$CI = L_j \mp t_{\alpha/2\,[k_j-1]} * S_j \tag{1}$$

where t is the two-tailed value from the Student's t-distribution at the critical level α and k_j is the number of pairwise comparisons in the j^{th}group. The mean effect size for each group (L_j) is considered to be statistically significant if its 95% confidence interval does not bracket zero38. For the purpose of interpretation we present the results as back-transformed response ratios R. We also assessed if statistically significant differences exist between the mean effect sizes for the groups through one-way analysis of variance (ANOVA).

Species Loss and Economic Trade-Offs

We considered 10 hypothetical forest management units (FMUs) operating with different timber producing management regimes in different world regions. For each case, we extracted the following variables from published literature: country, management type, timber species and their yields associated with that management type (m³/ha), market price of species (US$/m³), production costs (in $/ha or $/m³) and rotation cycle (in years).

To estimate the biodiversity cost of timber production in each case, we began with a hypothetical 10,000-ha of existing natural forest in a particular country, out of which 5,000-ha has been allocated to a commercial logging firm as a concession (hereafter referred to as forest management unit, FMU). We then estimated the fraction of regional species lost (f_{lost}) using the countryside species-area relationship (SAR) model37,53:

$$f_{lost} = \frac{S_{lost}}{S_{org}} = 1 - \left(\frac{A_{new} + h \cdot A_{mang}}{A_{org}}\right)^z \tag{2}$$

where S_{lost} is the species loss after forest management intervention;S_{lost} is the original number of species in the natural forest area; A_{org} is the original natural forest area with no habitat change; A_{mang} is the managed forest area (5000 ha); A_{new} is the remaining natural area ($=A_{org}-A_{mang}=5000$ ha); the exponent z is a constant indicating the rate of change in the number of species per unit area, which we set to be 0.25, as an overall mean for all taxa and continents; and h is a measure of affinity of a taxa to the altered habitat ($0<h<1$). We derived h values for each management regime from our meta-analysis, as $h = R^{1/z}$ here R is the response ratio for each continent-specific management type for all taxa combined (see Pereira *et al.*37 for details). For response ratios greater than 1, we assumed that no species loss would occur, and we set$R$ as 1. For selective logging case studies, we further adjusted R values by taking logging intensity into account. We did this by using an average coefficient, combined for mammals, birds, invertebrates and amphibians, from Burivalova *et al.*25 who examined the impact of selective logging intensity on species richness.

Note that countryside SAR predicts the *regional* species loss by considering FMU as a part of heterogeneous landscape surrounded by natural forest. The species loss estimates derived using our meta-analysis are plot-scale, *local* species richness reductions. From equation (2) we can observe that in the extreme case when all natural forest area is converted to a logging concession, the regional and local species loss estimates converge, i.e. when $A_{mang}=A_{org}$, $f_{lost}=1-R$.

We then calculated the net present value (NPV in US$/ha) from wood extraction in each FMU as follows:

$$NPV = \sum_{t=0}^{N} \frac{(P - C) \cdot Y_t}{(1 + r)^t} \tag{3}$$

where P and C are respectively the market price ($/m³) and production cost ($/m³) of timber species within FMU, r is the interest rate (assumed to be 10%) and t is the time of cash flow. N is the number of rotation cycles over a total of 75 year period and Y_t is the timber yield in a particular rotation cycle (m³/ha).

For FMUs operating with selective logging and reduced impact logging techniques we assumed a rotation cycle of 25 years and assumed that timber yields decline by 46% after the first harvest and remain at that level in subsequent rotation cycles9. For *Acacia Mangium* fuel plantation and *Tectona grandis* timber plantation FMUs in Malaysia –establishment, maintenance and harvesting costs along with timber yields and prices were taken from existing FAO report58 and the rotation cycle was assumed to be 8 years and 15 years respectively58. Here we assumed that the timber yields remain the same in each rotation cycle. Data for *Picea abies* FMUs in Sweden operating with clear-cut and retention regimes was imported from Hanell *et al.*59, assuming a rotation cycle of 75 years. We selected two *Aucoumea klaineiana* FMUs from Gabon - one operating with selective logging and another with RIL techniques. For these FMUs, production cost data was taken from Medjibe & Putz50, while yield data from Medjibe *et al.*60. For selectively logged FMUs in Malaysia and Indonesia, the price, yield and production cost data were taken from Fisher *et al.*55 and Ruslandi *et al.*51respectively. NPV was multiplied with total managed forest area (A_{mang} = 5000 ha) to obtain net profit (in million$) from each FMU over a 75 year period. Finally, we divided the fraction of species lost due to a particular management type in the FMU (equation 2) by the net profit from it. All FMUs were ranked according to their fraction of species lost per unit profit from worst (highest) to best.

ACKNOWLEDGEMENTS

Abhishek Chaudhary was funded within the National Research Programme «Resource Wood» by the Swiss National Science Foundation (NRP 66, project no. 136612). The authors thank Francis E. Putz and an anonymous reviewer for valuable comments.

AUTHOR CONTRIBUTIONS

A.C. conceived the study, compiled data from literature review, conducted meta-analysis and performed NPV calculations. Z.B. contributed to study design and drew plots. Both A.C. and Z.B. compiled case-studies, analyzed results and wrote the manuscript. L.P.K. provided comments on manuscript and S.H. discussed the work with A.C and provided comments on previous versions of the manuscript.

REFERENCES

1. Food and Agriculture Organization of the United Nations. Global forest resources assessment 2010: Main report. Food and Agriculture Organization of the United Nations (2010).
2. Laurance, W. F. et al. Averting biodiversity collapse in tropical forest protected areas. Nature 489, 290–294 (2012).
3. Hayes, T. & Ostrom, E. Conserving the World's Forests: Are Protected Areas the Only Way? Ind. L. J. 38, 595–617 (2005).
4. Edwards, D. P., Tobias, J. A., Sheil, D., Meijaard, E. & Laurance, W. F. Maintaining ecosystem function and services in logged tropical forests. Trends Ecol. Evol. 29, 511–520 (2014).
5. Hooper, D. U. et al. A global synthesis reveals biodiversity loss as a major driver of ecosystem change. Nature 486, 105–108 (2012).
6. Shearman, P., Bryan, J. & Laurance, W. F. Are we approaching "peak timber" in the tropics? Biol. Conserv. 151, 17–21 (2012).
7. Auld, G., Gulbrandsen, L. H. & McDermott, C. L. Certification Schemes and the Impacts on Forests and Forestry. Ann. Rev. Environ. Resour. 33, 187–211 (2008).
8. Lenzen, M. et al. International trade drives biodiversity threats in developing nations. Nature 486, 109–112 (2012).
9. Putz, F. E. et al. Sustaining conservation values in selectively logged tropical forests: the attained and the attainable. Conserv. Lett. 5, 296–303 (2012).
10. Putz, F. E. & Romero, C. Futures of Tropical Forests (sensu lato). Biotropica 46, 495–505 (2014).
11. Paillet, Y. et al. Biodiversity Differences between Managed and Unmanaged Forests: Meta-Analysis of Species Richness in Europe. Conserv. Biol. 24(1), 101–112 (2010).
12. Stork, N. E. et al. Vulnerability and resilience of tropical forest species to land-use change. Conserv. Biol. 23, 1438–1447 (2009).
13. Barlow, J. et al. Quantifying the biodiversity value of tropical primary, secondary, and plantation forests. PNAS 104(47), 18555–18560 (2007).
14. Putz, F. E., Sist, P., Fredericksen, T. & Dykstra, D. Reduced-impact logging: Challenges and opportunities. Forest Ecol. Manag. 256, 1427–1433 (2008).
15. Fedrowitz, K. et al. Can retention forestry help conserve biodiversity? A meta-analysis. J. Appl. Ecol. 51, 1669–1679 (2014).
16. Rosenvald, R. & Lohmus, A. For what, when, and where is green-tree retention better than clear-cutting? A review of the biodiversity aspects. For. Ecol. Manage. 255(1), 1–15 (2008).
17. Greenberg, C. H., Harris, L. D. & Neary, D. G. A comparison of bird communities in burned and salvage-logged, clearcut, and forested Florida sand pine scrub. Wilson Bull. 107, 40–54 (1995).
18. Gustafsson, L., Kouki, J. & Sverdrup-Thygeson, A. Tree retention as a conservation measure in clear-cut forests of northern Europe: a review of ecological consequences. Scand. J. For. Res. 25, 295–308 (2010).
19. Gustafsson, L. et al. Retention Forestry to Maintain Multifunctional Forests: A World Perspective. BioScience 62, 633–645 (2012).
20. Falk, K. J., Burke, D. M., Elliott, K. A. & Holmes, S. B. Effects of single-tree and group selection harvesting on the diversity and abundance of spring forest herbs in deciduous forests in southwestern Ontario. For. Ecol. Manage. 255, 2486–2494 (2008).
21. Puettmann, K. J. et al. Silvicultural alternatives to conventional even-aged forest management-what limits global adoption? For. Ecosyst. 2(1), 1–16 (2015).
22. Günter, S. M. et al. Silviculture in the tropics. Springer, Freisig (2011).
23. Putz, F. E., Redford, K. H., Robinson, J. G., Fimbel, R. & Blate, G. M. Biodiversity Conservation in the Context of Tropical Forest Management. The World Bank, Washington, DC (2000).

24. Pinard, M. A. & Putz, F. E. Retaining forest biomass by reducing logging damage. Biotropica 28, 278–295 (1996).
25. Burivalova, Z., Şekercioğlu, C. H. & Koh, L. P. Thresholds of logging intensity to maintain tropical forest biodiversity. Curr. Biol. 24, 1–6 (2014).
26. Obidzinski, K. & Chaudhury, M. Transition to timber plantation based forestry in Indonesia: towards a feasible new policy. Int. For. Rev. 11, 79–87 (2009).
27. Abood, S. A., Lee, J. S. H., Burivalova, Z., Garcia-Ulloa, J. & Koh, L. P. Relative contributions of the logging, fiber, oil palm, and mining industries to forest loss in Indonesia. Conserv. Lett. 8(1), 58–67 (2015).
28. Edwards, D. P. et al. Degraded lands worth protecting: the biological importance of Southeast Asia's repeatedly logged forests. Proc. R. Soc. B. 278(1702), 82–90 (2011).
29. Warren-Thomas, E., Dolman, P. M. & Edwards, D. P. Increasing Demand for Natural Rubber Necessitates a Robust Sustainability Initiative to Mitigate Impacts on Tropical Biodiversity. Conserv. Lett. 8(4), 230–241 (2015).
30. Faria, D., Laps, R. R., Baumgarten, J. & Cetra, M. Bat and bird assemblages from forests and shade cacao plantations in two contrasting landscapes in the Atlantic Forest of southern Bahia, Brazil. Biodivers. Conserv. 15(2), 587–612 (2006).
31. Russo, R. O. Agrosilvopastoral systems: A practical approach toward sustainable agriculture. J. Sustainable Agric. 7(4), 5–16 (1996).
32. Padoch, C. & Pinedo-Vasquez, M. Saving Slash-and-Burn to Save Biodiversity. Biotropica 42(5), 550–552 (2010).
33. Hedges, L. V., Gurevitch, J. & Curtis, P. S. The meta-analysis of response ratios in experimental ecology. Ecology 80, 1150–1156 (1999).
34. Gibson, L. et al. Primary forests are irreplaceable for sustaining tropical biodiversity. Nature 478, 378–381 (2011).
35. Duguid, M. C. & Ashton, M. S. A meta-analysis of the effect of forest management for timber on understory plant species diversity in temperate forests. For. Ecol. Manage. 303, 81–90 (2013).
36. Newbold, T. et al. Global effects of land use on local terrestrial biodiversity. Nature 520, 45–50 (2015).
37. Pereira, H. M., Ziv, G. & Miranda, M. Countryside Species–Area Relationship as a Valid Alternative to the Matrix-Calibrated Species–Area Model. Conserv Biol. 28(3), 874–876 (2014).
38. Englund, G., Sarnelle, O. & Cooper, S. D. The importance of data-selection criteria: meta-analyses of stream predation experiments. Ecology. 80(4), 1132–1141 (1999).
39. von Oheimb, G. & Härdtle, W. Selection harvest in temperate deciduous forests: impact on herb layer richness and composition. Biodivers Conserv. 18(2), 271–287 (2009).
40. Chambers, C. L., McComb, W. C. & Tappeiner, J. C. Breeding bird responses to three silvicultural treatments in the Oregon Coast Range. Ecol. Appl. 9(1), 171–185 (1999).
41. Martin, P. A., Newton, A. C., Pfeifer, M., Khoo, M. & Bullock, J. M. Impacts of tropical selective logging on carbon storage and tree species richness: A meta-analysis. For. Ecol. Manage. doi: 10.1016/j.foreco.2015.07.010 (2015).
42. Bicknell, J. & Peres, C. A. Vertebrate population responses to reduced-impact logging in a neotropical forest. For. Ecol. Manage. 259, 2267–2275 (2010).
43. Burivalova, Z. et al. Avian responses to selective logging shaped by species traits and logging practices. Proc. R. Soc. B. 282(1808), 20150164 (2015).
44. Dunn, R. R. Recovery of Faunal Communities during Tropical Forest Regeneration. Conserv. Biol. 18, 302–309 (2004).
45. Laurance, W. F. et al. The fate of Amazonian forest fragments: A 32-year investigation. Biol. Conserv. 144, 56–67 (2011).
46. Bhagwat, S. A., Willis, K. J., Birks, H. J. B. & Whittaker, R. J. Agroforestry: a refuge for tropical biodiversity? Trends Ecol. Evol. 23, 261–264 (2008).
47. Cowlishaw, G., Pettifor, R. A. & Isaac, N. J. B. High variability in patterns of population decline: the importance of local processes in species extinctions. Proc. R. Soc. B. 276, 63–9 (2009).
48. Lammertink, M. A multiple-site comparison of woodpecker communities in Bornean lowland and hill forests. Conserv. Biol. 18, 746–757 (2004).
49. Daily, G. C. et al. Countryside biogeography of Neotropical mammals: conservation opportunities in agricultural landscapes of Costa Rica. Conserv. Biol. 17, 1814–1826 (2003).
50. Medjibe, V. P. & Putz, F. E. Cost comparisons of reduced-impact and conventional logging in the tropics. J. For. Econ. 18(3), 242–256 (2012).
51. Ruslandi, X., Venter, O. & Putz, F. E. Overestimating conservation costs in Southeast Asia. Front. Ecol. Environ. 9, 542–544 (2011).
52. Sørensen, T. A method of establishing groups of equal amplitude in plant sociology based on similarity of species content. Biol. Skr. - K. Dan. Vidensk. Selsk. 5, 1–34 (1948).
53. Chaudhary, A., Verones, F., de Baan, L. & Hellweg, S. Quantifying Land Use Impacts on Biodiversity: Combining Species–Area Models and Vulnerability Indicators. Environ. Sci. Technol. 49(16), 9987–9995 (2015).
54. Edwards, D. P. et al. Land-sharing versus land-sparing logging: reconciling timber extraction with biodiversity conservation. Glob. Chang. Biol. 20(1), 183–191 (2014).
55. Fisher, B., Edwards, D. P., Giam, X. & Wilcove, D. S. The high costs of conserving Southeast Asia's lowland rainforests. Front. Ecol. Environ. 9(6), 329–334 (2011).
56. Ramage, B. S. et al. Pseudoreplication in tropical forests and the resulting effects on biodiversity conservation. Conserv. Biol. 27, 364–72 (2013).

57. Halme, P., Toivanen, T., Honkanen, M., Kotiaho, J. S., Mönkkönen, M. & Timonen, J. Flawed meta-analysis of biodiversity effects of forest management. Conserv. Biol. 24, 1154–1156 (2010).

58. Krishnapillay, D. B. & Varmola, M. Case Study of the Tropical Forest Plantations in Malaysia. Forest Plantations Thematic Papers. Working Paper (FAO) (2002).

59. Hånell, B., Nordfjell, T. & Eliasson, L. Productivity and costs in shelterwood harvesting. Scand. J. For. Res. 15(5), 561–569 (2000).

60. Medjibe, V. P., Putz, F. E. & Romero, C. Certified and uncertified logging concessions compared in Gabon: changes in stand structure, tree species, and biomass. Environ. Manage. 51(3), 524–540 (2013).

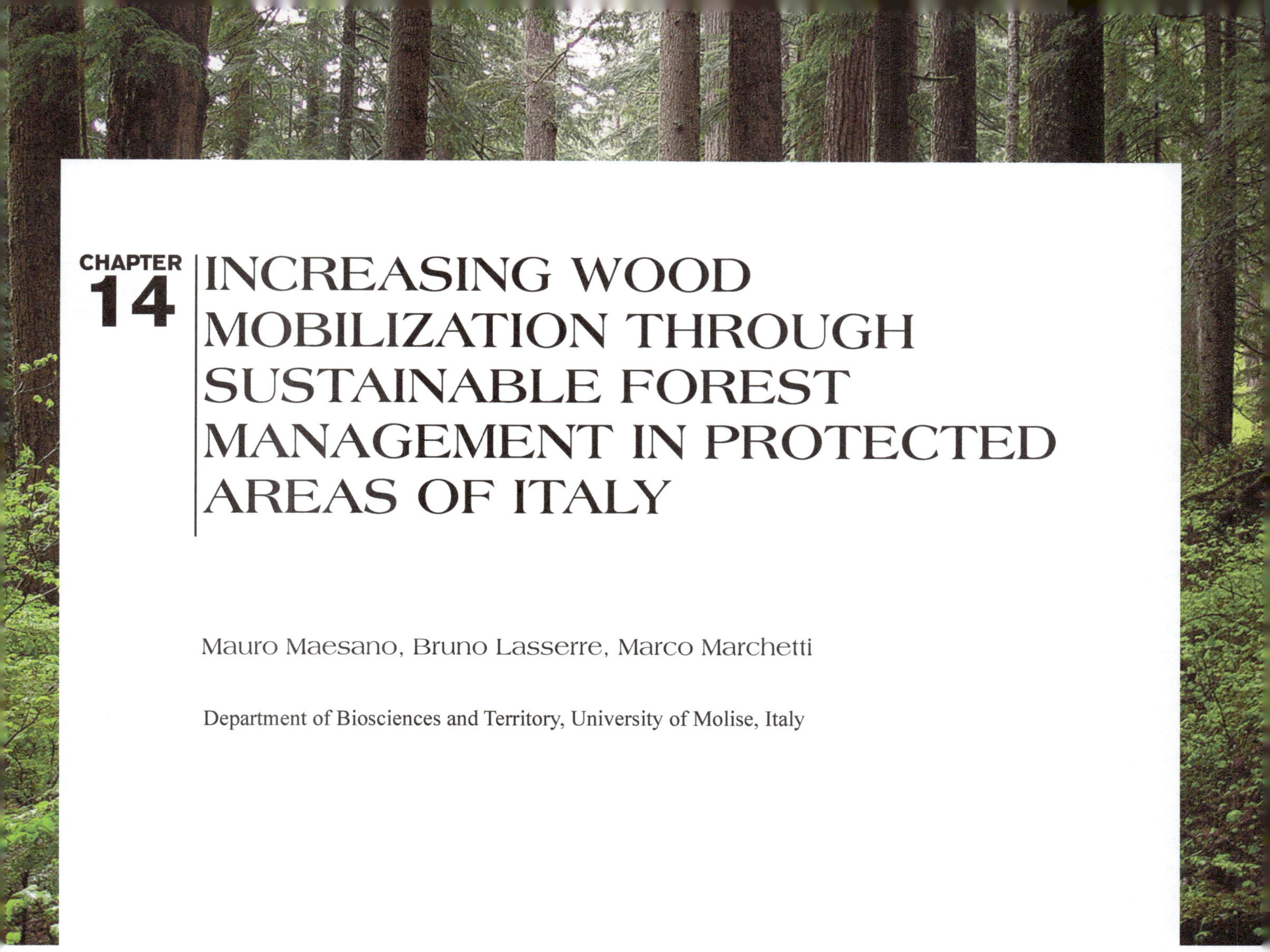

CHAPTER 14

INCREASING WOOD MOBILIZATION THROUGH SUSTAINABLE FOREST MANAGEMENT IN PROTECTED AREAS OF ITALY

Mauro Maesano, Bruno Lasserre, Marco Marchetti

Department of Biosciences and Territory, University of Molise, Italy

INTRODUCTION

In recent years, the use of forest biomass for energy purposes has gained outstanding importance due to its role in the reduction of fossil fuel dependence, the diversifi cation of the products deriving from forests (Lamers et al. 2013), the climate change mitigation and in providing job opportunities in rural areas. Within the 2020 strategy (http://ec.europa/europe2020/) for a smart, sustainable and responsible growth, forest biomasses play a crucial role in the achievement of objectives. The woody biomass accounts about half of Europe's renewable-energy consumption, in its various forms, from sticks to pellets to sawdust, as well as it is the more widespread renewable fuel used (Economist, April 6th 2013). In 2010, the total supply of all woody resources in the EU 27 was about one billion cubic meters where of almost 70% came from forest and 30% from outside the forest (Mantau et al. 2010). Currently, more than half of the wood harvested from European forests is used for industrial processing purposes and future increases in the demand for timber would require forest managers to increase future management intensity (Sohngen et al. 1999).

Citation: Mauro Maesano, Bruno Lasserre, Marco Marchetti, Increasing wood mobilization through Sustainable Forest Management in protected areas of Italy, http://cra-journals.cineca.it/index.php/asr/article/view/754.

The type of management within protected areas can infl uence the amount of increment that may be harvested. The choice between different forest management practices is a crucial step. A Forest Management Approach (FMA) provides a structure for decision-making, including a range of silvicultural operations throughout the stand development phases (PEER 2011). Dunker et al. (2012) proposed fi ve FMAs by placing the management goal and decisions along a gradient of intensity of intervention with the natural process. The strategic management choices of where to conserve nature, and where to produce wood are often done at the management unit level (PEER 2011).

Most of European forests are managed (MCPFE 2007) and thus offer the possibility to improve their adaptation ability by human intervention (Köhl et al. 2010). Nevertheless about 25% of the European forest area is subject to management constraints to secure the Ecosystem Services (ESs) such as nature conservation, soil protection, water supply or recreation (MCPFE 2003), but also timber production. In Italy forests account for about 10.9 million hectares corresponding to 37% of the land area and are mainly located in hill and mountain ranges. The forest area available for wood supply is about 8.08 million hectares (FOREST EUROPE, UNECE and FAO, 2011). According to the National Statistics Agency (Istituto Nazionale di Statistica - http://agri.istat.it - visited 07/2013), the national average harvested volume from forests amounted to almost 7.6 million cubic metres (21% of total annual volume increment) in 2001, of which about 67% was fuelwood.

Not all forest and Other Wooded Land (OWL) are available for wood supply, and this reflects the forests multifunctionality: as well as providing economic resources, the forests are important for both the environment and the community. To what extent Italian forests can satisfy an increased wood demand, without compromising the others ESs remains an open question. Main concerns are biodiversity protection and carbon neutrality (Muys et al. 2013). Maintaining biodiversity in forested areas can be achieved through an adequate network of Protected Areas (PAs) and by the implementation of a silvicultural management, which integrates conservation and production functions (Parviainen and Frank 2003). Establishing PAs is a common conservation strategy pursued for the aim of biodiversity conservation, but it is also an important opportunity to safeguard the others ESs provided by forests (Cash et al. 2006, Turner et al. 2007, Nelson et al. 2008, Egoh et al. 2009, Pettorelli et al. 2012, Willemen et al. 2013). Biological and landscape diversity protection measures can result in a range of possible impacts on the economic use of forest resources. These PAs are subject to national and/or regional legislation that may dictate some restrictions on forest management (Verkerk et al. 2008), for example, on the future land use options (Norton-Griffiths and Southey 1995).

In Italy, the Law (MATTM 2010) protects 27.5% of the forest area, with a higher incidence in some regions of central and southern Italy. The NATURA 2000 network sites (SCI and SPA) include 22.2% of the national forest area (INFC 2005). In this study, extensions and annual volume increment of the Italian forest area have been estimated with the aim of determining how PAs and others areas subject to different forms of protection weigh upon the annually available wood amount. The target was to identify concrete measures for wood mobilization in PAs. Extension and biomass volumes available for felling were estimated in three steps:

- assessment of forest extensions with different types of protection and silvicultural treatments through CLC 2006 (Corine Land Cover - 2006) (ISPRA 2010) and INFC (Italian National Forest Inventory - 2005) data;
- calculation of the annual increment of forest types through the correspondence between INFC and CLC 2006 data;
- assessment of the potential felling amounts in the network of PAs in Italy and determination of the annually available wood within the PAs supposing the implementation of an active management.

The volume of wood that cannot be harvested from these PAs was assessed estimating the unconstrained potentials for wood fellings (i.e. disregarding all forms of protection or limitations to mobilize these resources) and by reducing this potential by applying felling restrictions to the protected areas (Verkerk et al. 2008).

MATERIALS AND METHODS

General Approach

The paper is based on capturing a large set of data and information on Italian forests and elaborating it in a GIS (Geographic Information System) environment in the CLC 2006 resolution (1:100,000) at the IV level (ISPRA 2010). For this study OWL classes (322, 323, 324 CLC classes) were not considered. The analysis mainly focused on the CLC classes typically considered as "Forest" (311, 312, and 313). In the first step, the forest extensions were determined by overlapping the PAs network areas with CLC 2006 (ISPRA 2010). Each CLC 2006 forest polygon was classified according to the silvicultural system (coppice/non-coppice), as reported by the Italian National Forest Inventory 1985 (ISAFA 1988). A Protection Index (I_{PW}: Index of Protection Wood - ratio between protected forest area and total forest extension) was calculated at regional and national level for each type of protection, silvicultural system and CLC class.

In the next step, a correlation between the forestry categories listed within INFC (2003) and the IV level CLC forest classes was carried out to calculate the increments of each identified forest categories (Table 1) on the basis of INFC annual increment data. The correlation between INFC and CLC categories was performed considering the vegetation description of each category according to the inventory classification of forest vegetation (INFC 2003) and the description of the vegetation categories within CLC 2006 (http://sia.eionet.europa. eu/CLC2000/classes/index_html)

In the last step, the felling potential and restriction within the network of PAs were determined. Figure 1 reports the methodology applied to develop the objectives of the research.

Table 1: Correlation table between the forests categories INFI and IVth level CLC.

Inventory Category (INFC 2005)	CODE INFC	CLC 2006 CLASSES	CODE CLC 2006	Correspondence with INFC code
Larch and Stone pine (L. decidua, P. cembra)	1	Holm-oak and cork-oak forests	3111	15;16;17
Norway spruce (P. abies)	2	Deciduous oak forests	3112	9;10
Fir (A. alba)	3	Mesophilous broad-leaved forests	3113	12
Scots pine and mountain pine (P. sylvestris, P. mugo)	4	Chestnut forests	3114	11
Black pines (P. nigra, P. laricio, P. leucodermis)	5	Beech forests	3115	8
Mediterranean pines (P. domestica, P. marittima, P. halepensis)	6	Igrophilous forests	3116	13
Others coniferous forest	7	Non Native broad-leaved forests	3117	14;19;18
Beech (F. sylvatica)	8	Mediterranean pine forests	3121	6;7
Temperate oaks (Q. petrea, Q. pubescens, Q. robur)	9	Mountain and oromediterranean pine forests	3122	4;5
Mediterranean oaks (Q. cerrid, Q. frainetto, Q. trojana)	10	Silver fir and spruce forests	3123	2;3
Chesnut (C. sativa)	11	Larch and Arolla pine forests	3124	1
Hornbeam and Hophornbeam (Carpinu spp., Ostrya Carpinifolia)	12	Non Native coniferous forests	3125	20
Hygrophilous forests	13	Mixed forests dominated by Holm oak	31311	25%(15;16;17)+ 75%(6; 7; 9; 10; 14)
Others deciduos broadleaved forests	14	Mixed forests dominated by deciduous oak	31312	25%(9;10)+75% (6; 7; 14; 15; 16; 17)
Holm oak (Q. Ilex)	15	Mixed forests dominated by mesophilous broad-levead	31313	25%(12)+75%(5; 7; 14)
Coark oak (Q. suber)	16	Mixed forests dominated by chestnut	31314	25%(11)+75%(5; 7; 14)
Other evergreen broadleaved forests	17	Mixed forests dominated by beech	31315	25%(8)+ 75%(3; 4; 5; 7; 14)
Poplar plantations	18	Mixed forests dominated by igrophilous species	31316	25%(13)+75%(6; 7; 14)
Other broadleaved plantations	19	Mixed forests dominated by non native broad-leaved	31317	25%(14; 18; 19) +75%(6; 7; 14)
Coniferous plantations	20	Mixed forests dominated by mediterranean pine	31321	25%(6; 7)+ 75% (7; 9; 10; 14; 15; 16; 17)
		Mixed forests dominated by mountain and oromediterranean pines	31322	25%(4; 5)+ 75%(2; 3; 7; 8; 14)
		Mixed forests dominated by silver fir and spruce	31323	25%(2; 3)+ 75%(4; 5; 7; 8; 14)
		Mixed forests dominated by larch and Arolla pine	31324	25%(1)+ 75%(2; 3; 4; 5; 8; 14)
		Mixed forests dominated by non native coniferous	31325	25%(20)+ 75%(7; 14; 19)

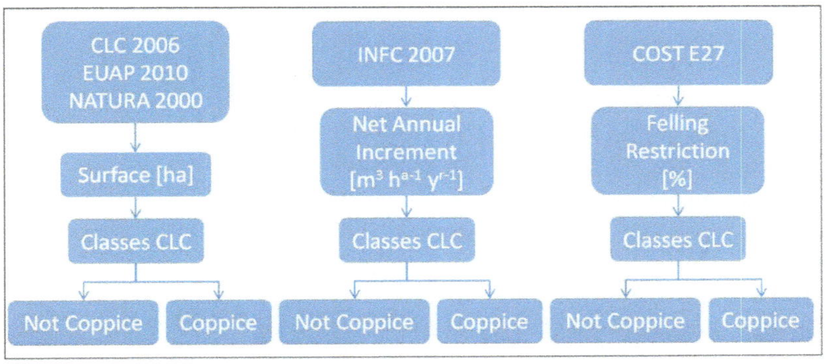

Figure 1: Schematic approach for determining the potential annual increment available and not.

Network of Protected Areas

The EUAP (Official List of the Protected Areas - MATTM 2010) areas and Natura 2000 sites made up the network of PAs used in this research. EUAP areas include 24 National Parks, 144 State Natural Reserves, 134 Regional Natural Parks, 365 Regional Natural Reserves and 171 Other Natural Protected Areas. Natura 2000 (Habitat directive 92/43/CEE) land sites (2011 Update - www.eea.europa.eu) consist in 2,549 Special Areas of Conservation (SAC) and include 2,269 Sites of Community Importance (SCI) and 600 Special Protection Areas (SPAs).

Table 2: Comparison between the MCPFE classes, EUAP, Natura 2000, EEA and the applicable felling restrictions

	MCPFE Classes	EUAP	Natura 2000	EEA	Restriction
Management Objective 1 "Biodiversity Conservation"	1.1 No Active Intervention	Zone A PNZ	-	A	1
	1.2 Minimum Intervention	-	-		A 2
	1.3 Conservation Through Active Management	Zone B-C-D PNZ, PNR, RNS, RNR, AANP	Natura 2000	A	3

The identified PAs network belong to the main classification of Protected Forest Areas (PFAs) at European level as they fulfilled the three general principles outlined in the MCPFE (Ministerial Conference on the Protection of Forests in Europe) guidelines (MCPFE 2002). The MCPFE guidelines distinguish two classes of forest management in the PFAs. The first one focuses on biodiversity and is split up in three subclasses; 1.1: no active intervention; 1.2: minimum intervention; and 1.3: conservation through active management (Table 2). The second one focuses on landscapes protection and specific natural elements. The PAs network areas are included in the first class with a differentiation of the sub-classes based on the parks zonation and/or management plans.

The EUAP areas include also the CDDA (Common Database on Designated Areas) list, coordinated by EEA (EEA 2007). Regarding National Parks within EUAP, the national law established as a planning tool, the zoning of territory (Law 394/91). Based on this law, Ciancio et al. (2002), define four classes different for management degree. Zone A: the goal is the preservation regardless of the naturalness degree; zone B: preservation is realized through systemic silviculture; zone C: in addition to systemic silviculture classical silviculture can be used; zone D: the choice of silvicultural system is broader. The extension of zone A in the national parks refers to INFC (2005) data, and being evident the conservation objective, we supposed a non-coppice system in these areas.

The partial overlapping between EUAP areas and NATURA 2000 sites was considered in data processing.

This network of PAs falls within the network of High Conservation Value Forest (HCVF) defined by the Forest Stewardship Council (FSC) that represent about 40 % of the total Italian forest extension (Maesano et al. 2014).

Felling Potential

The felling potential reflects the theoretical maximum volume that is available for harvesting. There are several approaches to estimate the maximum felling potential, but the most straightforward is the Net Annual Increment (NAI) method (Verkerk et al. 2008). NAI has to be considered an upper limit, because annual felling should not exceed the annual increment to ensure long-term sustainability (MCPFE 2003). The potential increment for the whole network of PAs was calculated for each forest type, administrative region and silvicultural system according to the INFC (2005) data of the NAI.

Felling Restrictions

Data on felling restrictions in forests protected for biodiversity aims were collected from the COST Action E27 study on PFAs in Europe (Frank et al. 2007). In COST Action E27 restriction are given on a scale of 1-4, where: 1- activity is strictly prohibited; 2 - activity is usually prohibited, but with some exceptions or conditions; 3 - activity is usually allowed, but with some exceptions or conditions; 4 - activity is allowed with no restrictions. Based on the restriction scale, the limitation levels were: 1 - 0% of the potential can be harvested; 2 - 33% of the potential can be harvested; 3 - 67% of the potential can be harvested; 4 - 100% of the potential can be harvested (Verkerk et al. 2008). Table 2 shows the felling restriction applied in the network of PAs.

RESULTS

Network of Protected Areas

The forest area inside the network of PAs was about 2.6 million hectares compared with 7.8 million hectares of Italian forest area (CLC data). This is equivalent to an IPW of 33.6% made up of 15.96% (about 1.23 million hectares) of coppice and 17.64% (about 1.37 million hectares) of non-coppice forests

Considering separately the two forms of protection that partially overlap, the EUAP areas represent an I_{PW} of 17.64% (about 1.37 million hectares) composed of 8.06% (about 0.63 million hectares) of coppice and 9.58% (about 0.74 million hectares) of non-coppice forests; while the NATURA 2000 sites represent an IPW of 29.72% (about 2.3 million hectares) with 14.33% (about 1.1 million hectares) of coppice and 15.39% (about 1.2 million hectares) of non-coppice forests. At regional level, the highest value of I_{PW} was in Campania region with 3.32%, while the smallest one was in Valle d'Aosta with 0.14%. Regarding silvicultural system, in coppice forests the I_{PC} (Index of Protection Coppice) varied from 2.25% in Campania to 0% in Valle d'Aosta, while in non-coppice forests the I_{PNC} (Index of Protection Non-Coppice) ranged from 1.74% in Calabria to 0.14% in Valle d'Aosta (Tab. 3).

Considering the CLC classes, the more representative was the deciduous oak forest (3112), which covers 0.55 million hectares, equal to I_{PW} of 7.14%. According to silvicultural system, the classes with the highest diffusion in coppice forests were the beech forests (3115) with an I_{PC} 4.49% (0.35 million hectares) while in non-coppice forests the deciduous oak forests (3112) with an I_{PC} 2.90 % (0.23 million hectares).

Felling Potential

At national level, the potential annual increment was about 35 million cubic meters, of which 42.42% derived from coppice and 57.58% from non-coppice forests. The highest annual increment of coppice forests was found in the Toscana region with 19.8%, equal to about 2.8 million cubic meters, while the Trentino Alto Adige represented the largest annual increment in non-coppice forests with 18.58% amounting to about 3.7 million cubic meters. The CLC classes with the highest annual increment were respectively the 3114 (chestnut forests) for coppice forests with 12.03% (about 4.1 million cubic meters) and the 3123 (silver fir and spruce forests) for noncoppice forests with 14.12% (about 4.9 million cubic meters) (Tab. 4).

The potential annual increment within the network of PAs amounted to about 11.7 million cubic meters, equivalent to 33.79% of the total annual increment divided in 13.53% for coppice forests (about 4.7 million cubic meters) and 20.26% for non-coppice forests (about 7 million cubic meters). The CLC classes with the highest annual increment were beech forests (3115) with 5.16% of the total annual increment (about 1.1 million cubic meters) in coppice forests, while silver fir and spruce forests (3123) with 4.17% (about 0.9 million cubic meters) in non-coppice forest (Tab. 4).

The potential annual increment distinguished by the form of protection and CLC classes was reported in Table 4.

Felling Restrictions and Available Wood

The availability of wood in the network of PAs depends on the type of protected area and on the silvicultural system. The felling restriction affected the Zone A of the national parks, involving 1.19% (0.09 million hectares) of the forest extension and 1.09% of the national annual increment. The annual increment unavailable within the network of PAs represents 11.88% (about 4.1 million cubic meters) of the total annual increment. However, the annual increment available through active management represents 21.9% of the total annual increment, amounting to about 7.6 million cubic meters (Tab. 5). At national level, beech forests (3115) class was the one with the higher available annual increment (5.6% - 1,940,000 cubic meters) (Fig. 2). This is easily explainable, because the beech forest in Italy is one of the most widespread forest type representing 12% of the national forests (INFC 2005 data). General overview of the results split up by silvicultural system showed that the annual increment was respectively 3.8 m^3 ha^{-1} in coppice forests and 5.1 m^3 ha^{-1} in non-coppice forests. Within the network of PAs the available annual average increment was 0.98 m3 ha-1 (0.81 m^3 ha^{-1} in coppice and 1.14 m^3 ha^{-1} in non-coppice forests). Considering the two forms of protection, the available annual average increment was 0.51 m3 ha-1 in EUAP areas (0.42 m^3 ha^{-1} in coppice, 0.60 m^3 ha^{-1} in non-coppice) while 0.9 m3 ha-1in Natura 2000 sites (0.74 m^3 ha^{-1} for coppice and 1.06 m^3 ha^{-1} for non-coppice).

Table 4: Share of the potential annual increment within the EUAP areas, Natura 2000 and CLC 2006. Legend: I_{CC}: Increment Current Coppice; I_{CNC}: Increment Current Non-Coppice; I_{CW}: Increment Current Wood.

CLC 2006 Classess	EUAP + NATURA 2000			EAUP			NATURA 2000			CLC 2006		
	Ic_C	Ic_{NC}	Ic_W	Ic_C	Ic_{NC}	Ic_W	Ic_C	Ic_{NC}	Ic_W	Ic_C	Ic_{NC}	Ic_W
3111	1.02%	0.90%	1.92%	0.54%	0.39%	0.93%	0.87%	0.70%	1.57%	2.22%	2.52%	4.74%
3112	2.46%	2.20%	4.66%	1.15%	1.37%	2.52%	2.23%	1.77%	4.00%	9.15%	7.49%	16.63%
3113	1.61%	1.02%	2.63%	0.70%	0.60%	1.30%	1.46%	0.94%	2.40%	6.00%	2.56%	8.56%
3114	2.60%	0.73%	3.33%	1.18%	0.49%	1.67%	2.29%	0.51%	2.80%	12.03%	2.42%	14.45%
3115	5.16%	3.36%	8.52%	3.09%	2.29%	5.38%	4.86%	3.12%	7.98%	9.86%	5.29%	15.15%
3116	0.03%	0.27%	0.31%	0.01%	0.12%	0.12%	0.03%	0.24%	0.27%	0.14%	1.16%	1.30%
3117	0.07%	0.16%	0.23%	0.05%	0.07%	0.11%	0.05%	0.15%	0.20%	0.37%	1.13%	1.50%
3121	0.00%	1.35%	1.35%	0.00%	0.68%	0.68%	0.00%	1.26%	1.26%	0.00%	2.99%	2.99%
3122	0.00%	1.06%	1.06%	0.00%	0.75%	0.75%	0.00%	0.78%	0.78%	0.00%	2.61%	2.61%
3123	0.00%	4.17%	4.17%	0.00%	1.65%	1.65%	0.00%	3.91%	3.91%	0.00%	14.12%	14.12%
3124	0.00%	1.08%	1.08%	0.00%	0.53%	0.53%	0.00%	1.05%	1.05%	0.00%	2.81%	2.81%
3125	0.00%	0.03%	0.03%	0.00%	0.01%	0.01%	0.00%	0.03%	0.03%	0.00%	0.19%	0.19%
31311	0.07%	0.20%	0.27%	0.05%	0.14%	0.18%	0.05%	0.14%	0.20%	0.14%	0.34%	0.48%
31312	0.10%	0.32%	0.42%	0.05%	0.19%	0.23%	0.08%	0.26%	0.34%	0.42%	1.08%	1.50%
31313	0.09%	0.17%	0.26%	0.03%	0.07%	0.11%	0.08%	0.15%	0.23%	0.73%	0.85%	1.57%
31314	0.08%	0.08%	0.16%	0.03%	0.04%	0.07%	0.06%	0.06%	0.12%	0.54%	0.49%	1.03%
31315	0.22%	0.64%	0.87%	0.13%	0.38%	0.50%	0.17%	0.59%	0.76%	0.71%	1.62%	2.32%
31316	0.00%	0.01%	0.01%	0.00%	0.00%	0.00%	0.00%	0.01%	0.01%	0.00%	0.02%	0.03%
31317	0.00%	0.00%	0.00%	0.00%	0.00%	0.00%	0.00%	0.00%	0.00%	0.04%	0.05%	0.09%
31321	0.00%	0.48%	0.48%	0.00%	0.25%	0.25%	0.00%	0.43%	0.43%	0.00%	1.65%	1.66%
31322	0.01%	0.96%	0.97%	0.01%	0.68%	0.69%	0.00%	0.69%	0.69%	0.02%	2.97%	2.99%
31323	0.01%	0.83%	0.84%	0.00%	0.30%	0.30%	0.01%	0.81%	0.81%	0.06%	2.55%	2.61%
31324	0.00%	0.21%	0.21%	0.00%	0.06%	0.06%	0.00%	0.21%	0.21%	0.00%	0.60%	0.60%
31325	0.00%	0.02%	0.02%	0.00%	0.01%	0.01%	0.00%	0.02%	0.02%	0.00%	0.07%	0.07%
Italy	13.53%	20.26%	33.78%	7.01%	11.04%	18.05%	12.26%	17.80%	30.06%	42.42%	57.58%	100.00%

Table 5: Annual increments available and not within the EUAP areas and Natura 2000.

	Wood Unavailable [m3]			Wood Unavailable [%]		
	Coppice	Not Coppice	Total	Coppice	Not Coppice	Total
EAUP (PNZ - Zone A)	0	380870	380870	0.0%	1.1%	1.1%
EUAP (Zone B-C-D PNZ, PNR, RNS, RNR, AANP)	806106	1144084	1950190	2.3%	3.3%	5.6%
EUAP (Total)	806106	1524954	2331060	2.3%	4.4%	6.7%
Natura 2000	1409644	2046279	3455923	4.0%	5.9%	9.9%
EUAP + Natura 2000	1555447	2584103	4139550	4.5%	7.4%	11.9%
	Wood Available [m3]			**Wood Available [%]**		
	Coppice	Not Coppice	Total	Coppice	Not Coppice	Total
EAUP (PNZ - Zone A)	0	0	0	0.0%	0.0%	0.0%
EUAP (Zone B-C-D PNZ, PNR, RNS, RNR, AANP)	1636640	2322836	3959476	4.7%	6.7%	11.4%
EUAP (Total)	1636640	2322836	3959476	4.7%	6.7%	11.4%
Natura 2000	2862004	4154566	7016570	8.2%	11.9%	20.1%
EUAP + Natura 2000	3158029	4473231	7631260	9.1%	12.8%	21.9%

DISCUSSION

At national level, 88.4% of the forest area was available for forest logging (Gasparini and Tabacchi 2011) and the share of net available annual increment was 39.5% in 1990, 33.1% in 2000 and 26.3% in 2005 (EUROSTAT 2012). The average increment potential available for felling at national level was 4.4 m^3 ha^{-1}.

According to EUROSTAT data (http://epp.eurostat.ec.europa.eu/), European countries have a large margin to increase logged wood volumes. Italy has a level of wood mobilization among the lowest ones in Europe. Considering the very low level of harvesting, a large majority of harvested volumes comes from production forests outside protected areas. Therefore, there is a potential to enhance wood supply in a sustainable way, for energy and raw materials for industry, encouraging wood mobilization (Muys et al. 2013). In contrast, due to the global economic crisis there was a rising interest in fuelwood, resulting in an increase of fuelwood imports of about 26% in Italy annually. Italy is the first importer country of firewood in Europe.

Increasing the consumption of the wood sustainably produced and boost the bio-energy in countries all over Europe could lead economic, environmental and geopolitical gains (FOREST EUROPE 2010). There are many credible reasons to move away from fossil fuels, including the reduction of the dependence on foreign petroleum, maintaining forest management infrastructures, and encouraging conservation of forest work (Gunn et al. 2012). Moreover, the biomass has several other well-recognized benefits such as improved security of supply, contributing to improve air quality and creation of new jobs and businesses - many of them in rural areas (Ragwitz et al. 2006).

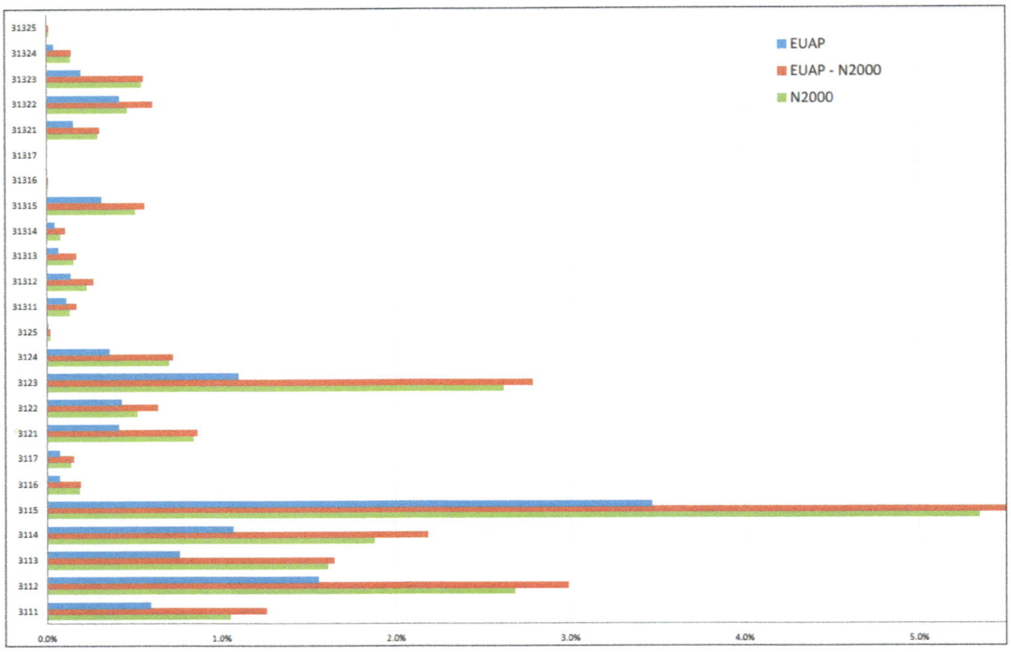

Figure 2: Annual increments available for each CLC class within the protected areas through an active management of the territory.

Forests represent a combination of important ecosystems that provide habitat for numerous species, regulate water cycles, clean air and provide timber for economic use (Maes et al. 2011). The forest growing stock, the increment and fellings could be considered directly ecosystem services indicators for sustainability of ecosystem services over time, to ensure that the long-term benefit flow of services is represented (Maes et al. 2011). Mobilizing more wood is an effort of the whole forest sector. There is a potential to enhance wood supply in a sustainable way, for energy and raw materials for industry. In line with the policy commitments all over Europe and in particular in light of the EU renewable energy targets, wood mobilization should be further encouraged in countries all over Europe (FOREST EUROPE 2010).

CONCLUSIONS

This paper assessed the amount of wood that is technically available in the protected areas by active management of the forest resources. To assess the felling potential in the network of PAs, the unbounded available wood was evaluated and divided into the two silvicultural systems (coppice, non-coppice). This upper limit does not take into account management restrictions due to biological and landscape diversity protection that may apply in PFAs and does not take into account other environmental, social and economic factors that may limit the felling potential. The research provides the dataset and suggests the rationale for wood mobilization in areas usually considered of close to strict environmental protection. Wood harvesting can be undertaken without compromising other forest functions if it is made in a sustainable way, in other words, without compacting soil, causing soil erosion, or disturbing the reproductive cycle of plants or animals.

REFERENCES

1. Cash D.W., Adger W., Berkes F., Garden P., Lebel L., Olsson P., Pritchard L., Young O. 2006 - Scale and cross-scale dynamics: governance and information in a multilevel world. Ecology and Society 11: 8.
2. Ciancio O., Corona P., Marchetti M., Nocentini S. 2002 - Linee guida per la gestione ecosostenibile delle risorse forestali e pastorali nei Pachi Nazionali. Ministero dell'Ambiente, Servizio conservazione della Natura, Accademia Italiana di Scienze Forestali, Firenze, 302 p. http://www.aisf.it/ monografie/linee%20guida%20 parchi/Linee%20Guida%20 Parchi.pdf.
3. Duncker P., Barreiro S., Hengeveld G.M., Lind T., Spiecker H., Mason B., Ambrozy S., Spiecker H. 2012 - Classification of forest management approaches: a new methodological framework and its applicability to European forestry. Ecology and Society 17 (4): 51.
4. EEA 2007 - Nationally designated areas (national-CDDA). Version 9. EEA, Copenhagen [Online] http:// www.eea. europa.eu/data-and-maps/data/nationally-designated-areasnational-cdda-5.1 [11 Oct 2011]

5. EUROSTAT 2012 - Forestry. Forest increment and fellings (tsdnr520). [Online] http://epp.eurostat.ec.europa.eu/portal/ page/portal/forestry/data/main_tables [2012]

6. FOREST EUROPE/UNECE/FAO 2010 - Information Document on Data Collection and Compiling the Statistics on Protected and Protective Forest and Other Wooded Land for Pan-European Reporting. [Online] http://www.unece.org/ fileadmin/DAM/timber/publications/soef2011-protectedforest.pdf [2012]

7. FOREST EUROPE 2010 - Good practice guidance on the sustainable mobilisation of wood in Europe. Publications Office of the European Union, ISBN: 978-92-79-13933-8, DOI: 10.2762/17910.

8. Frank G., Parviainen J., Vandekerkhove K., Latham J., Schuck A., Little D. (eds.). 2007 - COST Action E27 Protected forest areas in Europe-analysis and harmonization (PROFOR): results, conclusions and recommendations. Federal research and training centre for forests, natural hazards and landscape (BFW). Vienna. 211 p.

9. Gasparini P., Tabacchi G. 2011 - L'inventario Nazionale delle foreste e dei serbatoi forestali di carbonio INFC 2005. Secondo inventario forestale nazionale italiano. Metodi e risultati. Ministero delle Politiche Agricole, Alimentari e Forestali, Corpo Forestale dello Stato; Consiglio per la Ricerca e la sperimentazione in Agricoltura, Unità di ricerca per il monitoraggio e la pianificazione forestale. Edagricole, Milano, 653 p.

10. Gunn S.J., Ganz D.J., Keeton W. 2012 - Biogenic vs. geologic carbon emissions and forest biomass energy production. Global Change Biology 4: 239-242.

11. INFC 2005 - Inventario Nazionale delle Foreste e dei Serbatoi Forestali di Carbonio. Ministero delle Politiche Agricole Alimentari e Forestali, Ispettorato Generale - Corpo Forestale dello Stato. CRA - Istituto Sperimentale per l'Assestamento Forestale e per l'Alpicoltura, Trento. http://www.sian. it/inventarioforestale/jsp/home.jsp.

12. ISAFA 1988 - Inventario Forestale Nazionale Italiano 1985 (IFNI 85). Istituto sperimentale per l'Assestamento forestale e per l'Alpicoltura, Trento

13. ISPRA 2010 - La realizzazione in Italia del progetto europeo Corine Land Cover 2006. Rapporto 131/2010 ISPRA, Istituto Superiore per la Protezione e la Ricerca Ambientale. Roma, 50 p. ISBN 978-88-448-0477-0. http://www.isprambiente.gov. it/site/_contentfiles/00008300/8329_rapporto_131_2010.pdf.

14. Köhl M., Hildebrandt R., Olschofksy K., Köhler R., Rötzer T., Mette T., Pretzsch H., Köthke M., Dieter M., Abiy M., Makeschin F., Kenter B. 2010 - Combating the effects of climatic change on forests by mitigation strategies. Carbon Balance and Management 5: 8.

15. Maes J., Braat L., Jax K., Hutchins M., Furman E., Termansen M., Luque S., Paracchini M.L., Chauvin C., Williams R., Volk M., Lautenbach S., Kopperoinen L., Schelhaas M.J., Weinert J., Goossen M., Dumont E., Strauch M., Görg C., Dormann C., Katwinkel M., Zulian G., Varjopuro R., Ratamäki O., Hauck J., Forsius M., Hengeveld G., Perez-Soba M., Bouraoui F., Scholz M., Schulz-Zunkel C., Lepistö A., Polishchuk Y., Bidoglio G. 2011 - A spatial assessment of ecosystem services in Europe: methods, case studies and policy analysis - phase 1. PEER Report No 3. Ispra: Partnership for European Environmental Research. Italy, 148 p.

16. Mantau U., Saal U., Prins K., Steierer F., Lindner M., Verkerk H., Eggers J., Leek N., Oldenburger J., Asikainen A., Anttila P. 2010 - EUwood - real potential for changes in growth and use of EU forests. Final Report. University of Hamburg. Centre of Wood Science, Hamburg. http://ec.europa.eu/energy/ renewables/studies/doc/bioenergy/euwood final report.pdf.

17. MATTM 2010 - Elenco Ufficiale delle Aree naturali Protette (EUAP) 6° Aggiornamento 2010. Dipartimento per l'assetto dei valori ambientali del territorio, Direzione per la Conservazione della Natura, Ministero dell'Ambiente e della tutela del Territorio, Roma.

18. MCPFE 2002 - Assessment Guidelines for Protected and Protective Forest and Other Wooded Land in Europe. Annex 2 to Vienna Resolution 4. Fourth Ministerial Conference on the Protection of Forest in Europe, 28-30 April 2003, Vienna, Austria.

19. MCPFE 2003 - State of Europe's forests 2003. The MCPFE report on sustainable forest management in Europe. Jointly prepared by the MCPFE Liaison Unit Vienna and UNECE/ FAO, Vienna.

20. MCPFE 2007 - State of Europe's Forests 2007. http://www. foresteurope.org/filestore/foresteurope/Publications/pdf/ state_of_europes_forests_2007.pdf

21. Nelson E., Polasky S., Lewis D.J., Plantinga A.J., Lonsdorf E., White D., Bael D., Lawler J.J. 2008 - Efficiency of incentives to jointly increase carbon sequestration and species conservation on a landscape. In: Proceedings of the National Academy of Sciences 105: 9471-9476.

22. Norton-Griffiths M., Southey C. 1995 - The opportunity costs of biodiversity conservation in Kenya. Ecological Economics 12: 125-139.

23. Parviainen J., Frank G. 2003 - Protected forests in Europe approaches-harmonising the definitions for international comparison and forest policy making. Journal of Environmental Management 67: 27-36.

24. PEER 2011 - A spatial assessment of ecosystem services in Europe: Methods, case studies and policy analysis - phase I. European Commission Joint Research Centre Institute for Environment and Sustainability (JRC-IES).

25. Pettorelli N., Chauvenet A.L.M., Duffy J.P., Cornforth W.A., Meillere A., Baillie J.E.M 2012 - Tracking the effect of climate change on ecosystem functioning using protected areas: Africa as a case study. Ecological Indicators 20: 269-276.

26. Ragwitz M., Toro F., Resch G., Faber T., Haas R., Hoogwijk M., Voogt M., Rathmann M. 2006 - Economic analysis of reaching a 20 % share of renewable energy sources in 2020. European Commission, DG Environment. ENV.C.2/ SER/2005/0080r. http://ec.europa.eu/environment/enveco/ others/pdf/res2020_final_report.pdf

27. Sohngen B., Mendelsohn R., Sedjo R. 1999 - Forest Management, Conservation, and Global Timber Markets. American Journal of Agricultural Economics 81 (1): 1-13.

28. Turner W.R., Brandon K., Brooks T.M., Costanza R., da Fonseca G.A.B., Portela R. 2007 - Global Conservation of Biodiversity and Ecosystem Services. BioScience, 57: 868-873.

29. Verkerk P.J., Zanchi G., Lindner M. 2008 - Impacts of Biological and Landscape Diversity Protection on the Wood Supply in Europe. EFI Technical Report 27, European Forest Institute, 29 p.

30. Willemen L., Drakou E.G., Dunbar M.B., Mayaux P., Egoh B.N. 2013 - Safeguarding ecosystem services and livelihoods: Understanding the impact of conservation strategies on benefit flows to society. Ecosystem Services 4: 95-103.

CHAPTER 15

SUSTAINABLE FOREST MANAGEMENT:HOW CAN IT BE ACHIEVED THROUGH COMMUNITY FORESTRY?

OVERVIEW

Governments are under increased pressure to ensure their nation's forests deliver environmental, economic and social outcomes.

Community forestry allows governments to attain multiple benefits simultaneously due to its cross-cutting nature as an environmental, economic, social and cultural initiative.

The potential benefits of community forestry can be significantly and cost-effectively scaled up if governments: 1. increase the area of good quality forests under community management, 2. revise the regulations to enable communities to benefit from their tenure and 3. ensure that communities have access to information and technology that would improve the efficiency and effectiveness of forest management.

INTRODUCTION

Throughout the Asia-Pacific region, governments are under increased pressure to ensure their nation's forests deliver numerous benefits to its stakeholders. These include providing economic benefits including generating income for the national budget, environmental services such as water quality and biodiversity and social impacts such as poverty reduction, and providing subsistence products for many rural households. The challenges that governments face to manage these forests are numerous and seemingly overbearing. Those most adversely affected when the forests are not managed in a sustainable manner are those living in and around the forests, the same people that have the least capacity to adapt to the loss of forests. Community forestry provides a solution to these challenges, allowing governments to attain multiple benefits in an inclusive and sustainable manner.

This policy brief from RECOFTC - The Center for People and Forests discusses ways for governments to support community forestry, based on the understanding that it can provide numerous economic, environmental and social benefits. The brief also highlights the challenges facing community forestry throughout the region and then outlines recommendations for governments to further support community forestry as a pathway to inclusive development, where communities participate in decision-making and get equitable benefits, based on the principles of participation, non-discrimination and accountability. In many places throughout the region community forestry is the most appropriate option for the sustainable management of forest resources, as well as for addressing the wide set of economic, environmental and social objectives (box 1). Evidence in the recent report Community forestry in Asia and the Pacific: Pathway to inclusive development by RECOFTC shows that community forestry in the region:

- enhances livelihoods and covers subsistence needs of households (firewood, fodder, timber etc.);
- reduces deforestation and improves forest conditions and quality (wood volume, stem density, vegetation cover and biodiversity);
- strengthens governance and rural people's political rights.

By investing in community forestry governments can engage with local communities in inclusive development, based on participation, that goes beyond sustainable forest management (see box 2).

Box 1. What is community forestry, and why is it important?

Community forestry includes all aspects, initiatives, science, policies, institutions, and processes that increase the role of all local people – including women, youth and disadvantaged groups – in governing and managing forest resources.

Community forestry is important because it facilitates the incorporation of different interests and values in decision making for forest management. This is based on the understanding that those living in and around the forests are best placed to manage these forests in a sustainable manner, not least because they have the most to gain, such as escaping poverty, from the sustainable management of these forests, and most to lose, including loss of access to subsistence forest products. As those that are most affected, as well as knowing more about the forests, it is fundamental that they are included in decision making processes. This is underlined by the fact that forests are a safety net for many of the 450 million people that live in and around forests in the region.

Details of bemban mat edges. West Kalimantan, Indonesia
Photo by Ramadian Bachtiar/CIFOR

Box 2. Why should community forestry be central to government agendas?

Community forestry enables governments to:

- Respond to environmental objectives such as slowing deforestation, and conserving biodiversity.
- Contribute to decentralization, which is important as it gives local people more control over their lives with the result being that they are more able to take advantage of opportunities and deal with adversity.
- Meet global commitments such as on addressing illegal logging and reducing carbon emissions.
- Attain wider development goals through community forestry. Examples include climate change adaptation and mitigation, improved governance, recognition of indigenous peoples, gender equality, protection of human rights, development of grassroots democracy and poverty reduction.

SCALING UP THE BENEFITS OF COMMUNITY FORESTRY

The potential benefits of community forestry can be significantly and cost-effectively scaled up if more communities are given tenure to their forests. This is important as tenure provides security to the community, encouraging investment in the forests with many implications including for forest quality and the community's livelihoods.

Communities with tenure to their forests could increase the forest's contribution to their livelihoods if:

- Forest departments grant communities tenure rights to forests that can provide income. Currently communities are often granted tenure to highly degraded forests or barren forestland, severely limiting the potential of community forestry to contribute to poverty reduction.
- Regulations that govern communities' interaction with their forests are revised to ensure that the communities are able to make a living from the forests. The original intention of these regulations was to protect the forest, in fact many of these are counter-productive. Complex and rigid regulations for communities abound around timber harvesting, possessing or operating forestry equipment and machinery, transporting logs and processed timber, forest management planning, and technical and financial aspects. These barriers result in the loss of potential revenues for the state, as well as encouraging illegal logging as they make the legal logging of timber uncompetitive.

Further, developing the capacity of communities through improved access to information and technology could help improve the efficiency and effectiveness of forest management, enabling them to maximise the income from the forests in a sustainable manner.

RECOMMENDATIONS

Community forestry has been recognised in many places throughout the region as the most appropriate modality for sustainably managing forests, but it has not become the primary modality nor been fully mainstreamed in most countries. There is still significant potential for further development on which central governments can wield significant influence. When supported by governments together with other key stakeholders, community forestry can be a powerful pathway for inclusive development.

RECOFTC, together with our partners, call on governments to strengthen their commitment to community forestry through:

- Expanding, strengthening and respecting community tenure rights by making community forestry the main forest management modality, requiring that community rights are incorporated in all relevant legislation, that government staff are obliged and trained to support community forestry development, and through the institutionalisation of grievance mechanisms. This will lead to security for communities.
- Enabling communities to actively manage their forests. This will be achieved through the removal of regulations that greatly limit communities ability to profitably manage their forests.
- Making community forestry a foundation for all forest policy initiatives (e.g., Reducing Emissions from Deforestation and Forest Degradation (REDD+), Forest Law Enforcement, Governance and Trade (FLEGT), Payment for Environmental Services (PES)) and in wider development initiatives (e.g., on gender). This will ensure that community forestry is institutionalised.
- Coordinating activities of the forest sector by bringing together line ministries, departments, local governments and other stakeholders. By doing so this will reduce the contradiction and complexity in many of the laws and regulations that communities must adhere to in management of their forests.
- Signing up to relevant international treaties and ensuring that national policies and legislation are promulgated to implement them.
- Collaborating on an international level through renewing their commitment to regional networks such as the ASEAN Social Forestry Network (ASFN) and GREEN Mekong with the aim of promoting good policy and practices by sharing knowledge and experiences among key stakeholders.

CHAPTER 16

POLITIC CONSTELLATION OF FOREST MANAGEMENT ON CLIMATE CHANGE IN INDONESIA

Yayuk Siswiyanti [a], Dudung Darusman [b] , Hariadi Kartodihardjo [c] , Iin Ichwandi [d]

[a] Post Graduate School on Forest Policy, Forest Management Science, Bogor Agricultural University Academic Ring Road Campus IPB Dramaga, Bogor 16680 Indonesia

[b,c,d] Forest Management Science, Bogor Agricultural University, Academic Ring Road Campus IPB Dramaga, Bogor 16680 Indonesia

ABSTRACT

Indonesia's involvement in various international agreements expects to engage sustainable forest management. Indonesia adopted several international conventions, as well as United Nation Framework on Climate Change Convention. Nevertheless, forest management problems do not diminish. National policies in response to the Convention are constructed. Diverse interests are contested during raising of climate change policy processes, either in the forum of the Convention or the national policy. Actors within interests fabricated discourses and networks to establish certain narrative. This research highlights narrative of climate change that is deforestation and forest degradation would be able to reduce by incentive mechanism, while narrative of Indonesian forest management is by fairness tenure. The underlying problem of forest governance, particularly tenure governance does not being a prioritizing for forest management in Indonesia.

There is constellation magnitude of politic for policy change in regards economic interest and to eradicate climate disaster interest. Actors within interests play central role in the policy making process whereby works the bargaining power, circulate the knowledge, generate the networks, or coalesce discourse as power of politic within certain narrative.

INTRODUCTION

Policy-making influenced by global political power, particularly in the power of a global actor interest [1]. Facing conserve forest, since 1972 Indonesia has ratified various international conventions, to address national policies towards sustainable forest management. International agreements prioritize and benefit to national interests [2]. It has been more than four decades the ratification of international treaties. Sustainability of forest is facing many problems and conflicts. Illegal logging progressively more widespread [3] even illegally circulating in cross country [4] tenure conflicts of society increasingly wide open and some have involved international actors [5] reduce in biodiversity; rising endangered species [6] chance the illegal trade of endangered species in cross country [7] discourse of "the Forest First" dominates [8] forest communities remain poor and marginalized [9]; even deforestation and forest degradation persist to elevate [3].

Citation: Yayuk Siswiyanti, Dudung Darusman, Hariadi Kartodihardjo, Iin Ichwandi, Politic Constellation of Forest Management on Climate Change in Indonesia, ISSN 2307-4531.

In 1994, by the Law No. 6 of 1994, the President ratified the Convention on Climate Change. Ten years later, the President enforces the Kyoto Protocol through the Law No. 17 of 2004. Since the ratification, policies in the field of forestry progressing quickly respond to the Climate Change Convention. Associated with the ratification of the Convention on Climate Change, persistent issues of forest management, in the time of ratification of the Convention has been adopted since 1972; generate a question: would ratification be able to encourage national policies towards sustainable forest management? Furthermore this research to analyze the politic constellation of Indonesia forest management on Climate Change Convention. The purpose of this research are: (1) analyzing the politic of Indonesia state related to the Climate Change Convention (2) analyzing the ability of climate change convention to drive for policy changes to address the root problems of forest management in Indonesia.

METHODS

To answer the policy making is a political process [10, 11, 12] excavated by understanding the International Development Studies (IDS 2006) process of policy formulation. The debate on climate change negotiation process, extracted from the data and information covering the instrument discussion and debates that occur daily records, obtained from the secretariat of the Convention on Climate Change and the International Institute for Sustainable Development (IISD), that describe actors, interests, narrative, network and so politics which works onto. Similarly, associated with the Climate Change Indonesian policy, explored from the policies that is created by the Government of Indonesia to respond to the Convention, as well as documents regarding the drafting process. To deepen the excavation of the results of the findings, key informant interviews were conducted to clarify the findings of the analysis carried out by researchers. The key informant represent of the government and NGOs. Consistency and coherence reflects the authentication information and the reflectivity of the reality/ value research. This study was conducted over 24 months from May 2012 to April 2014.

RESULTS AND DISCUSSION

Contestation of Discourse

This research shows that international climate change constructs two different discourses in which play an important role in contested issues. These are green development discourse and conservationism discourse. Proponent of development discourse is developments/industries countries; while conservationism discourse is least developing/developing countries. These discourses have several argumentations as shown in Table 1. Both of discourses are constructed by economic frame, shaped by narrative of an economic towards an environment view, as storyline shown in table 2. The green development discourse argue that economic save an environment, obviously a stabilize atmosphere earth temperature from global warming. On the other hand the conservationism discourse assume that economic degrade an environment.

Central argument of green development discourse is that the forest natural resources as capital development to support economic growth for human welfare. Economic growth may increase human welfare, providing energy for humans to save the environment. Through Science and technology built by forest resources, the human be able to manage the environment. Conservationism discourse assumes that the forest is a buffer environmental quality of human life. Forest exploitation will damage the quality of the environment. Without a good quality of environment, the results of the development will be not working in vain because of the disasters caused damage to the environment.

Narrative of Forest Management in Indonesia

In Indonesia home country, the problem of deforestation and forest degradation in the scale of 1.17 million ha/year to 3.51 million ha/year [3, 13] is the failure indicator for forest management. The symptoms widespread tenure conflicts, not only access even the ownership, its more open; extend to move on the issue of human rights and latently. Tenure conflict is more escalating. Agricultural Development Commission (KPA) argues, that since the year 2009 until the year 2013 agrarian conflicts increase 3 times [14]. The forestry sector occupies the highest position, even an increase in land area as much 6 times compared to before year 2009. The most actors involved are private corporations or states corporations with the local community. The location of the conflict spread across 32 provinces. This happens from the accumulation of poor forest governance and land issues [15] dualistic scale of the map in the same location, the dualistic view between forestry and agriculture, dispute the authority of the central government and local governments, egosectoral [16] and corruption [17]. Legislation and management policies of land are not formulated clearly, disputably, overlap, even conflict between the policy; the policy deteriorate the other policies [17].

Tenure is associated with land access and ownership. Analysis of this research carried out on the level of rights to land, such as alienation rights, exclusion rights, management rights, the right to take the product and access rights [18].

Table 1: Discourses and Its argumentations in the Climate Change Conferences

	Green developmentalism discourse: Economic growth save an environment	Conservationism discourse: Economic growth degrade an environment
Central argumentation	The forest natural resources as capital development to support economic growth for human welfare. Economic growth may increase human welfare, providing energy for humans to save the environment particularly stabilize a climate. Through Science and technology built by forest resources, the human be able to manage the environment.	The forest is a buffer environmental quality of human life, particularly stabilization of climate. Forest exploitation will damage the quality of the environment. Without a good quality of environment, the results of the development will be not working in vain because of the disasters caused damage to the environment.
Arguments in the economic context	• Economic growth will damage the environment to a certain critical point, after that, growth actually increase the quality of the environment. • Social discount is negative, assuming the value of well-being in the time dimension	• Economic growth and the accumulation of wealth is the cause of environmental degradation. Economic growth is continuously damaging the environment • Social Discount is positive, assuming the uncertainty in the time dimension
Argumentations in the climate change context	• Heating /cooling of the Earth is a function of space and time • Science is currently not sufficiently able to infer events 100 years • A sharp rise in the temperature of the earth is not proven • Heating the Earth had no significant impact (negligible) to humans, because the human excellent adaptability. Evolution is the human defense system of the natural limitations. • Warming of the climate is the phenomenon of macro-cosmos (the geometry of the earth to the sun), solar radiation and the effect of turning on atmosphere. • Heating of the earth even extreme is a natural cycle of the planet (the evolution of the ice age cycles, growth century etc.) • The situation is local climate. In one region within certain limits do not affect other areas. At the same time, one can experience the cooling region, while in other regions warming.	• GHG emissions per capita is a function of time • With the knowledge of modeling and simulation, the state of the next 100 years can be predicted. • Earth heats up and sharpened since year 1900s • Global warming caused by anthropogenic (human) cause catastrophic • Heating of the earth's (scenario disaster / calamity, catastrophic scenario, alarmist scenarios) • Climate influence globally. The climate in the south may affect the climate in the north and vice versa, because the air is an integral atmosphere space.

Source: Primary Data Analysis, 2014; elaborate Wittmer dan Birner, 2006; Jacobs, 2012

In reference to the Schlager and Ostrom, the results of the analysis of narrative tenure briefly describe as Table 2. Narrative of forest management tenure revealed a weakening of the rights of the community to access, take benefit of forest products, or farming in forest areas. Policy makers of the Indonesian Government made efforts to deprive the rights of land [5]. Economic relations, social and historical relations were broken by the presence of concession [19, 20, 21]. The assemble narrative is a tenure narrative that reduces up to remove the rights of communities to forest products and forest areas, furthermore the tenure conflicts and deforestation are persevered. There are several attempts to provide space for the community, however, re-countered by following policy. Community resistance become latently arise [22]. The narrative consistently affirmed by key informants, revealed that nearly a century, there is a tendency of tenure policy maker actor has no desire to reformulate tenure policies in forest management in Indonesia. Policies actually allowed, somehow constructed overlapping, unclear, ambiguous, inconsistent, conflict with other policies that do not synergize to achieve forest sustainability. Thus, the issue of tenure is not a matter of priority in forest management in Indonesia

Table 2: Narrative of Indonesia Forest Management

Period	Narrative/storyline
1927-1956: Old Regime, Centralization, Exploitation of Forest Product	Dutch East Indies government/Indonesia government has the rights of alienation. Concession Permit Holders have the rights of exclusion. Communities have the rights of exclusion Analysis: Position of Concession Permit Holders rights and community rights grow land boundary conflict [18, 19, 21].
1957-1960: Old Regime, Decentralization, Exploitation of Forest Product	The Central Government has the rights of alienation. The Provincial Government has the rights of alienation. District Government has the rights of alienation. Concession Permit Holders have the rights of exclusion. Communities have the rights of alienation. Analysis: The position of community rights and Concession Permit Holders rights construct land boundary conflict. Likewise, the central government and local governments could potentially be a conflict of authority. [18, 19, 21].
1960-1970: New Regime, Centralization, Exploitation of Forest Products	The Central Government also has the right of alienation in outside Java. The Concession Permit Holders have the rights of exclusion. The community rights is not clear. Analysis: Position of Concession Permit Holders rights rise, the community rights threatened to eradicate [18, 19, 21].
1970-1990: The New Regime, Centralization, Forest Land Use Arrangement (TGHK)	Central Government has the alienation rights in outside Java. The Concession Permit Holders have the right of exclusion. The community rights is still not clear (hazy). Analysis: Position of Concession Permit Holders rights rise, the community rights in forest areas eradicate. There is enormous potentially conflict between the community and Concession Permit Holders [18, 19, 21].
1990-2000: Reformation, Decentralization, Limited exploitation of Forest Product, Region spatial planning/Provincial Spatial Planning/ synchronization	The Central Government has the rights of alienation. The Provincial Government has the rights of alienation. District Government has the rights of alienation. The concession permit holder has the rights of exclusion. The community has a right to manage. Analysis: The rights of the central government and local government, provincial governments and the district have the potentially conflicts of authority. Position of the Concession Permit Holders slightly deteriorates. The rights of people in forest areas strengthened. There is still a potential conflict of land boundaries [18, 19, 21].
2000-2011: Reformation, Decentralization, Appointment of Forest Management Unit	Central Government (Ministries related to Forestry) has the rights of alienation. Ministry of Agriculture, Ministry of mining has the rights of alienation. District Government has the rights of alienation. The concession permit holder has the rights of exclusion. Community rights eliminated. Analysis: The rights of the central government and local government, provincial government and district government have potentially conflict of authority. There are potential conflicts among sectors with respect to land use. Position concession rights rebound, the rights of community in forest areas removed. There is a potential conflict of boundaries between that community and concessionaires. [18, 19, 21].
2012-2014: Reformation, Decentralization, Conflict Resolution	There are the same conditions as the period 2000-2011, but the central government alienation Rights (Ministry of Forestry related) begin to be restricted. The rights of communities to use/manage the forest begin to be recognized, but not guaranteed. [18, 19, 21].

Source: Primary Data Analysis, 2014

This reflects the policy conditions such as regulatory and policy functions of the legislation is not reached [23] which is regulation successfully organize and gain public legitimacy [21] not achieved; policy which has fairness force, failed [17]. Policy functions only on mere administration, obviously the administration approaches. Narrative tends to land tenure centrally controlled by the government, more over land use policies tend to be approached by means of licensing, or protection of utilization, which negates the benefit of community [24]. On the other hand, the earth atmosphere temperature heats up due to a sharp increase in the concentration of greenhouse gases. Global warming causes a natural disaster. Forest has a role to contribute in a reduction of greenhouse gases emissions concentration [25] by which reduction of deforestation and forest degradation. Dealing with facing deforestation and forest degradation, table 3 briefly describes how the narrative of the existence forests roles towards greenhouse gas concentration in the climate change conference. Analysis performed within two periods, with the base of following the achievement of emission restriction commitments for the countries of the Conference of the Parties of the Kyoto Protocol in 1997 and Amendments to the Kyoto Protocol in 2012.

Table 3: Narrative of Climate Change Negotiation Process

Period	Narrative/storyline
1962-1997 Commitment I, Towards the Kyoto Protocol, Annex I emission limitation	Countries aware the destruction of nature, the atmosphere temperature of earth heats up 2 ° C per century. GHGs are anthropogenic gases, cumulative and long life (even CO_2 up to 115 years). GHG produce radiative forcing, thus increasing the level of heating. There is the role of forests, which contributes 17.4% of global emissions. Resulting CO2 emissions from deforestation and decay of biomass accounted for 17.3% of global emissions. To reduce GHG emissions, insist a role of multilateral, including the commitment of developed countries. Non-Annex I countries bear the negative impact (natural disasters, social disaster) because warming of the earth atmosphere temperature. Necessary restrictions on emissions for developed countries (Annex I). Developing countries (non-Annex I Parties as members) are not required to reduce emissions. Countries with tropical forests are almost entirely not mandatory emissions reduction, whereas in the management of tropical forests, the IPCC acknowledges the substantial contribution to greenhouse gas concentrations. Contestation of developed countries versus developing countries, more powerful than the emission production state versus emissions reduction state. Annex I countries can reduce their own emissions or compensate for the negative impact of the provision of services or compensate for emissions reduction. Analysis: Forest plays a role in reducing GHG. Annex I Parties shall provide incentives for states that forest owners shown to reduce emissions. Annex I Parties are also necessary to help the forest management to reduce deforestation and forest degradation through capacity building and technology transfer, which in turn provides incentive for countries that are proven to reduce emissions.
1998-2014 Commitment II, Addendum Towards the Kyoto Protocol, Restricting emission limits of Annex I	Deepened global warming reaches 5 ° C per century. If no deep cuts action, bear bigger and/or faster disaster occurred. Fire of forest and decay of biomass are a high emitter, as well as deforestation and forest degradation. Costs of reduction due to emissions incurred 5 times compared to the costs for GHG reduction actions. Increasingly powerful role of forests: calculations based on the principle of additional of carbon, containing the release of emissions, emission absorption and storage of emissions (emission, sink, stock; ESS), the estimated afforestation, reforestation and deforestation. By reduce emissions, benefit for conservation. Emissions of Annex I Parties between 10.3% -19.4%, with a population of 19.7% of the total world population. While emissions of non-Annex countries, which has a population of 80.3%, accounted for 7.8% -17.3% emissions. Strictly necessary emission limitations for developed countries (Annex I). Tropical forest owners are given the opportunity to voluntarily reduce their emissions. Provide incentives for countries with tropical forests be able to reduce deforestation, with additional principles. Analysis: GHG reduction action more imperative. Insist Annex I countries to immediately reveal commitment to provide incentives to owners of tropical forests for reducing deforestation and forest degradation.

Source: Primary Data Analysis, 2014

By Table 3, the Climate Change Convention negotiations construct a narrative that is the problem of deforestation and forest degradation in countries with tropical forests can be suppressed by the incentive mechanism of GHG emission reductions by developed countries. Almost all products of Indonesian policy response is concerning onto the Climate Change Convention mechanism, obviously the launch of the implementation of the Climate Change Convention does not imply for policy changes regarding to the root of the problem of forest management in Indonesia. Narrative convention contested by Indonesia state is a convention mechanism, which contesting about the commitment of developed countries, solidarity with the affected countries, compliance, and methodology of calculation of emission reductions. The underlying cause of persistence of forest deforestation and forest degradation is failure of tenure governance, as a basic of land management. This issue is poorly contested. On the other side, the preparation of Indonesia in the Convention on climate change negotiations tends to incentive mechanism oriented, as well as administration. It is confirmed that the governance of tenure problems is considered as a matter of home country

Interests of Actors

In the forum of the Convention are approximately 195 state actors. There are six groups, i.e. developed countries, developing countries, Least Developed Countries, Small Developing States, (SIDS), Non Governmental Organization (NGO) and multilateral international organizations. By the analysis of the national communication documents of states parties to the Convention are identified at least nine interests, among others (1) fossil fuel production (2) the industry (3) negatively impact that are vulnerable countries to global warming; (4) positively affected, (5) struggle for sovereignty and the fate of the impact of colonialism, release from dependence on developed countries (6) combating environmental and regional solidarity; promoting regional peace and stability through respect for justice and the rule of law, (7) FFP consumption, (8) improving economic conditions (9) zone of peace, freedom and neutrality and nuclear-free zone.

Strong interest is the reducing of deforestation and forest degradation, fairness tenure, biodiversity and conservation, law enforcement. Powerful actors are forestry ministry, the ministry of Energy, Ministry of agriculture, and local government. Powerless actors are NGOs, while community and private sector absence from policy making process. Somehow, NGOs allocate their time resource between participating on forum of climate change convention and fundraising, which attracts private donations [26].

Powering Politic

Discourse is particularly pertinent to the study of climate change because it provides a framework that is sensitive to the political construction and use of scientific knowledge [27]. The degree of success of policy change by joint knowledge depends on the actors involved, contents of dominant discourses, presence of rules and the availability of resources [28]. In Climate Change Conference, both of Green development discourse and conservationism discourse constitute in the economic framework, wherein the danger and the dispossession of the atmosphere [29]. The economic perspective of the green development discourse is different from the perspective of conservationism discourse. The green development narrative is counter-narrative for conservationism narrative and vice versa. Green development discourse promoted by industrialized countries, while conservationism discourse promoted by countries where in threatened by climate disasters. Countries building a knowledge to reach the own interest, that is an industry/economic benefit of certain countries, similarly a conservation interest of others. The strong 'materialization' of discursive shifts in forest management imply policy change as well as sustainability effects on the ground. Green development discourse itself is a coalition discourse build by developmentalism discourse and conservationism discourse [30]. Sustainable development discourse opened the door to coalitions, which succeeded in combining economic interests with ecological aims [31].

Although within the same framework of economic, industrial interests prioritize a profit and market, while the conservationists prioritize stabilization with preservation of environmental quality and protection. Mission of industry interests encompasses developing GHG disaster adaptation systems, the impact of the compensation system, incentive system (mechanism of environmental services) and the internalization of environmental costs. Option for industrialized countries is to prioritize the incentive system, the system of compensation and contesting mechanism for environmental services. Incentives system and compensation systems designed by industrialized countries, to offer incentives to the parties that provide reduction of greenhouse gas concentrations. The principle of compensation is constructed to provide compensation for disaster borne by certain parties. Thus, argument to strengthen adaptation to climate change is an orientation of the mission. It is opposed by countries threatened disaster prone. The drought and threat of sea level rise of African countries, small island coastal state more urgent more than others countries, these are related to the survival of the nation [32]. Compensation system and incentive systems are considered still provide space industry practices regardless of the environment. Conservationists insist to immediately stop adding of greenhouse gas concentrations, i.e. reducing GHG emissions, such as by the action of deep cuts. Conservationists promote an additional principles and to strengthen mitigation actions. This is disputed by the industrialized countries, because it would threaten the survival of the industry. By reducing GHG will deteriorate their industrial activities, thereby ultimately weaken the economic situation. On the other side, the acceleration of the Asian economies become increasingly however strengthened its own concerns for the countries the United States and Europe, which is threaten the United States and European markets. The taken approach is oriented south-north negotiating compensation [33, 34, 35] also donor-recipient negotiation [26].

Negotiations then move to the initiation to internalizing environmental costs. Waste treatment policy will add up a total operating costs and profit industries as such affecting the selling price of industrial products. That is closely associated with the purchasing power of consumers and ultimately to the market demand. Environmental costs are also not cheap [36] likewise also environmental tax policy. It is still not getting a deal in the negotiations of the Convention. Negotiations with respect to them is often become a dead lock, because it does not obtain the agreement of the parties that threatened its interests.

Powerful actors are developed countries/industries. Powerless actors are poor countries and/or vulnerable to climate disasters, i.e. islanders. Islanders face is not so much the hazard of climate per se, but the reason why SIDS communities often do not have the resources or options to resolve climate change and other development challenges themselves. Indicate the actor has bargaining power which is associated with the Convention's financial contribution borne by the developed countries/industries. Developed countries/industries withdraw financial flows when the results of the negotiations of the Convention are not relevant to their interests, even they do not provide incentives or financial assistance from developed countries (Annex I, annex II) to the non Annex I, obviously, leave from the commitment to the Conventions. The bargaining position encourages countries to the next negotiation. The main contestation of climate change meeting related to economic issues, also issues of controlling GHG disaster. The problem of forest governance is considered as the internal affairs of home countries, therefore poor governance of forest tenure is Indonesia's domestic affairs.

To achieve its interests, actors affiliate into a network, require to effective implementation of climate change convention [37, 38]. Economic/industrial networks are OECD, EU, and OPEC, while G77 + China solidarity to fight the former imperialism. AOSIS and CFRN struggle a climate disaster. Network is used to strengthen the bargaining position of the intervention in the climate change convention negotiations, either verbal intervention or written intervention [32] with a joint proposal or common position. This is due to the parties that each country has one vote in the intervention; therefore the greater networks create for the stronger position [38], such as sustain network [40] The network enable accelerate the implementation of the Convention on Climate Change in Indonesia [41]. Every country contributes to GHG emissions, however, developed countries/industries are mandatory reduce emission, while developing countries voluntarily. Emissions of Annex I Parties between 10.3% -19.4%, with a population of 19.7% of the total population of the world; while emissions of non-Annex countries, which has a population of 80.3%, accounted for 7.8% -17.3% of emissions, so that emissions of developed countries should be restricted. That is political network done by developing countries (G77 + China), insist to developed countries to mandatory reduce emissions.

Narrative of forest management Indonesia that is deforestation and forest degradation failed reduce by the presence of community alienation. Policymaker actor reluctant to reforms the current policy. It is indicate that the actor associate with policy makers take a part of free riding and rent-seeking, safety player person, skepticism, and the presence of the historical of the policy.

Such the situation indicates that strongly narrative contested, have no guarantee to be adopted. It is depend on magnitude of actor interests, in which form a bargaining position [42, 43] by construct a discourse and a network both. These findings draw the negotiation are still dominated by powerful business-as-usual interests, reflected in some countries reluctance to undertake the larger policy reforms that would enable the required change and effectiveness in tackling the often underlying causes of deforestation and forest degradation.

In engaging the interests of actors, decision-making mechanism becomes important [23] The political decisionmaking is not controlled by one (group) actor, but rather the process of consensus [44] thus providing power space for the benefit of the actors whose have high magnitude of interest [32, 45, 46, 47, 48]. Obviously politics is constructed by actors become critical aspect [49].

CONCLUSION

Climate Change Convention able to encourage the production of policies associated with incentive mechanisms, such as a proposing procedure for forest management, inventory and monitoring of GHG of Indonesia, and its institutions. Climate Change Convention enforces international cooperation related to research programs, technology transfer, human resource development and institutional capacity building associated with incentive mechanism. Such a situation, benefit for the government, NGOs, research institutions and the owner of the concession, while the people ignored even become sufferers of a system of incentives that will be built. Climate Change Convention does not promote for policy changes toward fairness tenure because it has a different focus to the issue of forest management in Indonesia. On the other hand, in the home country, tenure issues have not been a priority in Indonesia's forest management yet.

ACKNOWLEDGEMENTS

Authors cordially would like to thank Dr. Harry Santoso and Dr. Agus Sarsito for their most inspiring lens on this article, several of which were adopted into the text. Authors also thank Prof. Wasmen Manalu for his support for this achievement of scientific writing. Authors thank anonymous reviewers for their comments on the manuscript. Despite the many constructive contributions that authors have received during the research and writing, authors alone bear the responsibility for the analysis made in the article.

REFERENCES

1. R. Peet, P. Robbins, MJ. Watts. Global Political Ecology. New York: Routledge, 2011.
2. U.A. Cowie, Schneider, L. Montanarella. Potential synergies between existing multilateral environmental agreements in the implementation of land use, land-use change and forestry activities. Environmental Science and Policy 10: 335– 352, 2007. http://dx.doi.org/10.1016/j.envsci.2007.03.002.
3. Forest Watch Indonesia. Protret Keadaan Hutan Indonesia Periode 2009-2013. Jakarta: FWI, 2014.
4. World Wide Fund for Nature. Illegal Logging & The EU an Analysis of The EU Export & Import Market of Illegal Wood and Related Products. Jakarta:WWF Indonesia, 2008.
5. Epistema Institute. Menuju Kepastian dan Keadilan Tenurial. Jakarta: Epistema Institute, 2011
6. M.T. Brooks , RA. Mittermier. Habitat Loss and Extinction in the Hotspot of Biodiversity. Washington: Center for Applied Science, 2002
7. ProFauna Indonesia. Perdagangan 2012: Maraknya Perdagangan Satwa Langka Via Online. Malang: ProFauna Indonesia, 2012.
8. A. Khan, H. Kartodihardjo, S. Soedomo, D. Darusman. Kebijakan usaha kehutanan: sebuah analisis diskursus. Jurnal Manajemen Hutan (JMHT) 16 (2), 2010.
9. Center for International Forestry Research. Why are forest relevant to reduce poverty in Indonesia. Bogor, Indonesia: CIFOR, 2000
10. R. Sutton. Policy Process: An Overview. Working Paper 118. Overseas Development Instititut. London SW1E 5 DP: Portland House. Stag Place, 1999. http://hdl.handle.net/10068/576355.
11. Institute Of Development Studies. Understanding Policy Processes: a Review of IDS Research on The Environment. Institute Of Development Studies University Of Sussex. UK, 2006.

12. W. de Jong, B. Arts, M. Krott. Political theory in forest policy science. Forest Policy Economics 16: 7-13, 2012. http://dx.doi.org/10.1016/j.forpol.2011.07.001.

13. Kementerian Kehutanan

14. Ministry of Forestry.. Peta jalan Sektor Kehutanan Indonesia menuju Tahun 2030. Jakarta: Kementerian Kehutanan, 2010.

15. Konsorsium Pembangunan Agraria. Membangun kekuatan politik agraria. Land Reform. Vol II-Desember 2014.

16. A.M. Larson. Forest tenure reform in the age of climate change: Lessons for REDD+. Global Environmental Change 21: 540–549,2011. http://dx.doi.org/10.1016/j.gloenvcha.2010.11.008.

17. Koyuncu, R. Yilmaz. The Impact of corruption on deforestation: a cross country evidence. Developing Areas 42: 213-222, 2009. http://dx.doi.org/10.1353/jda.0.0010.

18. Komisi Pemberantasan Korupsi. Program Indonesia Memantau Hutan (IMH). Jakarta: KPK, 2012.

19. E. Schlager, E. Ostrom. Property rights regime and natural resources: a conceptual analysis. Land Economics Journal 68: 249-262, 1992. http://www.jstor.org/stable/3146375.

20. N.L. Peluso. A political Ecology of Violence and Territory in West Kalimantan. Asia Pasific Viewpoints 49: 48–67, 2008. http://dx.doi.org/10.1111/j.1467-8373.2008.00360.

21. L.B. Raymond, A. Paniaguab, T. Kizosc. Conceptualising 'shadow landscape' in political ecology and rural studies. Land use Policy 28:460–471, 2011. http://dx.doi.org/10.1016/j.landusepol.2010.09.005.

22. S. Adiwibowo, M. Shohibuddin, H. Kartodihardjo. Kontestasi Devolusi: Ekologi Politik Pengelolaan Hutan Berbasis Masyarakat dalam Kembali ke Jalan Lurus. Kartodihardjo H, editor, Yogyakarta: Forci Development & Tanah Air Beta, 2013

23. P. Maring. Bagaimana Kekuasaan Bekerja: di Balik Perlawanan dan Kolaborasi, sebuah Sudut Pandang Antropologi tentang Perebutan Sumberdaya Ekologi. Jakarta: Lembaga Pengkajian Atropologi Kekuasaan Indonesia, 2010

24. W. Dunn. Public Policy Analysis: an Introduction. New Jersey: Prentice Hall, Inc, 1994

25. C. Luttrell, L. Loft, M.F. Gebara, D. Kweka, M. Brockhaus, A. Angelsen, WD. Sunderlin. Who should benefit from REDD+? Rationales and realities. Ecology and Society 18: 52, 2013. http://dx.doi.org/10.5751/ES-05834-180452.

26. Intergovernmental Panel on Climate Change. The Fourth Assessment Report (AR4). Bonn: UNFCCC Secretariat, 2007G.

27. Aldashev, T. Verdier. Goodwill bazaar: NGO competition and giving to development. Development Economics. 9:48– 63, 2010. http://dx.doi.org/10.1016/j.jdeveco.2008.11.010.

28. A. Arnall, U. Kothari, I. Kelman, Introduction to politics of climate change: discourses of policy and practice in developing countries. Geographical Journal 8: 98–10, 2013. http://dx.doi.org/201410.1111/geoj.12054.

29. D. Hegger , M. Lamers, A.C. Van Zeijl-Rozema, Dieperink. Conceptualising joint knowledge production in regional climate change adaptation projects: success conditions and levers for action. Environmental Science & Policy 18: 52–65, 2012. http://dx.doi.org 10.1016/j.envsci.2012.01.002.

30. D.M. Liverman. Conventions of climate change: constructions of danger and the dispossession of the atmosphere. Journal of Historical Geography 35: 279–296, 2009. http://dx.doi.org/10.1016/ j.jhg.2008.08.008.

31. M. Jacobs. Green Growth: Economic Theory and Political Discourse. Centre for Climate Change Economics and Policy Working Paper No. 108 and Grantham Research Institute on Climate Change and the Environment Working Paper No. 92., 2012

32. B. Arts, M. Buizer. Forest, discourse, institutions: a discursive-institutional analysis of global forest governance. Forest Policy And Economics 11: 340–347, 2009. http://dx.doi.org/10.1016/ j.forpol.2008.10.004.

33. C. Betzold, P. Castro , F. Weiler. AOSIS in the UNFCCC Negotiations: from Unity to Fragmentation. Zurich: Center for Comparative and International Studies (CIS), 2011.

34. D.R. Fisher, J.F. Green. Understanding disenfranchisement: civil society and developing countries' influence and participation in global governance for sustainable development. Global Environmental Politics 4:65-84, 2004. http://dx.doi.org/10.1162/1526380041748047.

35. A. Najam. Developing countries and global environmental governance: from contestation to participation to engagement. International Environmental Agreements 5: 303-321, 2005. http://dx.doi.org/10.1007/s10784-005-3807-6.

36. L.E. Pedersen. Climate Change Negotiations and Their Implications For International Development Cooperation. Denmark: Danish Institute for International Studies (DIIS), 2011. http://hdl.handle.net/10419/59850.

37. A. Giddens. The Politics Of Climate Change National Responses To The Challenge Of Global Warming. London: the Centre for the Study of Global Governance, 2008.

38. K. McComas, J. Shanahan. Telling stories about global climate change: measuring the impact of narratives on issue cycles. Communication Research 26: 30-57, 2009. http://dx.doi.org/10.1177/009365099026001003.

39. S. Lester, K. Neuhoff. Policy targets: lessons for effective implementation of climate actions. Climate policy 9: 464-480, 2009. http://dx.doi.org/10.3763/cpol.2009.0633.

40. D. Katz, I. Fischhcendler. Spatial and temporal dynamics of linkage strategies in arab israeli water negotiations. Political Geography 30: 13–24, 2011. http://dx.doi.org/10.1016/j.polgeo.2010.12.002.

41. M. Ingram, HR. Ingram, Lejano. What's the story? Creating and sustaining environmental networks.

Environmental Politics 1-19, 2014. http://dx.doi.org/10.1080/09644016.2014.919717.

42. A. Angelsen. The 3 REDD 'I's. Forest Economics 16: 253-256, 2010. http://dx.doi.org/10.1016/j.jfe.2010.10.001.

43. A. Sandstrom, L. Carlsson. The performance of policy networks: the relation between network structure and network performance. Policy Studies 36: 497–524, 2008. http://dx.doi.org/10.1111/j.1541- 0072.2008.00281.

44. j. Ayers, D. Dodman. (2010). Climate change adaptation and development I: the state of the debate. Progress in Development Studies 10: 161-168. http://dx.doi.org/10.1177/146499340901000205.

45. S. Oberthur. Interplay management: enhancing environmental policy integration among international institutions. International Environmental Agreements 9:371-391, 2009. http://dx.doi.org/10.1007/s10784-009-9109-7.

46. Gupta. Transparency under scrutiny: information disclosure in global environmental governance. Global Environmental Politics Journal 8: 1-7, 2008. http://dx.doi.org/10.1162/glep.2008.8.2.1.

47. M. Sippel, K. Neuhoff. A history of conditionality: lessons for international cooperation on climate policy. Climate policy 9: 481-494, 2009. http://dx.doi.org/10.3763/cpol.2009.0634.

48. H. Schroeder. Agency in international climate negotiations: the case of indigenous peoples and a) voided deforestation. International Environmental Agreements. Politics, Law and Economics 9: 58-78, 2010. http://dx.doi.org/10.1162/glep.2009.9.1.58.

49. D.R. Dalton, D. Catherine. Trips and tips for negotiation self-defense: Forewarned is forearmed. Business Horizons 54: 63–72, 2011. http://dx.doi.org/10.1016/j.bushor.2010.09.002.

50. M. Brockhaus, D.M Gregoria, R. Carmenta. REDD+ Policy Networks: Exploring Actors and Power Structures in an Emerging Policy Domain. Ecology and Society 19: 29, 2014. http://dx.doi.org/10.5751/ES-07098-190429.

INDEX

L

M

N

O

P

R

S

T

U

W